# Health, Hygiene and Eugenics in Southeastern Europe to 1945

# CEU Press Studies in the History of Medicine

## Volume II

Series Editor: Marius Turda

PUBLISHED IN THE SERIES:

**Svetla Baloutzova**

**Demography and Nation**

Social Legislation and Population Policy in Bulgaria, 1918–1944

# Health, Hygiene and Eugenics in Southeastern Europe to 1945

Edited by

**Christian Promitzer**
**Sevasti Trubeta**
**Marius Turda**

CEU PRESS

Central European University Press
Budapest—New York

Published in 2011 by
Central European University Press
An imprint of the
Central European University Share Company
Nádor utca 11, H-1051 Budapest, Hungary
*Tel:* +36-1-327-3138 or 327-3000
*Fax*: +36-1-327-3183
*E-mail*: ceupress@ceu.hu
*Website*: www.ceupress.com

400 West 59th Street, New York NY 10019, USA
*Tel:* +1-212-547-6932
*Fax*: +1-646-557-2416
*E-mail*: mgreenwald@sorosny.org



ISBN 978-963-9776-82-1 cloth

**Library of Congress Cataloging-in-Publication Data**

Health, hygiene, and eugenics in southeastern Europe to 1945 / edited by Christian Promitzer, Sevasti Trubeta, Marius Turda.
p. ; cm. -- (CEU Press studies in the history of medicine ; v. 2)
Includes bibliographical references and index.
ISBN 978-9639776821 (cloth)
1. Eugenics--Balkan Peninsula--History. 2. Public health--Balkan Peninsula--History. I. Promitzer, Christian. II. Troumpeta, Sevaste. III. Turda, Marius. IV. Central European University. V. Series: CEU Press studies in the history of medicine ; v. 2.
[DNLM: 1. Public Health--history--Europe, Eastern. 2. Public Health--history--Greece. 3. Eugenics--history--Europe, Eastern. 4. Eugenics--history--Greece. 5. Health Policy--history--Europe, Eastern. 6. Health Policy--history--Greece. 7. History, 19th Century--Europe, Eastern. 8. History, 19th Century--Greece. 9. History, 20th Century--Europe, Eastern. 10. History, 20th Century--Greece. 11. Social Medicine--history--Europe, Eastern. 12. Social Medicine--history--Greece. WA 11 GA3]
HQ755.5.B28H43 2010
362.109496'09041--dc22
2010036918

Printed in Hungary by Akaprint Kft., Budapest

# CONTENTS

# ACKNOWLEDGEMENTS

The idea for this volume originated with the conference on "Hygiene—Health Politics—Eugenics: Engineering Society in Twentieth Century Southeastern Europe," held between 31 May and 2 June 2007 at the Institute of East European Studies at the Free University of Berlin. Participants were historians and historians of medicine, social anthropologists and sociologists, all with a research interest in social history generally and the social history of medicine in particular. Their different approaches point towards the interdisciplinary scholarship attempted in this volume.

This volume's publication is possible due to the support of several colleagues and institutions to which we are much obliged. First, we would like to thank the Fritz Thyssen Foundation for facilitating the organization of the conference. A grant from the Institute for Historical and Cultural Research at Oxford Brookes University enabled Matt Feldman and Paula Bowles to provide the first rounds of copy-editing. We would like to thank Elisabeth Jay for her support. Another grant from the University of Graz helped to bring the manuscript to its final form due to Andreas Hemming's meticulous final editing.

We would also like to thank Tanja Petrović, Gisela Bock and Roger Griffin for many helpful conversations during the preparation of this volume as well as the anonymous reviewers for their comments and suggestions on an early version of the manuscript. Last but not least, we would like to express our gratitude to Central European University Press and especially to Linda Kunos, István Bart and Péter Inkei for welcoming and supporting the publication of this volume.

# INTRODUCTION

## FRAMING ISSUES OF HEALTH, HYGIENE AND EUGENICS IN SOUTHEASTERN EUROPE

*Christian Promitzer · Sevasti Trubeta · Marius Turda*

In a programmatic text on "Health Politics" penned immediately after the end of the First World War, the pioneer of social medicine in interwar Yugoslavia, Andrija Štampar (1888–1958), identified the emergence of a "national and social renaissance," which he insisted was "at the same time a health renaissance."[1] Guided by this vision, this volume engages with developments in the history of health, hygiene and eugenics in Southeastern Europe and the national contexts within which these developments took place. The geographic scope of what is usually understood as Southeastern Europe (Albania, Bulgaria, Greece, the former Yugoslavia and Romania) inevitably carries with it various conceptual problems, including asymmetrical comparisons between Central and Western Europe and the countries in the region. This juxtaposition of different, and often antagonistic, perceptions of European symbolic geographies presents a picture of intellectual and cultural history that is characterized by more complex processes of scientific appropriation and knowledge transfer across European countries during the first half of the twentieth century than was previously assumed.

As this volume aims to show, various developments in health, hygiene and eugenics in Southeastern Europe have a number of overlapping regional patterns. Two of them are particularly revealing: first, the idea

---

1 Andrija Štampar, "On Health Politics" (1919), republished in *Serving the Cause of Public Health. Selected Papers of Andrija Štampar*, ed. by M. D. Grmek (Zagreb: Medical Faculty of the University of Zagreb, 1966), 62.

of the national community figures prominently in leading medical discourses from the late nineteenth century onwards. Annexed to various forms of nationalism and concepts of society, health and hygiene became central components of larger biopolitical agendas; that is, the creation of a eugenically healthy nation and society. During the interwar period this agenda served as the progressive vehicle for transmitting a biological message transcending both specific political differences and generally opposed ideological camps. Second, health, hygiene and eugenics were complexly linked to political processes and state policies. Like elsewhere at the time, in Southeastern Europe the health of the collective body (defined either as the nation or society) became synonymous with the health of the state, and this political concern with health and hygiene, in turn, took several forms. One was the dominant role the state was asked to play in the expansion and administration of the collective body. New regimes of health and hygiene were instituted, especially after 1918, within which state intervention was normalized. It was assumed that with the help of experts, society and the state could be shaped and controlled by means of systematic regulation. This form of scientific management through medical education and schemes of hygiene and public health became central to the professionalization of medicine in Southeastern Europe, a process that only matured during the 1930s and 1940s. Comparing and contrasting various case studies in this region therefore provides the opportunity to delve into the archives in an attempt to redefine the wider debates on health, hygiene and eugenics within the broader European context of the late nineteenth and twentieth centuries.[2]

This volume also aims to explore the possible integration of the social history of medicine and health into broader Southeastern European studies. It thus raises questions about common regional patterns in various developments in health, hygiene and eugenics in Southeastern Europe. Of particular interest are the ways in which national states and emerging professional elites addressed (Ottoman and Habsburg) imperial legacies while at the same time contributing to the creation of new political and national systems based on the principle of national uniformity and centralism.

---

[2] For a discussion of recent historiography, see Marius Turda, "Focus on Social History of Medicine in Central and Eastern Europe," *Social History of Medicine* 21, 2 (2008): 395–401 and "Focus on Austria and Germany," *Social History of Medicine* 23, 2 (2010): 408-412

Two interrelated assumptions therefore guide this volume's conceptual unity: first, health, hygiene and eugenic policies in their various permutations were instruments of state building as expressed in professional networks and encompassing various social and political agendas. Second, in such agendas the idea of the nation was particularly salient, and, crucially, such nationalist narratives relied on biological metaphors of belonging and identification. It hardly needs emphasizing that these debates about health, hygiene and eugenics were conterminous with the wide-ranging effort to consolidate the Southeastern Europe nation-states and their societies. Once these dimensions have been fully appreciated, we can turn to examine how health, hygiene and eugenic projects were implemented and understood in their local contexts.

## Health and Hygiene in a 'Backward' Region

Before 1945, Southeastern Europe was frequently viewed as a "backward" region. For example, one encounters arguments over unhygienic conditions in commentaries written by visitors to the region during the late nineteenth and early twentieth century. In 1909, the English traveler Mary Edith Durham (1863–1944) depicted the situation in Albania (then still under Ottoman rule) as dreadful: "No words can tell the misery of the sick in these lands, who, swarming with lice, rot helpless on a heap of ferns or filthy rags in a dark corner till death releases them. No doctor has penetrated these wilds, nor any teacher save the Franciscans, whose medical knowledge is usually of the slightest."[3] Paul Dudley White (1886–1973), the prominent cardiologist working for the American Red Cross, alerted his readers to the fact that: "On the old battle-scarred valleys and mountains of Eastern Macedonia lurk some of the most dangerous of the world's diseases. The wars, which have been waged over this territory almost constantly since 1912, have exposed to broad day the menace of this land to the world's health."[4] Local authorities in Southeastern Europe were equally unflattering. A Bulgarian sanitary report describes the poor health of the

3 Edith Durham, *High Albania* (London: Edward Arnold, 1909), 88.

4 Paul Dudley White, "Public Health in Eastern Macedonia," *American Journal of Public Health* 10, 1 (1920): 14.

population in the countryside of the eastern Balkans in the early twentieth century thus: "Our rural population, which is busy with the everyday worries of the difficult provision of its subsistence, has had no time left to think about its appearance. It has neglected the cleanliness of the body. Consequently, among peasants several different parasites breed that are the vectors for various infectious diseases."[5]

The above commentaries were written at a time when Southeastern Europe was plagued by wars and population exchanges—all accompanying the transition from empire to nation-states. But these views also endorse the image of Southeastern Europe as a place of enduring otherness. As Maria Todorova has convincingly argued with respect to the Balkans, this region belongs to those sub-divisions of Europe's symbolic geography that "are locked in a hierarchical matrix."[6] Following a similar pattern, various "Balkanisms" and "Nesting Orientalisms" had developed within Southeastern Europe, appropriating Western European representations of this region in order to create its own cultural gradations.[7] In many cases, therefore, developments in Southeastern European health and hygiene reproduced ideological representations of this region as intrinsically "diseased," in a manner similar to the description of the Ottoman Empire as "the sick man of Europe."[8] And such representations did not simply provide stereotypes; they performed a wide variety of further functions, many of which were allegedly on behalf of Southeastern European countries themselves. At the same time, portrayals of this region as "backward" long included as

5 Toshko Petrov, *Raport za sanitarnoto sastoyanie na II-y Inspektorski Sanitaren rayon (Burgaski, Varnenski, Rusenski i Shumenski okrazi) prez 1909 godina* (Sofia: Direktsia za opazvane obshtestvenoto zdrave 1912), 120.

6 Maria Todorova, "Spacing Europe: What is a Historical Region?" *East Central Europe* 32, 1-2 (2005): 63. See also her earlier *Imagining the Balkans* (Oxford: Oxford University Press, 1997). For an extensive discussion of Eastern Europe as the "Other" Europe for the philosophers of the Enlightenment, see Larry Wolff, *Inventing Eastern Europe: The Map of Civilization on the Mind of the Enlightenment* (Stanford: Stanford University Press, 1994). A more specific analysis of how Southeastern Europe was represented in German debates on symbolic geography can be found in Dietmar Müller, "Southeastern Europe as a Historical Meso-Region: Constructing Space in Twentieth-Century German Historiography," *European Review of History* 10, 2 (2003): 393–408.

7 Milica Bakic-Hayden, "Nesting Orientalisms: The Case of Former Yugoslavia," Slavic Review 54, 4 (1995): 917–931. See also Slobodan Markovich, *British Perceptions of Serbia and the Balkans, 1903–1906* (Paris: Dialogue, 2000); and Andrew Hammond, ed., *The Balkans and the West: Constructing the European Other, 1945–2003* (Aldershot: Ashgate, 2004).

8 As early as 1720, Montesquieu described the Ottoman Empire in the following terms: "I have seen with astonishment the weakness of the Osmanli empire. This sick body does not sustain itself with a mild and temperate regimen, but by violent remedies that continually exhaust and undermine it." See Charles de Montesquieu, *The Persian Letters* (Indianapolis: Hackett Publishing, 1999 [1720]), 36 (Letter XIX).

much cultural rhetoric as a grounds for practical intervention in the fields of health and hygiene.

Some other features of this symbolic medical geography can be tentatively enumerated. Prior to the early twentieth century, the Great Powers regarded the Ottoman Empire's European possessions to be an area where sanitary interventions were necessary to protect the general health of Europe. Various quarantines were initiated to protect Europe against epidemics such as bubonic plague and cholera. These two diseases were thought to have originated in the Middle East, and local authorities were called upon to collaborate with international agencies in applying multilateral conventions concerning sanitary protection.[9] Indeed, a system of quarantines was established in the Ottoman Empire in 1838 by Sultan Mahmud II (1785–1839) with the specific purpose of containing and protecting against epidemics. This system consisted of thirteen sanitary offices in the Ottoman Empire's European possessions and forty in Asia Minor. They were controlled by a *Conseil supérieur de santé* representing, as equal partners, the Ottoman government and the European powers. Initially this program was meant only for the European part of the Ottoman Empire, where the Christian population was purportedly yearning for measures similar to those in the neighboring countries. It was assumed that Southeastern Europe, especially the Danubian Principalities and Serbia, where quarantines had been effective since 1832, and the Habsburg Empire, where quarantines had been in effect since the mid-eighteenth century, would protect the rest of Europe from the Oriental plague by means of a "partition wall."[10] Ten years later, after the bubonic plague was exterminated in the European part of the Ottoman Empire, one of the experts on quarantine at the time, the Austrian physician Karl Ludwig Sigmund (1810–1883) believed that quarantines could be lifted in this region in order to promote it economically as a result of free-trade and social mobility.[11] These examples point to

9 Erna Lesky, "Die österreichische Pestfront an der k.k. Militärgrenze," *Saeculum* 8 (1957): 82–104; Daniel Panzac, *La peste dans l'Empire ottoman: 1700–1850* (Leuven: Peeters, 1985); *Quarantaines et lazarets: l'Europe et la peste d'Orient (XVIIe–XXe siècles)* (Aix-en-Provence: Édisud, 1986); and *Population et santé dans l'Empire ottoman: (XVIIIe–XXe siècles)* (Istanbul: Éd. Isis, 1996); and Nadya Manolova-Nikolova, *Chumavite vremena (1700–1850)* (Sofia: "IF-94", 2004).

10 [Friedrich August] Vetter, "Nachrichten neuester Beobachter über die Pest auszugsweise nach dem von Hrn Dr Bulard herausgegebenen Journale la Peste bearbeitet," C. W. *Hufeland's Journal der practischen Heilkunde* 86, 3 (1838): 91–108, esp. 96–7.

11 Karl Ludwig Sigmund, *Die Quarantäne-Reform und die Pestfrage* (Vienna: Wilhelm Braumüller, 1850), 56 and 62.

the intermediary position of Southeastern Europe vis-à-vis the centre of the continent; the discussion was mainly about the inclusion or exclusion of this region by means of a *cordon sanitaire*. Such debates are not specific to medical history alone; rather, they confirm a general construction of the Balkans as an "incomplete self" (according to Todorova)[12] of Central and Western Europe vis-à-vis the "Oriental other" which, in terms of symbolic geography, commences on "the other side" of the Bosporus.

Whether or not accompanied by stereotypical Orientalist representations, there is no doubt that Southeastern Europe was characterized by a number of features that indicated a certain structural difference between this region and Central and Western Europe and which in turn shaped the development of health, hygiene and eugenics after 1900. The sanitary conditions in particular, which were typical for predominantly rural societies, came under the critical evaluation of hygienists and physicians who followed precepts elaborated in industrialized societies. They criticized, for example, the precarious hygienic conditions among the population, malnutrition, the spread of social diseases (such as alcoholism) and venereal diseases (syphilis in particular), high levels of infant mortality, the rejection of modern medicine and a persistence of traditional methods of healing.[13] The junction of these features indexes a variety of different historical and cultural phenomena that are associated with the protracted process of establishing modern medicine. Other elements of this process are revealed in two areas in which national and international interests were constantly negotiated: health campaigns and the establishment of health and hygiene institutions.

## International Health Campaigns

There is no doubt that Southeastern Europe constituted a critical zone of interaction between local and international health agencies during the prolonged collapse of the Ottoman Empire. That became evident during vari-

12 Todorova, *Imagining the Balkans*, 18.

13 See, for example, Aida Brenko, Željko Dugac, Mirjana Randić, *Narodna medicina – Folk Medicine* (Zagreb: Etnografski muzej Zagreb, 2001); Filip Šoć, *Narodna i naučna medicina u Crnoj Gori: istorijski pregled do 1916*

ous International Sanitary Conferences organized between 1851 and 1911 that focused on cholera in the Ottoman Empire and its threat to the rest of Europe as well as on the sanitary control of European Muslim pilgrims to Mecca.[14] Western medical intervention also occurred during the Russo-Turkish War of 1877/78 and the two Balkan Wars of 1912/13. Indeed, right from the start of these conflicts, Western medical organizations were present at the front in Macedonia, Thrace, Albania and Bulgaria.[15] All of these conflicts challenged military medicine and, at the same time, provided physicians with new opportunities for medical research: field physicians studied the healing of wounds caused by new weaponry and the containment of war epidemics, of which cholera consistently ranked among the most devastating.[16]

(Podgorica: Institut za zdravlje Crne Gore, 2002); Tomislav Mijović, ed., *Za zdravlje: iz istorije narodne medicine i zdravstvene kulture. VII naučni skup Rajačke pivnice, 2000* (Zaječar: Narodni muzej – Zavod za zaštitu zdravlja "Timok", 2001); Dobrivoje Janković, Živko Slavković, *Etnomedicina balkanskih naroda: načini narodnog lečenja iz Srbije, Srpske, Crne Gore, Makedonije, Bugarske i severne Grčke* (Kraljevo: Slovo, 1997); Milorad Dragić, *Etnomedicina*, 2 vol. (Belgrade: Odeljenje za etnologiju Filozofskog fakulteta, 1991); and Mincho Georgiev, ed., *Balgarska narodna medicina. Enciklopediya* (Sofia: Izdatelska kashta "Petar Beron," 1999).

14 Norman Howard Jones, *The Scientific Background of the International Sanitary Conferences 1851–1938* (Geneva: World Health Organization, 1975); Peter Baldwin, *Contagion and the State in Europe, 1830–1930* (Cambridge: Cambridge University Press, 1999), 123–243; Mariko Ogawa, "Uneasy Bedfellows: Science and Politics in the Refutation of Koch's Bacterial Theory of Cholera," *Bulletin of the History of Medicine* 74, 4 (2000): 671–707; and Valeska Huber, "The Unification of the Globe by Disease? The International Sanitary Conferences on Cholera, 1851–1894," *The Historical Journal* 49, 2 (2006): 453–476.

15 Paul J. Weindling, *Epidemics and Genocide in Eastern Europe, 1890–1945* (Oxford: Oxford University Press, 2000), 73–74; Charles S. Ryan and John Sandes, *Under the Red Crescent. Adventures of an English Surgeon with the Turkish Army at Plevna and Erzeroum*, 1877–1878 (New York: Charles Scribner's Sons, 1897); Alexander Jacob Schem, *The War in the East. An Illustrated History of the Conflict between Russia and Turkey with a Review of the Eastern Question* (New York: H. S. Goodspeed & Co., 1878), 472–482.

16 In 1913/14, an impressive number of articles on combating cholera in Bulgarian, Ottoman, Serbian and Romanian armies was published in the German and Austrian medical press, for example: [Johannes] Eckert, "Die Rolle der Kontaktinfektion in der Epidemiologie der Cholera (Nach in Bulgarien gesammelten Erfahrungen)," *Berliner klinische Wochenschrift* 50, 50 (1913): 2326–2328; Rudolf Kraus, "Über Maßnahmen zur Bekämpfung der Cholera auf dem bulgarischen Kriegsschauplatz," *Wiener klinische Wochenschrift* 26, 7 (1913): 241–247; Franz Rosenthal, "Medizinische Eindrücke von einer Expedition nach Bulgarien, speziell ein Beitrag zur Diagnose und Therapie der Cholera asiatica," *Berliner klinische Wochenschrift* 51, 8 (1914): 342–344; K. J. Schopper, "Erfahrungen über die Cholera in Ostrumelien während des Balkankrieges 1912," *Wiener klinische Wochenschrift* 26, 10 (1913): 366–370; Rudolf Strisower, "Meine Erfahrungen aus der Choleraepidemie in Serbien im Sommer 1913," *Wiener klinische Wochenschrift* 26, 50 (1913): 2078–2081; Julius Stumpf, "Über Cholerabehandlung und Choleraprophylaxe auf Grund meiner Erfahrungen in Nisch und Belgrad," *Münchener medizinische Wochenschrift* 61, 14 (1914): 759–763; [August] Aumann, "Welche Bedeutung kommt dem Kontakt bei der Verbreitung der Cholera in Serbien 1913 zu?," *Berliner klinische Wochenschrift* 51, 2 (1914): 62–64 and "Über Maßnahmen bei der Bekämpfung der Cholera in Serbien 1913," *Berliner klinische Wochenschrift* 51, 13 (1914): 589–592. With respect to cholera in the Romanian army, see Victor Babes, "Studien über cholerabekämpfung," *Zeitschrift für Hygiene und Infektionskrankheiten* 77 (1914): 501–533.

The typhus pandemic in Serbia in spring 1915, for instance, brought a British Military Sanitary Mission as well as various relief organizations from the United States into the country.[17] Following the Serbian defeat in 1915, military medicine continued to play a prominent role in containing malaria among the Entente troops disembarking near Salonica.[18] Concurrently, German and Austro-Hungarian armies established a harsh system of military medicine and anti-epidemic measures in occupied Serbia, Montenegro and northern Albania.[19] The German expert in tropical medicine Peter Mühlens (1874–1943) contributed to establishing military hygiene in various parts of the Ottoman Empire and Bulgaria during World War I.[20]

Direct medical assistance and control by the Great Powers ended with World War I; international organizations, however, advised the application of bacterology and various methods of disinfection in the region in order to combat infectious diseases. Furthermore, the regional medical network established during the war provided the basis for hygiene and public health projects that were later established to aid the civilian population—especially those carried out by international organizations like the *Office International d'Hygiene publique*, which was created in 1907 with Romania as a founding member, the Health Organization of the League of Nations and the Rockefeller Foundation.[21] The Epidemic Commission of the League of Nations was similarly important in Greece in 1922/23, when more than a million refugees arrived from Asia Minor in the wake of Greece's defeat by

17 M. R. Smallman-Raynor and A. D. Cliff, *War Epidemics: An Historical Geography of Infectious Diseases in Military Conflict and Civil Strife, 1850–2000* (Oxford: Oxford University Press, 2006), 657–664; Vojislav Soubbotitch, "A Pandemic of Typhus in Serbia in 1914 and 1915," *Proceedings of the Royal Society of Medicine* 11, 5 (1918): 31–39; William Hunter, *The Serbian Epidemics of Typhus and Relapsing Fever in 1915* (London: John Bale, 1920); and Richard P. Strong et al., *Typhus Fever with Particular Reference to the Serbian Epidemic* (Cambridge, Mass.: Harvard University Press, 1920).

18 Arthur G. Phear, "Medical Experiences in Macedonia and the Caucasus," *Proceedings of the Royal Society of Medicine* 13, 5 (1920): 57–100.

19 See *Sanitärer Wiederaufbau Serbiens. Festschrift anläßlich des einjährigen Bestehens des k. und k. Militär-General-Gouvernements in Serbien* (Vienna: K. und k. Hofbuchhandlung, 1917); *Militärärztliche Arbeiten aus dem k. u. k. Militär-Generalgouvernement in Montenegro* (Vienna: K. und k. Hofbuchhandlung, 1918); and Max Eugling, *Über Malariabekämpfung. Beobachtungen und Untersuchungen aus dem albanischen Malaria-Gebiet* (Leipzig: Johann Ambrosius Barth, 1921).

20 Stefan Wulf, *Jerusalem - Aleppo - Konstantinopel: der Hamburger Tropenmediziner Peter Mühlens im Osmanischen Reich am Vorabend und zu Beginn des Ersten Weltkriegs* (Münster: LIT Verlag, 2005); Peter Mühlens, "Kriegshygienische Erinnerungen," *Archiv für Schiffs- und Tropenhygiene* 43, 12 (1939): 531–561.

21 Paul J. Weindling, ed., *International Health Organisations and Movements 1918–1939* (Cambridge: Cambridge University Press, 1995).

Turkey.[22] Many refugees were diagnosed with malaria just a few months after their arrival in Macedonia or Thrace.[23] Physicians played an important role in these operations. In 1924, for instance, the German expert on tropical medicine Bernhard Nocht (1857–1945) was entrusted by the head of the Hygiene Commission of the League of Nations Ludwik Rajchman (1881–1965) with investigating malaria in Yugoslavia, Greece, Bulgaria, Romania and the Soviet Union.[24]

External funding and expertise was provided to Southeastern European countries throughout this protracted process of implementation of modern ideals of health and hygiene. Indeed, it was international organizations such as the Rockefeller Foundation, the Health Commission of the League of Nations and the International Committee of the Red Cross that offered financial aid and assisted local governments.[25] But the medical expertise and financial support received from international organizations also shed light on the mutual interaction between modern ideas of health and hygiene and the local cultures in Southeastern Europe. The connection between health, hygiene and eugenics was strongly mediated by means of medical metaphors, devised to express the wide-spread feeling of optimism about social and scientific progress and its effect on the intellectual, moral and material welfare of society. In Southeastern Europe, it was demanded that the state take the leading role in disseminating modern ideas on hygiene and technologies of health. The professional elites in the region relied heavily on the state for funding, and particularly so during the interwar period when local governments repeatedly acknowledged the need for improvement in health and hygiene and attempted to achieve it by establishing institutes of research and nation-wide programs of sanitation and preventive medicine.

22 See Elisabeth Kontogiorgi, *Population Exchange in Greek Macedonia. The Rural Settlement of Refugees 1922–1930* (Oxford: Oxford University Press, 2006).

23 Bruce-Chwatt and de Zulueta, *Malaria in Europe*, 36; Marta Alexandra Balińska, "Assistance and Not Mere Relief: the Epidemic Commission of the League of Nations, 1920–1923," in Weindling, ed., *International Health Organizations*, 99–100.

24 Stefan Wulf, *Das Hamburger Tropeninstitut 1919 bis 1945. Auswärtige Kulturpolitik und Kolonialrevisionismus nach Versailles* (Berlin: Dietrich Reimer Verlag, 1994), 67–68; League of Nations, Health Organisation, Malaria Commission, *Report on its Tour of Investigation in certain European Countries in 1924* (doc C. H. 273) (Geneva, 1925).

25 John Farley, *To Cast Out Disease. A History of the International Health Division of Rockefeller Foundation (1913–1951)* (Oxford: Oxford University Press, 2003).

## Institutionalizing Health and Hygiene

The process of embracing modern theories of health and hygiene did not occur simultaneously across Southeastern Europe, nor was the spread of medical education—let alone the extent of medical institutionalization—similar in the individual countries. Under the Ottoman Empire, a medical school and a teaching hospital had been opened in the state shipyard in 1805, the result of a private initiative aimed solely at the military. A school for surgeons followed in 1828. These two schools were fused in 1839 as the newly-established Imperial Medical School.[26] The first medical faculty in Southeastern Europe not belonging to the military, however, was founded in 1837 in Athens, and another was established in Bucharest twenty years later. Yet the rest of the countries in the region had to wait until the twentieth century: in Bulgaria the first medical faculty was only founded in 1918, in Serbia in 1920 and in Albania only in 1952. Until these institutions were established there was a chronic shortage of physicians in these countries. Those looking to join this profession needed to be trained abroad, and many could only afford this elite education with the help of scholarships. Right up to the end of World War II, numerous medical students from Southeastern Europe benefited from this transfer of knowledge by studying with leading specialists of various medical disciplines at various European universities.[27] These medical students later joined initiatives for combating diseases and epidemics and contributed to building the national health systems. To some extent, the values of Western medicine thus became the standard against which the advance of hygiene and preventive medicine in Southeastern Europe was assessed.

Ministries of Health were also founded in Southeastern European countries in this time. In November 1918, the last Serbian government founded

26 Arslan Terzioglu and Erwin Lucius, eds., *Mekteb-i Tibbiye-i Adliye-i Şahane ve bizde modern tip eğitiminin gelişmesine katkıları – Die hohe Medizinschule Galatasaray und ihre Bedeutung für die moderne türkische Medizin* (Istanbul: Akeoloji ve Sanat Yayınları, 1993) and Oya Dağlar, *War, Epidemics and Medicine in the Late Ottoman Empire (1912–1918)* (Haarlem: SOTA 2008).

27 Richard Georg Plaschka and Karlheinz Mack, eds., *Wegenetz Europäischen Geistes 1: Wissenschaftszentren und geistige Wechselbeziehungen zwischen Mittel- und Südosteuropa vom Ende des 18. Jahrhunderts bis zum Ersten Weltkrieg* (Vienna: Verlag für Geschichte und Politik, 1983); and Richard Georg Plaschka and Karlheinz Mack, eds., *Wegenetz Europäischen Geistes 2: Universitäten und Studenten: die Bedeutung studentischer Migrationen in Mittel- und Südosteuropa vom 18. bis zum 20. Jh.* (Vienna: Verlag für Geschichte und Politik, 1987).

a Ministry of Public Health, which became the Ministry of National Health in the first Yugoslav government only a month later. The republican government in Turkey established a Ministry of Health and Social Care in 1921, while Romania established one in 1922. Greece passed a law establishing such a ministry in the same year. Bulgaria and Albania only established their Ministries of Health in 1944 and 1946 respectively.[28] Two factors had a significant impact on the development of public health systems across Southeastern Europe: the need to establish professional health care in urban centers and the necessity of coping with the consequences of industrialized warfare.[29]

Following the watershed of World War I, countries that benefited territorially from the ensuing peace treaties, like Romania and Yugoslavia, had to address regional disparities and different institutional traditions in these areas. This disparity existed, for instance, between the Romanian Old Kingdom, which developed a health system as an independent nation-state, and Transylvania, or between independent Serbia and Croatia, Bosnia and Slovenia, which had been a part of the Habsburg Empire prior to November 1918. Yet this historical discrepancy within the new states was compensated by sustained institutionalization and centralization. Any inventory of the elements that constitute the culture of health and hygiene in Southeastern Europe thus cannot avoid considering the key role of the state.

But even if there is no doubt that the state played the most significant role in the development of health and hygiene projects, the influence of outstanding individuals is not to be completely neglected. Leading health reformers like the Croat Andrija Štampar (1888–1958), the Greeks Constantinos Savvas (1861–1929) and Emmanuel Lampadarios (1882–1943), the Bulgarian Toshko Petrov (1872–1942) and the Romanians Iuliu Moldovan (1882–1966) and Gheorghe Banu (1889–1957) played decisive roles in creating centralized systems for health and hygiene. Their concepts

28 See Karl Kaser, *Handbuch der Regierungen Südosteuropas (1833–1980)* vol 1 (Graz: Institut für Geschichte, Abteilung für Südosteuropäische Geschichte, 1981), 31, 124–125, 310–311; *Handbuch der Regierungen Südosteuropas (1833–1980)* vol. 2 (Graz: Institut für Geschichte, Abteilung für Südosteuropäische Geschichte, 1982), 4, 258, 520; and *Handbuch der Regierungen Südosteuropas (1833–1980)* vol. 3 (Graz: Institut für Geschichte, Abteilung für Südosteuropäische Geschichte, 1984), 92.

29 On the case of Vardar Macedonia, see Patrick Zylberman, "Mosquitos and the Komitadjis: Malaria and Borders in Macedonia (1919–1938)," in I. Borowy and W. D. Gruner, eds., *Facing Illness in Troubled Times: Health in Europe in the Interwar years 1918–1939* (Frankfurt am Main: Peter Lang, 2005), 305–343.

of nationalized hygiene and health systems became paramount in the interwar years, when these individuals held important ministerial positions. And although Štampar was forced to resign in the 1940s to assume a role as a leading expert for the Health Organization of the League of Nations, his ideas remained prominent in Europe, most notably among Bulgarian experts on public health.[30]

Central to theories of health and hygiene developed during the interwar period was the idea that the biological conditions of communities could also be improved upon with the help of external factors such as education and a controlled environment, including both the prevention and eradication of contagious diseases and parasites as well as sanitation and housing. The Rockefeller Foundation in particular favored a "program of positive health based on prevention medicine,"[31] which dovetailed with the institutionalization of health and hygiene. Most public health institutions and institutes of hygiene in Southeastern Europe were established with external funding. Although departments of hygiene existed at universities in Southeastern Europe since the late nineteenth century (a department of hygiene and microbiology was, for instance, established at the University of Athens in 1900) many public health institutions and institutes of hygiene were established in the 1930s with external funding. The most important of them were, in Greece, the School of Hygiene (est. 1930), the School of Public Health Nursing (est. 1931) and the Athens Health Centre (est. 1939); in Romania, the Institute of Hygiene and Social Hygiene in Cluj (est. 1919), the Institute of Hygiene and Public Health in Bucharest (est. 1927) and the Institute of Hygiene and Public Health in Iaşi (est. 1930); in Bulgaria, the Institute for National Health in Sofia (est. 1929); and in Yugoslavia, the Central Institute for Hygiene in Belgrade (est. 1926) and the Institute of Hygiene and the School of Public Health in Zagreb (est. 1927).[32] The intention was in each case to create a modern infrastructure for the national promotion of health and hygiene.

30 Patrick Zylberman, "Fewer Parallels than Antitheses: René Sand and Andrija Stampar on Social Medicine, 1919–1955," *Social History of Medicine* 17, 1 (2004): 77–92 and Zahari Bochev, "Yugoslavyanska zdravna politika," *Balgarski higienski pregled* 2, 3 (1932): 206–213.

31 Paul Weindling, "Public Health and Political Stabilisation: The Rockefeller Foundation in Central and Eastern Europe between the Two World Wars," *Minerva* 31, 3 (1993): 254.

32 On Yugoslavia, see Željko Dugac, *Protiv bolesti i neznanja: Rockefellerova fondacija u međuratnoj Jugoslaviji* (Zagreb: Srednja Europa, 2005); on Bulgaria, see Milena Angelova, "Rokfelerovata fondatsiya i amerikanskata

The other phase of institutionalization was characterized by active social and national politics, such as training in modern health techniques, the introduction of hygiene and eugenic education into the school curriculum, hygienic courses for adults and programs of rural sanitation. It also meant that suitable health legislation be adopted by these countries, which included the National Popular Health Education Act in Yugoslavia (1928), the Public Health Law in Bulgaria (1929) and the Sanitary Law in Romania (1930). On the basis of these legislative acts, health and hygiene became institutionally defined and centralized.

Moreover, issues of health and disease gradually became a matter of the political and national economy.[33] Statistical records on birth, fertility, mortality and morbidity rates provided a scientific basis for population and eugenic policies, which intensified during the 1930s and 1940s. A paramount goal was the reduction of infant mortality by means of various projects on child protection and puericulture.[34] A number of prominent physicians were active in this field at both the national and international levels. Yet the effectiveness of such projects depended on specific infrastructural conditions. Infant mortality in Bulgaria, Romania and Yugoslavia declined in the interwar period, although not at a rate comparable with that of other European countries like Austria, France, Germany or Switzerland.[35] On the other hand, the fact that Southeastern European countries had higher birthrates than Western European countries was considered a biological advantage as well as a particular feature of the nation's healthy racial qualities. Fertility and ethnicity were understood to be mutually supportive.

Another aspect that deserves highlighting is the international role of health reformers from Southeastern Europe. If in the field of public health it was Andrija Štampar, who was highly regarded by organizations and officials around the world, in the field of child protection it was the work of

blizkoiztochna fondatsiya v Balgariya – initsiativi v poleto na sotsialnata rabota, 20-30te godini na XX vek," in Kristina Popova and Milena Angelova, eds., *Obshtestveno podpomagane i sotsialna rabota v Balgariya. Istoriya, institutsii, ideologii, imena* (Blagoevgrad: Yugozapaden universitet "Neofit Rilski" Blagoevgrad, 2005), 112–125; on Greece, see Vassiliki Theodoru and Despina Karakatsani, "Health Policy in Interwar Greece: The Intervention by the League of Nations Health Organisations," *Dynamis* 28 (2008): 53–75.

33 Thus, the 1933 report by the Bulgarian Central Directorate of Public Health regarding the national state of health started with the words: "A healthy population, i.e. physically and mentally apt to work, indeed is the very base for the understanding of nation and state." Central State Archive Sofia (f-372k, op-1, ae-541).

34 Kristina Popova, *Natsionalnoto dete. Blagotvoritelnata i prosvetna deynost na Sajuza za zakrila na detsata v Balgariya 1925–1944* (Sofia: LIK, 1999).

the Romanian social hygienist and eugenicist Gheorghe Banu that received wide international recognition. In the 1930s Banu became actively involved in the International Association for the Protection of Infancy as seen in his numerous contributions to the *Bulletin International de la Protection de l'Enfance* and his leading role in organizing various conferences on child protection, like those in Liège in 1930 and Paris in 1933. Together with Emmanuel Lampadarios, the Head of the Office for the School of Hygiene in Athens, Banu counterbalanced the influence exercised by French and Belgian activists and child reformers within international organizations.[36] The First and the Second Balkan Congress on the Protection of Children organized in Athens in 1936 and Belgrade in 1938 respectively illustrate Southeastern Europe's growing importance in the field of child protection and public health.[37] To this one must add the emergence of local models of health and hygiene, like theories of rural biology and peasant universities in Yugoslavia and Romania.[38]

These sanitized versions of local rural life, the ideal peasant village as a repository of specific national values and traditions, were not only an essential component of the new health policies developed after 1918 but were similarly incorporated into the emerging eugenic discourses. Projects to improve public health were perceived as a matter of politics and science. This is because health posed a socialized and socializing resource for those technologies of power employed by the state to control, supervise and discipline its subjects.[39] The social and national pressure put on women is illustrative. Women were predominantly perceived as "mothers of the nation" in both nationalist and eugenic discourses on reproduction. One of the key areas, then, in which the relationship between health, hygiene and eugenics was conveyed were the debates surrounding mar-

35 Milan Hauner, "Human Resources," in M. C. Kaser and E. A. Radice, eds., *The Economic History of Eastern Europe, 1919–1975: Economic Structure and Performance Between the Two Wars*, vol. 1 (Oxford: Claredon Press, 1986), 98.

36 In 1933 Banu published *L'Hygiène sociale de l'Enfance*, 2 vols. (Paris: Masson, 1933) a book which established him as an international authority on the subject.

37 *Premier Congrès Balkanique de la Protection de l'Enfance. Compte-Rendu* (Geneva : Union Internationale de Secours aux Enfants, 1936) and *Deuxième Congrès Balkanique de la protection de l'Enfance* (Belgrade: n.p. 1938).

38 See Andrija Štampar, *Public Health in Yugoslavia* (London: School of Slavonic and East European Studies, 1938) and G. Banu, *Mari probleme de medicină socială* (Bucharest: n.p., 1938).

39 Michel Foucault, "*Society Must be Defended*": *Lectures at the Collège de France, 1975–1976* (New York: Picador, 2003), 239–264.

riage certificates and sterilization. Within this biopolitical transformation of the relationship between states and populations, eugenics was "applied to both the body and the population, both the organism and biological processes" for "both disciplinary effects and regulatory effects."[40] Indeed, on the one hand, health politics aimed to improve living conditions, individual lives and the nation's welfare; on the other, physicians intervened in the lives of these individuals, sometimes to the point of challenging their right to life itself. The aspiration for a "healthy population" thus emerged as a reflection of the ambivalence inherent to the alleged egalitarian character of modern health care.

This process echoes developments elsewhere at the time. By the 1920s, physicians and health reformers in Southeastern Europe too would have been convinced of the general connections between health, hygiene and nationalism. Since the establishment of nation-states in the region, physicians had sought to protect national health through medical education, health reform and large-scale schemes of preventive medicine.[41] One of the most important corollaries to this development was the physician's extensive social and national involvement: a physician was now more than just a medical doctor caring for patients. He (and increasingly she) gradually became an instrument of state politics while medicine became a medium for addressing moral and ethical questions pertaining to the health of the nation and society.

A host of further questions derive from this transformation, such as to the links between theories of national health, political philosophies and social policies or how best to care for sick individuals in the wake of an increased demand for a healthy society and nation. One might also ask in how far visions of national decline and fragmentation so vividly portrayed in nationalist discourses throughout Southeastern Europe in the 1930s and 1940s shaped or were shaped by the wider debate on health, hygiene

40 Michel Foucault, "*Society Must be Defended*", 252.

41 For a convincing discussion of these two aspects of eugenics (medical and social) see Peter Weingart, "Eugenics—Medical or Social Sciences," *Science in Context* 8, 1 (1995): 197–207. More recently, the issue has been addressed by Lene Koch, "How Eugenic was Eugenics? Reproductive Politics in the Past and the Present," in Regina Wecker et. al, *Wie nationalsozialistisch ist die Eugenik? What is National Socialist about Eugenics? Internationale Debatten zur Geschichte der Eugenik im 20. Jahrhundert/ International Debates on the History of Eugenics in the 20th Century* (Vienna: Böhlau Verlag, 2009), 39–63 and Lene Koch, Regina Wecker, "How Eugenic is Eugenics? A dialogue between Lene Koch and Regina Wecker," idem, 65–72.

and eugenic welfare? What was the connection between endeavors during World War II to bring eugenics and racial hygiene to bear on the understanding of "undesirable" ethnic minorities and political debates on the Jewish question? To what degree did medical discussions on the health of the population, for instance, amplify contemporaneous social debates about demography, including natality, fertility and mortality rates? One cannot pursue these questions regarding modern technologies of health and hygiene without also addressing theories of human improvement or the eugenic expression of modern biopolitics. As actively politicized eugenic discourses developed in Southeastern Europe during the 1930s and 1940s, ideas of health and hygiene became part of a new political language used to describe and justify new models of national belonging.

## Southeastern European Eugenics in Context

Long before the history of eugenics in Southeastern Europe became a subject of scholarly research,[42] the historian of medicine Mark B. Adams pointed out the remarkable diversity characterizing international eugenic movements. Simultaneously, he placed the development of eugenic ideas in a wider, international context, reacting against those approaching eugenics as a "single, coherent, principally Anglo-American movement with a specifiable set of common goals and beliefs."[43] While research largely continues to limit its focus to individual countries—sometimes extending to particular regions—today there is also a historiographical movement toward highlighting the international character of eugenics, particularly in

42 See Maria Bucur, *Eugenics and Modernization in Interwar Romania* (Pittsburgh: Pittsburgh University Press, 2002); Turda and Weindling, eds. *Blood and Homeland*; Marius Turda, "'The Nation as Object: Race, Blood and Biopolitics in Interwar Romania," *Slavic Review* 66, 3 (2007): 413–441; Marius Turda, "'To End the Degeneration of a Nation': Debates on Eugenic Sterilization in Interwar Romania", *Medical History* 53, 1 (2009); Ayça Alemdaroğlu, "Politics of the Body and Eugenic Discourse in Early Republican Turkey," *Body and Society* 11, 3 (2005): 61–76; Murat Erkin, "Biometrics and Anthropometrics: The Twins of Turkish Modernity," *Patterns of Prejudice* 42, 3 (2008): 281–304; Παναγιώτα Καζολέα-Καβολάρη, "Ψυχολογία και Ευγονική στην Ελλάδα", *Τετράδια Ψυχιατρικής* 78 (2002): 53–64; and Gergana Mircheva, "Balgarskiyat evgenichen proekt ot 20-te i 30-te godini na minaliya vek i normativniyat kod na 'rodnoto'," *Kritika i humanizam* 17, 1 (2004): 207–221.

43 Mark B. Adams, "Toward a Comparative History of Eugenics," in Mark B. Adams, ed., *The Wellborn Science: Eugenics in Germany, France, Brazil, and Russia* (Oxford: Oxford University Press, 1990), 217. For recent developments, see Alison Bashford and Philippa Levine, eds., *The Oxford Handbook of the History of Eugenics* (New York: Oxford University Press, 2010).

terms of international networks emerging at the beginning of the twentieth century.[44]

The consequent challenge of these approaches involves recognizing and interpreting eugenics' complex character as a movement, both nationally and internationally. To this end, one way to integrate eugenics into a discussion of health and hygiene in Southeastern Europe is to contextualize it historically.[45] This volume thus aims to introduce a version of the history of eugenics in that region that accentuates its national and political affinities and explores its possible connections with more racially-oriented discourses on human improvement, like those developed in Germany after 1933. Yet, contrary to many of the universal values it shared and foreign models it sought to emulate, Southeastern European eugenics had distinctive national overtones, differentiated by each country's individual culture and social context. Again, compared to other regions in Europe, the increasing ties to a racialized politics that one can observe during the 1940s in countries like Romania, Bulgaria and in the Independent State of Croatia should not be allowed to obscure the fact that biological determinism never became the basis of official policy in Southeastern Europe. Whether it was articulated in the more specialized idioms of eugenics, or in the vernacular paradigm of culture, the dominant discourse about the nation in these countries rarely conjured fantasies of racial purity. That, for example, no institute for eugenics or racial research was established in Southeastern Europe before or during World War II, was not only due to financial constraints. Governments were skeptical about the introduction of practical eugenic measures, deeming their societies insufficiently prepared to accept them. Instead, resources were directed towards the improvement of sanitary and hygienic conditions and nation-wide vaccination campaigns.

Another argument why eugenics failed to become a politically sanctioned practice pertains to its belated institutionalization. One distinguishing characteristic of Southeastern European eugenics is its late appearance on the international map of eugenic movements. Compared to other regions in Europe and the United states, no eugenics society existed in Southeast-

44 Stefan Kühl, *Die Internationale der Rassisten. Aufstieg und Niedergang der internationalen Bewegung für Eugenik und Rassenhygiene im 20. Jahrhundert* (Frankfurt am Main: Campus Verlag, 1997). See also Weindling, "International Eugenics: Swedish Sterilisation in Context," *Scandinavian Journal of History* 24, 2 (1999): 179–197.

45 Maria Todorova, "Introduction: Learning Memory, Remembering Identity," in Maria Todorova, ed., *Balkan Identities. Nation and Memory* (London: Hurst, 2004), 12–13.

ern Europe before the late 1920s (the first being an informal "Society for the Study of Racial Hygiene" that was established in Bulgaria in 1926). Moreover, with the exception of Romania, where the Eugenics and Biopolitical Section of the "Astra" Association in Transylvania functioned uninterrupted between 1927 and 1947, eugenics societies in Southeastern Europe were ephemeral and without a broad geographical membership. If we can speak of a specific "Scandinavian model" for eugenics, "Latin" eugenics, or a particular German *Rassenhygiene,* can we then also argue for the existence of a specific Southeastern European eugenics model? Some scholars have argued that perhaps Romania should be included in the "Latin eugenic" movement based on Romanian eugenicists participating in the First International Congress of Latin Eugenics held in Paris in 1937.[46] But is this claim warranted? Existing research indicates that Romanians, like other Southeastern European eugenicists, were conceptually syncretic, fusing various eugenic models and ultimately producing an assorted eugenic philosophy tailored to their national milieu and professional concerns.[47] As much as their discourse on environmental factors, natality and puericulture evoked eugenic tropes suffusing French and Italian eugenics, Romanian eugenicists also believed in the specter of racial degeneration and lurking physical and mental defects threatening to incapacitate the race—those eugenic phantoms persistently trumpeted by German racial hygienists. Yet what differentiated Romanian eugenicists—and similarly engaged professionals and adherents of eugenic ideas from other Southeastern European countries—from their French and German counterparts was how they treated their own multi-ethnic societies as the center of eugenic reform.

Raising the issue of ethnic minorities has an additional significance. Historians of the region have not appreciated the importance of the minority eugenic traditions that were constituted against the dominant majority, like the Saxon eugenics in Transylvania. But in recognizing the significance of these biologized minority cultures in Southeastern Europe, one has a better chance to understand the majority's biopolitical anxieties about the future

46 This line of argumentation is offered by Nancy L. Stepan, *"The Hour of Eugenics": Race, Gender, and Nation in Latin America* (Cornell: Cornell University Press, 1996) and William H. Schneider, *Quality and Quantity: The Quest for Biological Regeneration in Twentieth-Century France* (Cambridge: Cambridge University Press, 2002). For a discussion, see Turda, "'To End the Degeneration of a Nation.'"

47 Marius Turda, *Eugenism şi antropologie rasială în România* (Bucharest: Cuvântul, 2008).

of their own nations. During the interwar years, the biopolitical meanings of eugenics were easily re-articulated into fears of absorption by (politically and numerically stronger) neighboring nation-states and by ominously growing minority enclaves. However, minorities preoccupied eugenicists to different degrees in each individual Southeastern European country. While in Bulgaria and Greece ethnic minorities were only a marginal concern for eugenic projects, if at all, the Romanian case is exemplary for developing eugenic agendas by both the Romanian majority and the ethnic minorities. Similarly, initial eugenic projects developed by Serb, Croat and Slovene scientists and politicians aiming to improve a common "Yugoslav race" were—especially in the late 1930s and 1940s—replaced by preoccupations with regionally specific agendas attached to their respective national centers: Belgrade, Zagreb and Ljubljana. As a general tendency, however, the entire population of the nation-state rather than minorities in particular was the focus of eugenic concerns.

The population was consistently portrayed in eugenic discourses as a biological entity whose natality, mortality, longevity and morbidity needed both regulation and supervision. Evincing a sense of an enduring threat, eugenic rhetoric ultimately produced an asymmetrical relationship between the image of the nation as existing through centuries and its members' biological finitude. But eugenic ideas of human improvement not only augmented forms of nationalism and anti-Semitism; more broadly they endorsed the modernization of state and society in Southeastern Europe. More research is needed to explore how health, social and population policies overlapped in eugenic projects in nationally specific cases or how several medical initiatives in the fields of population policy and eugenics gradually penetrated larger intellectual circles and associations, becoming in some cases a matter of national policy. In practice this meant that the eugenic importance of reproduction and heredity was often discussed when health laws were debated in parliaments or were taught in schools established for nurses or social workers.[48]

Existing scholarship on Southeastern European eugenics has concentrated on a very thin segment of the educated professional and political elite.

48 Dagmar Schule, "The History of Social Work in Eight Eastern European Countries from 1900–1960: An Overview," in Sabine Hering and Berteke Waaldijk, *Guardians of the Poor—Custodians of the Public. Welfare History in Eastern Europe, 1900–1960* (Opladen: Barbara Budrich Publishers, 2006), 121.

But while eugenic solutions to the nation's alleged biological deterioration were in fact initially advanced by the elites and later discussed by much of the urban middle class in Southeastern Europe, in the interwar period eugenic arguments were extended, to be drawn from low birth rates in urban centers when compared with birth rates in rural areas, from the growing numbers of workers and their precarious living conditions and in some cases, like in Romania and Yugoslavia, from the existence of ethnic minorities and political scapegoating, including racism and anti-Semitism.[49] Even when they did not profess it openly, by the late 1930s, most health reformers and hygienists in Southeastern Europe became fluent in the language of eugenics and the "racial sciences." These occurrences should, of course, be placed in the context of domestic and international instability, the cultural propaganda of the Third Reich in Southeastern Europe since 1933, and, not least, a World War. It bears repetition that all countries discussed in this volume were either allied to or occupied by Nazi Germany between 1940 and 1944.

Finally, much still remains to be investigated about the political ramifications of interwar and wartime relations between health, hygiene and eugenics. This volume aims to answer only some of the questions pertaining to these difficult topics. To approach health, hygiene and eugenics as international movements with specific regional and national characteristics invites us to examine the means by which these ideas were mobilized in Southeastern Europe. The chapters included to this volume suggest that if framing health, hygiene and eugenics in transnational and regional settings indeed needs to be a major aspect of research, exploring the complexity of these phenomena within each national society can reveal specific social, cultural and political dynamics. What this volume ultimately proposes as a future research agenda is that scholars should avoid the dichotomy "East" and "West" in their approach to the local context in Southeastern Europe. The case studies provided here help reconstruct precisely how such an international diffusion of health, hygiene and eugenic ideas and their implementation became possible and in some instances how geographical boundaries were challenged and ultimately rejected.

49 This general framework was shared by other European countries, as Peter Weingart observed in his comparison of eugenics in Sweden and Germany; see Peter Weingart "Science and Political Culture: Eugenics in Comparative Perspective," *Scandinavian Journal of History* 24, 2 (1999): 163–177. See also Marius Turda, *Modernism and Eugenics* (Basingstoke: Palgrave Macmillan, 2010).

## The Contributions to this Volume

Based on a wide rage of primary sources, the authors of this volume show that processes of health institutionalization were usually accompanied by attempts to educate the population in the values and principles of modern hygiene and eugenics. Furthermore, these contributions address the crucial problems of the porous border between care for the population and the repression of their individual rights. In particular during the interwar period, this antinomy assumes its clearest manifestation in visions and projects of a eugenically healthy nation and society. In order to reflect this dual quality in the health discourse, the volume is divided into two sections, one dealing with hygiene and health policies, the other with case studies of eugenic control over reproduction. Using different methodologies, the authors attempt to situate these issues within their regional and respective national context, hoping to enlighten issues related to the transfer of ideas as well as the correlation and interaction of international, national and local agencies.

The influence of German racial hygiene is addressed by Paul Weindling. This opening chapter is designed to serve as a point of reference for the eugenic paradigms developed in Southeastern Europe. Furthermore, the chapter places emphasis on the professional networks between experts from Germany and those in Southeastern Europe. In particular, Weindling evaluates how the multiethnic setting of the Habsburg Empire and its Southeastern European neighbors contributed to the foundation of German racial hygiene and how the latter radiated back into this broader geographical framework.

Some contributions in the first section on health politics and hygiene direct their attention to the intermediary position of Southeastern European health and hygiene between norms of Western medicine and the Balkan and/or Orientalist imaginations of 'backwardness', ethnic fragmentation and lack of hygiene. "Balkanism" and "Orientalism" are key notions in Brigitte Fuchs' and Christian Promitzer's chapters, which examine questions of hygiene and health politics in the case of the Muslim populations in Bosnia-Herzegovina and Bulgaria respectively. Fuchs integrates the Orientalist theme in a social history of medicine and gender relations by examining the extent to which the Bosnian population, and in particular its Muslim element (including women), was largely perceived by Habsburg administrators as "neurasthenic," "degenerated" and prone to syphilis. Promitzer deals with manifestations of "medical Orientalism" in the case of the Muslim minorities in Bulgaria. He looks at the way in which Turks, Roma and Pomaks were perceived as unwelcome reminders of the Ottoman legacy in Bulgaria and analyzes how typhus shifted from a war epidemic affecting Bulgaria between 1915 and 1920 to be perceived in the interwar period as a disease endemic to the Muslim population.

The other chapters in the first section show how cumbersome the relationship between diagnosed dissemination of disease and the establishment of a public health system in Southeastern European countries was in the early twentieth century. These case studies bring to light different strategies pursued by state politics, professional groups, international organizations or other agencies involved in the fight against epidemics, the establishment of welfare institutions and educating the population to adopt hygienic norms. In her chapter KATERINA GARDIKAS discusses these issues with respect to malaria treatment in Greece. Although state initiatives were set up with the assistance of international agencies, they were not able to replace the private provision that had spearheaded the fight against malaria since the nineteenth century, long before the public health reforms of the 1930s. KRISTINA POPOVA examines how medical efforts to reduce infant mortality in Bulgaria were fueled by the example of initiatives in Western and Central European countries and led to the development of national pediatrics. She also considers regional and national networks and initiatives emerging after World War I. As opposed to the other contributions, which mainly deal with medical issues in rural areas, LEDA PAPASTEFANAKI's chapter refers to urban centers in early twentieth century Greece. The author demonstrates how the health of the growing working class and especially the fight against tuberculosis increasingly became a matter of public and occupational health policies in Greece.

Apart from institutionalization processes, the contribution of outstanding physicians to the establishment of public health systems needs also to be assessed. Hence, in addition to a discussion of various health institutions established during the 1920s, the "Golden Era" of public health in Yugoslavia, ŽELJKO DUGAC examines the massive influence of the Croatian physician Andrija Štampar on the development of preventive and social medicine in the Kingdom of Serbs, Croats and Slovenes. As the author argues, Štampar's career serves as a paramount example to test the success of the "Yugoslav idea" in the sphere of public health in the interwar period.

The second section of this volume is dedicated to eugenic projects. The contributors examine various attempts to implement eugenic measures in several Southeastern European countries in order to control the reproduction of population. These took place in the spheres of marriage regulation, pro-natalist propaganda and child care. In doing so, they point at a multitude of agencies that were instrumental in articulating eugenic projects. Gergana Mircheva addresses the issue of marital health and eugenics in Bulgaria between 1878 and 1940. She outlines the normative guidelines on illnesses that formed the legal base for impediments to marriage and for divorce under Church law and Church judicial practice. These rules were later adopted in various eugenic approaches in Bulgarian health legislation. With respect to the Greek case, Sevasti Trubeta examines how the international rhetoric about 'overpopulation' and the perceived threat of "social diseases" were translated into the Greek context, leading to heated discussions on the implementation of eugenic measures among physicians, feminists, the Orthodox Church, politicians and even the legislative body. In contrast to Bulgaria, eugenic ideas in Greece were addressed primarily (although not exclusively) by liberal intellectuals and politicians. Greek eugenics is also discussed by Vassiliki Theodorou and Despina Karakatsani but from the perspective of childcare and pediatrics. Their main argument is that a form of eugenic "puericulture" developed in Greece that was influenced by French eugenics and embraced a "mild eugenic project" oriented towards "positive" rather than "negative eugenics." Far more radical forms of biological engineering in the form of "negative eugenics" in interwar Romania are discussed by Marius Turda. The author argues that controlling the national body during the interwar period was an ideal that Romanian intellectuals and politicians found attractive in their quest for a nation purified of both dysgenic individuals and any ethnic minorities deemed degenerate. Further cases of radical eugenic projects connected with fascist ideologies are considered by Rory Yeomans and Tudor Georgescu. Yeomans discusses the influence of eugenic ideals and pronatalism in the Independent State of Croatia (1941–1945) with respect to the drastic anti-abortion legislation passed by the Ustasha regime. He argues that even if negative eugenic policies regarding natalism were rejected by the regime, other forms of eugenics were certainly not excluded, as long as they were in accordance with the Catholic doctrine and could reverse the "foreign infiltration" in Croatian villages purportedly caused by the preceding Yugoslav regimes. Georgescu likewise focuses on fascist ideology and religion in his case study on the Transylvanian Saxon eugenic discourse in interwar Romania and on the activities of Alfred Csallner. Georgescu excavates the ideological and methodological interfaces between the German minority's Church institutions and the fascist experiment of its "Self-Help" or "Movement for Renewal" with national regeneration.

Despite their specific examinations of various case studies, all contributions provide a common framework for examining the transfer of medical ideas via local, national and international agencies. As Maria Bucur points out in her concluding chapter, this volume forces the reader to think comparatively, revising outdated assumptions about health, hygiene and eugenics in one of Europe's most frequently misunderstood regions. She stresses in particular the importance of regional political, cultural, social and economic contexts in understanding how ideas about health became translated into politics in Southeastern Europe, while she also keeps in mind the need for conceptual plurality necessary in order to frame the relationship between nation-state, society, modernity and the biologization of different forms of collective belonging—a relationship dominating regimes of health, hygiene and eugenics in Southeastern Europe until 1945.

# PART I
# GERMAN EUGENIC PARADIGMS

# RACIAL EXPERTISE AND GERMAN EUGENIC STRATEGIES FOR SOUTHEASTERN EUROPE

*Paul Weindling*

Until the 1980s eugenics was seen as historically marginal to mainstream German history. Race and reproduction were absent from discussions about the social consequences of Germany's rapid industrialization and its disturbed international relations. Since then, eugenics and racial hygiene have moved from being interpreted as only a facet of right-wing, *völkisch* extremism to being identified as central to twentieth century German history. Issues concerned with population policy, health and welfare policies, as well as the role of science and the professions emerged as part of a new wave of gender and social history. All these contentious topics arose in Germany's rapid social restructuring before 1945, as eugenicists tried to predict and reshape the profound consequences of modernization for their emergent nation. This chapter examines the broader consequences for European history, given German eugenics' move from the margins to mainstream German history.

## Historiographic Reflections

During the 1980s social historians resurrected Eckart Kehr's thesis of a primacy of domestic policy [*Primat der Innenpolitik*] in an effort to shift attention away from traditional national histories shaped by foreign policy and military concerns.[1] The point was that social questions, gender

1 Eckhardt Kehr, *Economic Interest, Militarism, and Foreign Policy: Essays on German History* (Berkeley: University of California Press, 1977).

and class appeared to be more important than power politics as determinants of power relations in Imperial Germany. In this context, a new interest arose in the politics of German public health and population policy before 1945. This reversed an earlier, wholly positive view on the development of German social welfare and public health provision—following the introduction of sickness insurance under Bismarck in 1883—as benign and inclusive. Eugenics, as linked to public health and population policy, was interpreted as a means for resolving socio-political tensions in Imperial Germany that arose from social deprivation, economic inequalities and rapid industrialization. World War I saw the family, reproduction and sexuality move to the center of state policy. The Weimar Republic's welfare state focused on reproductive controls for the eugenically "unfit" in addition to incentives for "child rich" families. The eugenic welfare state rapidly gave way to the Nazi "racial state", with selective welfare benefits for racially elite groups and the later extermination of the racially undesirable from 1939. In the postwar period of "the two Germanies," family politics remained center stage, albeit with a contrast between the conservative moral frame of Adenauer's Federal Republic and the socialist focus on youth sexuality—and having children young while working—in the German Democratic Republic.

Generally speaking, the study of the history of eugenics was transferred across to Germany from studies of Anglo-American Social Darwinism and eugenics. Social interests in scientific knowledge—indeed, the social construction of that knowledge base—and in professional power were key issues. By the 1970s, historians became interested in civil rights for the oppressed and in gender politics. By contrast, German historians were interested in issues like mass political mobilization or the corporate interests of the military, industry and labor, meaning that the role of professional power and the knowledge base of the industrializing economy appeared very marginal to them. In 1983, I demonstrated the process by which eugenic policies permeated state public health in Prussia between 1905 and 1933.[2] Yet only in the early 1990s did readings of Michel Foucault

2 Paul Weindling, "Die Preussische Medizinalverwaltung und die 'Rassenhygiene'," in A. Thom and H. Spaar, eds., *Medizin und Faschismus*, new edn. (Berlin: Verlag Volk und Gesundheit, 1985), 48–56. Extended version published as "Die Preussische Medizinalverwaltung und die 'Rassenhygiene'. Anmerkungen zur Gesundheitspolitik der Jahre 1905–1933," *Zeitschrift für Sozialreform*, 30 (1984): 675–687.

(taken up by Detlev Peukert) and the critique of state welfare—evident in the work of Michael Burleigh and Wolfgang Wippermann—fuel writings on "the racial state," which has now become an accepted paradigm for historical studies of German eugenics.[3] The problem, though, remains how to conceptualize and interpret features of German eugenics that fall outside the frame of state social policy—not least the culturally distinctive concerns of German racial scientists.

The extent of influence by German eugenics beyond the borders of the Reich remains similarly unresolved. On the one hand, we have the issue of Germany's place in Southeastern Europe; on the other, that of a threat from the East in terms of "alien races and their diseases." Whether or not German eugenic models were transferred to ethnic German communities at this time or whether each geographical context was distinctive in terms of differing eugenic models remains an open question. The prevailing *kleindeutsch* approach is nevertheless too restricted, particularly in its failing to take account of the broader vision of a Greater German racial hygiene regime espoused by its many proponents. Protagonists at the time certainly refused to accept the restrictions of national borders. Here the nation-state as a conventional historical category obscures significant transnational issues, not least the analysis of *völkisch* national associations and local initiatives in public health provision for *Volksdeutsche*. These continue essentially to be seen as pressure groups extending Imperial German state policies into the spheres of welfare, physical and mental fitness as well as reproductive health and fertility. Studies of *völkisch* ideologues, such as of the earlier Pan-German League agitating for imperial expansion, do not adequately allow for the racial-political and scientific networks among German eugenicists.[4] We therefore need to look carefully at the place of eugenicists among the politics of mass public health organizations, notably in combating alcoholism, for preventing sexually transmitted diseases and in promoting infant welfare and maternal health. From a racial-biological perspective, the domestic/foreign policy divide is a wholly artificial one.

3 Detlev Peukert, "The Genesis of the 'Final Solution' from the Spirit of Science," in Thomas Childers and Jane Caplan, eds., *Reevaluating the Third Reich* (New York: Holmes & Meier, 1994), 234–252. Originally published as "Die Genesis der "Endlösung' aus dem Geist der Wissenschaft," in Detlev Peukert, ed., *Max Webers Diagnose der Moderne* (Goettingen: Vandenhoeck & Ruprecht, 1989), 102–121.

4 Roger Chickering, *We Men who Feel Most German. A Cultural Study of the Pan-German League 1886–1914* (London: George Allen and Unwin, 1984).

At the time, it was perceived as necessary, within the paradigm of defending racial health, to sustain the scattered German communities in Central and Southeastern Europe.[5] To achieve this, there were broadly two conceptions. One was inherited from nineteenth century Austrian ideas of a Central European confederation. Here, Germany would work alongside other states and ethnic groups. Thus the liberal Friedrich Naumann (1860–1919) favored the scheme of a German-led confederation, but one including Austria-Hungary and the South Slavs. The second idea was that Germany should dominate and control *Mitteleuropa*. This appealed to ultranationalist annexationists in World War I, who saw the opportunity for a colonial Empire in the East. Indeed, *Mitteleuropa* underpinned the ideas of German *Lebensraum*.

As a term, *Mitteleuropa* was inherited from Austrian social thinking, and it encompassed a variety of geographical definitions. For his part, Adolf Hitler adopted a Germanic view, rejecting Habsburg federalism in favor of "living space" for Germans in the area to Germany's east, from Poland to the Ukraine. Other conceptions were oriented toward the Austro-Hungarian sphere—either north of the Danube or, potentially, including not only the Adriatic ports but stretching south-eastwards, covering the Balkans and reaching to the Black Sea. During World War I, both annexationists and advocates of a Germanic confederation saw *Mitteleuropa* as extending to Southeastern Europe. *Mitteleuropa* consequently should not be interpreted too narrowly under either the confederate or annexationist models.

For racial hygienists, there was consensus that the borders of the Reich were too narrowly set. By the 1920s, the key issues were:

- Should German-speaking territories (such as German Austria) be added to a Third Reich?
- Should contiguous lands lost under the Treaty of Versailles be restored?
- Should lands with Slav populations be annexed under German domination (as advocated in World War I)?

---

[5] J. Brechtefeld, *Mitteleuropa and German Politics, 1848 to the Present* (London: Macmillan, 1996). W. O. Henderson, *Friedrich List: Economist and Visionary 1789–1846* (London: Frank Cass, 1983); and F. List, *Das nationale System der politischen Ökonomie* (Stuttgart: Cotta, 1841).

- Should some form of confederate model be developed (as liberal/center-left eugenic thinkers preferred) with the successor states? Here one might examine the Austrian sociologist Rudolf Goldscheid (1870–1931) and center-left figures in Germany like the advocate of social gynecology Max Hirsch (1877–1948).

These questions existed alongside considerations of settlement policy and community organization, geopolitics and living space. Varying evolutionary and hereditary biological theories underpinned the geopolitical conceptual armory. The Darwinian zoologist Ernst Haeckel's (1834–1919) ideas of the superior racial qualities of the scientific outlook, for example, were ambivalent: what if racially non-Germans (not least Jews) espoused these superior cultural values?[6] Likewise, Wilhelm Schallmayer's (1857–1919) idea of a *Volkseugenik* was a sociological model for a eugenic state, leaving it open as to whether he was writing about Germany or other so-called *Kulturvölker*.[7] Here he related "cultural level" to decreasing infant mortality, an issue bearing directly upon Central and Southeastern Europe, where infant mortality rates were relatively high.[8]

Naumann prioritized the superior virtues of German cultural and economic values, investing them with the right to provide an integrating structure for Central and Southeastern Europe. For instance, he noted the positive qualities of the scattered German communities in Transylvania. Such eugenicists saw a variety of alternatives, from supporting these German exclaves with advice and resources to bringing back these populations, whose German identity was in part fragile, within the fold of the Reich. Expanding the Reich's borders was only an option when taken together with an aggressive militarism that would, ironically, mean sacrificing the nation's best breeding stock. Alfred Ploetz (1860–1940), a key medical advocate of racial hygiene, spoke about regenerating the fitness of youth and seeking to restore its primitive vigor.[9]

---

6 Ernst Haeckel, *Naturliche Schopfungsgeschichte* (Berlin: Reimer, 1909), 807–812.

7 Wilhelm Schallmayer, *Vererbung und Auslese* (Jena: Fischer, 1910), 174–178.

8 Schallmayer, *Vererbung*, 241–252.

9 Paul Weindling, *Health, Race and German Politics between National Unification and Nazism 1870-1945* (Cambridge: Cambridge University Press, 1989).

Ploetz saw this form of regeneration of the primal racial stock as palingenetic renewal. The term was originated by Ernst Haeckel to denote a recapitulation of primal forms, but has recently been deployed by Roger Griffin and Marius Turda in defining fascism and certain tendencies in modernism and eugenics.[10] Yet Ploetz until ca. 1900 advocated socialism on a biological basis. Haeckel, in turn, was a National Liberal in opposition to democratic liberals like Rudolf Virchow (1821–1902), who advocated a centralizing state. By the time of Haeckel's death in 1919, he was a supporter of annexationist rights, rallying behind Erich Ludendorff (1865–1937) and the powerful military High Command. They hoped for the acquisition of Eastern and Southeastern territories, but instead found themselves deploying the *Freikorps* to protect the defeated Reich.

Eugenics thus has a political ambivalence: while renewal through recovery of the primal vitality of a race was a central characteristic of generic fascism, eugenics and racial hygiene appealed across a broad political spectrum. If Ploetz spoke of socialism on a biological basis, Wilhelm Schallmayer advocated a system of medical surveillance with the physician acting as state official. Clearly, the palingenetic idea of rebirth and renewal was not in itself fascist, although it could easily be aligned with incipient National Socialism. Indeed, between 1900 and the early 1920s, Ploetz kept racial hygiene at a distance from *völkisch* racial schemes. What was clearly shared was a community-building ethic, one where the basis of the community and its political structuring were to be defined by a set of biological qualities. Yet, the inner Nordic convictions held by Ploetz and Fritz Lenz (1887–1976) meant a reserve towards those of Jewish, Greek, Latin or Slav ancestry in the inner core of German racial hygienists.

Eugenicists presciently diagnosed the problems of modernity, which was seen in fatalistic terms, although with uncertain social and demographic consequences. There was also the declining birth-rate and the shift of attention toward chronic degenerative diseases. While eugenicists accepted the desirability of fewer children of better quality, they saw the modern woman's demands for control over her own reproductive autonomy as a threat to the race. The birth-rate problem involved inculcating a sense of moth-

10 Roger Griffin, *Modernism and Fascism: The Sense of a New Beginning under Mussolini and Hitler* (Basingstoke: Palgrave Macmillan, 2007) and Marius Turda, *Modernism and Eugenics* (Basingstoke: Palgrave Macmillan, 2010).

erhood in women of sound, child-bearing stock. The focus upon chronic diseases similarly involved a set of disturbed social conditions with degenerative physical consequences. As a result, combating of syphilis and alcohol as "racial poisons" became priorities.[11] France offered a model for such changes, given that country's rapid decline in fertility. A renowned German public health physician, Hans Harmsen (1899–1989), also studied this problem in some detail.[12] Eugenicists proposed models applicable to the scattered, ethnic German communities in Eastern and Southeastern Europe, as well as to domestic welfare policies more generally.

The German welfare state in its relations to family policy has been the main context for analyzing the impact of eugenics. The historical concept of a racial state has conveniently been the framing and defining paradigm for eugenics. As noted earlier, the authors of the racial state paradigm, Michael Burleigh and Wolfgang Wippermann, were well aware of the German sense of a historic mission to conquer the East, but considered this more a case of stamping German national identity on Eastern territories.[13] The transnational history of race as a cultural concept is under-explored, although there are a few pointers in this interpretative direction. Niall Ferguson in his *The War of the World* claims that racial politics is the central engine of twentieth-century history. Ferguson rightly observes the tension between areas of culturally and ethnically mixed settlements on the one hand and the nation-state established on the basis of national self-determination on the other. But Ferguson glosses over how race was defined as a category, not to mention the role of eugenic public health in the machinery of state-building. Hitler perceived the Jews as a "biological threat," but Ferguson leaves the methods and modes of this definition ill-defined, not appreciating how even biological anti-Semitism contained varying formulations in its poisonous ideology.[14] Here we need to contextualize varying conceptions of race and to develop interpretations based on professional expertise, scientific networks, medical power and the welfare state,

11 Ploetz Papers, Max-Planck-Institut für Psychiatrie, Munich, 'Denkschrift über die Gründung einer Internationalen Gesellschaft für Rassen-Hygiene'.

12 Hans Harmsen, *Bevölkerungsprobleme Frankreichs, unter besonderer Berücksichtigung des Geburtenrückganges* (Berlin: Vowinckel, 1927).

13 Michael Burleigh, *Germany Turns Eastwards. A Study of Ostforschung in the Third Reich* (New York: Cambridge University Press, 1988).

14 Niall Ferguson, *War of the World: History's Hatred of Revenge* (Harmondsworth: Penguin Books, 2007), 256.

as against the popularization of hygienic ideals as part of the politics of national identity.

In the 1980s, the history of Social Darwinism and eugenics reached German history less as an interest in the origins of Nazism and the implementation of the Holocaust than as a counterpart to studies into Anglo-Saxon eugenics. Historians of science were engaging with eugenics as an extension of Social Darwinism and focusing on genetics and its social implications in addition to the social construction of scientific representations. The first wave of social historians of medicine added eugenics and welfare to this list of considerations. Yet the link to mainstream Holocaust history has been—and remains—problematic since eugenics and Nazism had to be brought into alignment despite their very different social origins. A breakthrough came when Gisela Bock showed how compulsory sterilization became a Nazi measure, prompting studies of euthanasia, which began to gather momentum from the mid-1980s.[15] Nevertheless, attempts to tightly link Social Darwinism to Nazi race policy remain problematic.[16] Again, as part of Nazi propaganda and occupation policies, the idea of race requires a fuller consideration.

Approaches to German eugenics need a fuller integration into its proper, international context. As such, a number of dimensions come into play:

1. German eugenics as a response to American, British and Scandinavian influences. A critical issue is that of the transfer of sterilization techniques, notably the Californian and Danish precedents for Germany's Sterilization Law of 1933.
2. Nordic racial networks linking groups of eugenicists in Germany, Scandinavia and the United States. Here membership in the International Society for Racial Hygiene is a key source for the period prior to 1914.
3. Germany's sanitary defense of its borders against Eastern European Slav and Jewish migrants.

---

15 Gisela Bock, *Zwangssterilisation im Nationalsozialismus* (Opladen: Westdeutscher Verlag, 1986).

16 Paul Weindling, "Dissecting German Social Darwinism: Historicizing the Biology of the Organic State," *Science in Context* 11 (1998): 619–637.

4. German revulsion against decadent French modernism. Here, the specter of a declining birth-rate threatened ideals of the "child rich" German family.
5. Far less investigated than Nordic racism is how "Greater German" racial hygiene and eugenics worked across lands bordering Germany, including Austria-Hungary and Switzerland, the border territory of Alsace-Lorraine and ethnically mixed communities such as Danzig/Gdansk.
6. Sustaining expatriate German communities in Eastern and Southeastern Europe.
7. German overseas colonial territories and areas of settlement.
8. The question of scientific/medical transfer: were German eugenic models (as opposed to, for example, Latin models), adopted by national-racial, eugenic and ethnic movements in Central and Southeastern European nation-states? This perspective allows for the regional autonomy of national and ethnic eugenic movements as well as for diverse and competing strands. Such a view also accords with what is known about Austrian, Czechoslovak, Hungarian and Romanian eugenicists.

The above dimensions, however, remain relatively unexplored. There is a real need to delve into the national and local histories of eugenics—like in Austria in general, in addition to specific localities like Vienna and Graz. Swiss cantons likewise offer a history of a complex set of eugenic measures. Here, Stavros Zurukzoglu (1896–1966), the Greek advocate of sterilization based at the University of Basel, is highly relevant as advocating sterilization in a modernist social framework. Often, localities were formative, such as in Zurich, where Ploetz and Ernst Rüdin (1874–1952) studied and where the psychiatrist August Forel (1848–1931) applied heredity to understanding psychoses. These peripheral phenomena raise questions about the history of German eugenics more generally. Eugenic and racial issues clearly operated in conjunction with *völkisch* German aspirations; but that said, while the *völkisch*-national and the eugenic approaches may overlap, they have their own dynamics. Rather than eugenics being regarded as done "to" or "against" other Central and Southeastern European countries after 1933, German racial/social eugenics emerges most strongly from the peripheries. We shall now advance from the situation in German historiography 30 years ago (more or less predating the work of historian George Mosse), when racial policy was seen exclusively in terms of a movement of the ultra-right. These broader parameters raise central issues both in Southeastern European history and international history more generally, which should also be borne in mind.

## Origins

The *Gesellschaft für Rassenhygiene*—the first eugenics society worldwide—was founded in Berlin on 22 June 1905 in the Steglitz apartment of the ethnologist Richard Thurnwald (1869–1954) located in the Fichtestrasse 47/III. Thurnwald was born in Vienna and studied social institutions in Bosnia in the 1890s. He maintained Viennese links through his correspondence, particularly when researching in German New Guinea or when, in the 1920s, he lived in the United States, where he keenly observed racial segregation.[17] He remained secretary of the Racial Hygiene Society until resigning in 1910.[18] Another key Austrian figure in this Berlin nucleus was the distinguished anthropologist Felix von Luschan (1854–1924), who joined the nascent Racial Hygiene Society early in 1907; by 1909 he was named chair of anthropology in Berlin.

The *Gesellschaft für Rassenhygiene* was broadly *Grossdeutsch* in orientation in so far as it involved Swiss nationals like Ernst Rüdin, Austrians like the Munich professor of hygiene, Max von Gruber (1853–1927) and the Lamarckian hygienist, Ignaz Kaup (1870–1944), who admired Georg von Schönerer (1842–1921) of the Austrian ultra-right. The nascent Society also had international pretensions, seeking to develop federated societies in France and Sweden. Thurnwald and the Viennese anthropologist Rudolf Pöch (1870–1921) had come to know Ploetz through the anti-alcohol movement.[19]

17 Ploetz Papers, Diary. Marion Melk-Koch, *Auf der Suche nach der menschlichen Gesellschaft: Richard Thurnwald* (Berlin: Museum für Völkerkunde, 1989), 44–7. Ploetz Papers. Diary. Thurnwald's address was Fichtestrasse 47/III.

18 Luschan Papers, Ploetz to Luschan 21 January 1910.

19 Rudolf Pöch, "Rassenhygiene und ärztliche Beobachtungen aus Neu-Guinea," *Archiv für Rassen- und Gesellschaftsbiologie* 5 (1908), 46–66 and Heinrich Reichel, "Rudolf Pöch als Arzt," *Wiener klinische Wochenschrift* 14 (1933) (cited as offprint).

Medical and biological concerns among temperance campaigners thus linked German, Austrian and Swiss eugenicists at the turn of the century. The founding of the *Gesellschaft für Rassenhygiene* therefore reveals a deeper Greater German context: not only Swiss but also strongly Imperial Austrian components. Anthropological and lifestyle-reforming elements fused within early "German" racial hygiene; moreover, these connections take us to the Eastern fringes of Europe, as well as to so-called "primitive" cultures overseas.[20]

The foundation of the *Gesellschaft*—renamed the *Internationale Gesellschaft für Rassenhygiene* in 1907—raises important issues as to the meanings and influence of racial hygiene in Central and Southeastern Europe. The composite term *Rassenhygiene* first involved the component of "race" as the intersection of anthropology with varieties of race theory: physical, psychological and cultural. The second part of this double-barreled term, "hygiene," primarily referred to the science of public health and prevention of pathogens. Ploetz took as his reference point the prestigious medical discipline of hygiene, equating the bacteriologist Robert Koch's prevention of infectious pathogens with measures to stop the spread of degenerative hereditary factors. Ploetz's point was that racial hygiene should be an integral part of preventive medicine; it should have scientific status and prestige equivalent to that of bacteriology. Ploetz further stressed that the dissemination of hygiene provided a potential basis for including racial hygiene as a sub-discipline within it.

Assorted members of the International Society included various Swiss and Austrian nationals. Although there was no Austrian or Swiss chapter, the Racial Hygiene Society unmistakably had an influence in these countries. Ploetz was careful to keep the Society open to a range of eugenic strategies and scientific agendas. In 1909 the Racial Hygiene Society rejected a proposal to exclude all but "Nordic" racial members, preferring instead a more inclusive policy for other "white races."[21] That meant that Jews, Slavs and Southern Europeans could find a place in the federated International Society.

Yet the situation remained more complex since Ploetz was forced to steer a course between his Nordic inclinations and the ideal of an inclu-

20 Paul Weindling, "A City Regenerated: Eugenics, Race and Welfare in Interwar Vienna," in Deborah Holmes and Lisa Silverman, eds., *Interwar Vienna: Culture between Tradition and Modernity* (New York: Camden House, 2009), 81–113.

21 Ploetz Papers 4. Bericht der Internationalen Gesellschaft vom 3. bis 18 März 1909.

sive science. As a result, Vienna turned out to be a problematic situation. In May 1909 Ploetz wrote to the British eugenicist Karl Pearson (1857–1936), indicating that an Austrian would be joining the committee of the International Society.[22] Count Hans Wilczek (1837–1922)—a member of the Society since April 1908—had agreed to found an Austrian group, along with the economist Michael Hainisch (1858–1940) and the eminent professor of pathology and fervent temperance campaigner, Anton Weichselbaum (1845–1920). The latter three joined the Society on 15 February 1909.[23] Hainisch was already a member of the *Wiener Fabier Gesellschaft* (modeled on the Fabian Society, but with a pronounced national focus). As *Bundespräsident* of the new Austrian Republic from 1920–1929, Hainisch offered a point of entry for eugenics into the interwar welfare state.[24] Rudolf Goldscheid further linked eugenics to social science, developing a theory of "human economy." But then matters spun out of control. Goldscheid established a section (really, an embryonic society) for Social Biology and Eugenics in the Sociological Society in Vienna, founded in 1909. The secretary was Paul Kammerer (1880–1926), the Lamarckian zoologist and socialist.[25] Julius Tandler (1869–1936), newly-appointed professor of anatomy in Vienna, was the chair of this society. On 20 March 1914, the society met to discuss Tandler's views on "Constitution and Condition."[26] For Ploetz the problem was that the nascent Viennese society was becoming dominated by Jewish intellectuals—the sort who might join a group for Sexual Science and "Eugenetics" as organized by Max Hirsch of the *Archiv für Frauenkunde und Eugenetik* in Berlin.

On 12 February 1909, Ploetz had a six hour discussion with Goldscheid in Vienna.[27] Goldscheid was keen to maintain relations with Ploetz but—and this is the puzzle—Ploetz withdrew from Vienna and instead chose to cultivate links to Nordic Scandinavia. One explanation is that the Viennese academic milieu appeared overly Jewish to Ploetz in terms

22 Ploetz Papers. Ploetz to Karl Pearson May 1909.

23 Ploetz Papers, Mitglieder der G.f.R.H., [1905–1907]; Internationale Gesellschaft für Rassen-Hygiene. Mitgliederliste vom 10 December 1909. Ploetz Diary 1909. See also Anton Weichselbaum, *Die gesundheitsschädlichen Wirkungen des Alkoholgenusses* (Vienna: Philipp, 1905).

24 Michael Hainisch, *75 Jahre aus bewegter Zeit* (Vienna: Böhlau Verlag, 1978), 109–110.

25 Albrecht Hirschmüller, "Paul Kammerer und die Vererbung erworbener Eigenschaften," *Medizinhistorisches Journal* 26 (1991) 26–77.

26 Staatsbibliothek Preussischer Kulturbesitz Berlin, Max Hirsch M 6084.

27 Ploetz Papers, Diary. 12 February 1909.

of the racial background of intellectuals attracted to eugenics. The *Eugenics Review* recorded the founding of an Austrian Society, but Ploetz wished only to work with a select number of Viennese eugenicists.[28] Nonetheless, Goldscheid's grouping amounted to the first Vienna-based eugenics association, and while it was limited to discussion at a conceptual level, the Society nonetheless acted as a focalization point. Expectations for the founding of an Austrian eugenics society were—as Marius Turda observes—in and of themselves influential in the Dual Monarchy.[29] By 1914 the *Deutsch-Österreichische Beratungsstelle für Volkswohlfahrt* [German-Austrian advisory centre for welfare] (in the Lagergasse) had become a centre for eugenic information.[30]

Potentially confronted by the specter of too many Jewish eugenicists, Ploetz balked at founding a Vienna chapter. Instead, the Austrian group developed into a Society for Population Policy. Thereafter we find a number of individuals with a Moravian background active in eugenics, not least Hugo Iltis (1882–1952), the Mendel biographer based in Brunn/Brno. For Ploetz the risk was that Viennese eugenics would take a left-leaning and Jewish stance. Instead, Ploetz worked selectively with non-Jewish Austrians like the bacteriologist Heinrich Reichel (1876–1943), who dedicated his life's work to eugenics despite never joining the Nazi Party. Significantly, Iltis had aligned with Kammerer in developing a socialist Central European eugenics in Vienna and Brno. Ploetz was always concerned to maintain the breadth of the International Society, trying to avoid restricting it to any one Nordic or *völkisch* grouping. This meant that, although the Eastern and Southeastern profile was less pronounced, there was still a scattering of such elements within the Society. By way of example, Géza von Hoffmann (1885–1921) represented the Berlin group at a meeting of the German Society for Racial Hygiene on 29 July 1917 at the Munich Hygiene Institute.[31]

28 "Eugenics in Austria," *Eugenics Review* 5 (1913/14): 387

29 Marius Turda, "The First Debates on Eugenics in Hungary, 1910-1918", in Marius Turda and Paul Weindling, eds., *'Blood and Homeland': Eugenics and Racial Nationalism in Central and Southeast Europe, 1900–1940* (Budapest: Central European University Press, 2007), 185–221.

30 Géza von Hoffmann to Harry Laughlin, 26 May, 1914. Available at www.eugenicsarchive.org (Last accessed on 1 July 2008)

31 Lenz Papers [copies made available to author by Widukind Lenz], 6. Also "Aus der Gesellschaft für Rassenhygiene," *Archiv für Rassen- und Gesellschaftsbiologie* 12–13 (1916/17): 403–404.

Another element of relevance here is the concern with German colonies and the building of an Empire on a racial basis. Eugen Fischer (1874–1967) and Rudolf Pöch were among the anthropologists who studied primitive peoples. The Racial Hygiene Society whose members had to be from "white races," had several colonial medical officers as members and favored colonies of high eugenic worth. Colonial race theorists—notably Ludwig Külz (1875–1938) and Philaletes Kuhn (1870–1937)—however, were forced to return to Germany in the 1920s. The loss of the colonies prompted a new stress on *Lebensraum* and racial health, entailing an orientation towards the East in interwar Germany.

We also see efforts to mount a scientific critique of predominant trends in bacteriological hygiene by pointing to environmental and racial factors in susceptibility to disease.[32] The dissident bacteriologist Ferdinand Hueppe (1852–1938) is a prime example. His environmental-based views on infection went hand in hand with an interest in race; for him Greeks had descended from the Aryans. At the same time, as a sense of the Germanic race, its history, racial features and psychology began to be outlined, a biomedical program evolved. The centripetal problem was racial survival in the face of modernity, which was soon compounded by the lost First World War and interwar traumas. Indeed, the way in which the racial and sociobiological were interlinked has yet to find an adequate historical reconstruction.

Currently, dominant historiography stresses that ideas of German racial superiority were imposed on other European countries by the Third Reich. It is all too easy to identify the expansion of race theory with the expansion of Nazism or to argue that more peripheral locations had no substantive interest in eugenics.[33] And even if one sees eugenic movements as developing in parallel in cities like Vienna and Zurich, the assumption is always that a standardized German model of racial hygiene was replicated. In reality, race and the racialization of hygiene, health and medicine were contested and open to a multiplicity of interpretations. Eugenics was central to welfare and social care from the late nineteenth century onwards and

32 Paul Weindling, "Typhus and Scientific Racism: A Social Constructionist Approach," in W. Ernst and B. Harris, *Race and Modern Medicine* (London: Routledge, 1999), 218–234.

33 Dieter Kuntz and Susan D. Bachrach, eds., *Deadly Medicine: Creating the Master Race* (Washington, D.C.: United States Holocaust Memorial Museum, 2004).

racial hygienists were often prescient in their analysis of social, medical and population trends. They foresaw the shift to chronic diseases and small families—even though they attempted to ameliorate undesirable aspects through medical intervention. New sources have become accessible, and there have been some excellent case studies, various overviews and increasing recognition of the importance given to eugenics and race. Yet the assimilation of eugenics into wider historiographic discourse has been at the cost of inaccuracy, oversimplification and distortion. This has implications for interpreting race in the whole of Southeastern Europe and—I will argue—taking account of the peripheries has serious implications for an understanding of German eugenics.

A broader international awareness shaped the views of German eugenicists. This may be illustrated by the case of veteran eugenicist, Hans Harmsen, who was affiliated with the Protestant welfare organization, *Innere Mission.*[34] French population policy was, for Harmsen, the situation that Germany had to avoid. He advocated the German Mothers' Day in the Weimar Republic. Much later, in the 1950s, he took a close interest in Eastern European health and welfare while administering the *Pro Familia* birth control organization. Harmsen demonstrates the strong eugenic continuities through the upheavals of twentieth-century German history.

In reaching such conclusions, a seismic shift occurred in 1991 thanks to the introduction of the concept of the "racial state" by Burleigh and Wippermann.[35] Race had replaced class as the key issue in studies of Nazi policy. The prevailing historical approach to Hitler's rule—on officials, on the mobilization of the Nazi vote and on the political economy of the *Volksgemeinschaft*—was exposed as overly restricted in its focus. A new synthesis covered a range of racially persecuted groups including not only Jews but also Sinti and Roma and policies against homosexuals and single mothers in addition to the targeting of "anti-socials." Added to the ranks of victims were forced laborers, the "work-shy" and those deemed to be hereditary criminal.

These two former mediaevalists and scholars on the Teutonic Knights saw the close linkage between Nazi social policies and Germanization pol-

34 Sabine Schleiermacher, *Sozialethik im Spannungsfeld von Sozial- und Rassenhygiene: Der Mediziner Hans Harmsen im Centralausschuss für die Innere Mission* (Husum: Matthiesen Verlag, 1998).

35 Michael Burleigh and Wolfgang Wippermann, *The Racial State: Germany 1933–1945* (Cambridge: Cambridge University Press, 1991).

icies in the occupied East. Burleigh and Wippermann's lucid and sophisticated path-breaking analysis provided a new framework for the understanding of genocidal impulses in twentieth century medicine. Yet trends in the modern historiography of the Third Reich and the realization of its racial utopia continue to blandly describe racial hygiene as an overwhelmingly German and *völkisch* movement. Instead, what we have is a very literal instrumentalization of scientific and medical categories by political elites.

The production, forms and application of racial, eugenic and hygienic knowledge—and not least their institutional manifestations—represent a potential key to this area of enquiry. A dense network of institutions focused on local population and racial studies, notably anthropometric and serological research centers, arose across interwar Europe. Yet the agencies and agents of Nazi race policy, in addition to the ways in which race was a contested category under National Socialism, have only been studied superficially. Despite excellent studies of the *Rassepolitisches Amt, SS Ahnenerbe,* and in particular, Heinemann's important study of the SS *Rasse- und Siedlungshauptamt* [RSHA], available analyses do not probe the medical and scientific rationales to any great extent.[36]

Yet the racial hygienist and human geneticist Fritz Lenz advised the SS on criteria for camp selection and RSHA officers took courses at the *Kaiser-Wilhelm-Institut für Anthropologie, menschliche Erblehre und Eugenik* [KWI for Anthropology, Human Heredity and Eugenics]. And if we examine how one SS RHSA officer, Hans Poppendiek/Poppendick—who attended the KWI course and remained close to Lenz—came to be tried at Nuremberg for his eugenic and racial activities, the details of his actions within this organization remain shadowy and unexplored. How social/racial anthropology developed within international networks, and the overall politics of anthropological knowledge produced by these networks, remain under-researched. The human economy of forced labor and ghettoization could fit with the Nazi ideals of promoting Germanic racial fitness. While some rhetoric of racial hygiene, cleansing and so on is invoked, these motives have rarely been

36 Michael Kater, *Das 'Ahnenerbe' der SS 1935-1945: Ein Beitrag zur Kulturpolitik des Dritten Reiches* (Stuttgart: Deutsche Verlags-Anstalt, 1974); Isabel Heinemann, 'Rasse, Siedlung, deutsches Blut'. *Das Rasse- und Siedlungshauptamt der SS und die rassenpolitische Neuordnung Europas* (Goettingen: Wallstein Verlag, 2003); and Claudia Koonz, *The Nazi Conscience* (Cambridge: Harvard University Press, 2003).

seriously examined in recent historiography. One would at least like to know what the criteria of a "*rassische Auslese*" as practiced by RSHA were.

What restricts analyses of the racial state paradigm is that the state is the central, determining social entity, while race remains of secondary importance. Thus race has remained immune from rigorous historical scrutiny and its mobilizing power and social effects continue to be largely obscured. That race was theoretically unstable and politically contested between a range of factions in Nazi Germany is seen as only of limited, marginal interest to esoteric coteries of intellectual and medical historians.

More generally non-medical historians tend to:

- Instrumentalize science and its associated cultural formations.
- Neglect their meaning and incentive, as well as the parameters and dynamics underpinning the varieties of racial theories and policies.
- Neglect disciplines like anthropology, genetics and serology as technical and insignificant in relation to broader political and ideological issues.
- Take no interest in processes of scientific and medical research as socially significant—despite large-scale state funding for groups like the *Kaiser Wilhelm Gesellschaft* (KWG) and the *Deutsche Forschungsgemeinschaft* (DFG) under National Socialism.
- Ignore eugenic policies towards to the *Auslandsdeutsche* and the place of racial studies in the *völkische Wissenschaft*.

The ends become the explanation and those parts that do not fit are omitted. If we take the example of sterilization, the emphasis has been on operations carried out in Germany under the 1933 legislation. Illegal operations, X-ray sterilization and sterilization in annexed German territories, however, are far less documented. The figures shaping and transmitting knowledge of American sterilization policies require attention. Thus, Géza von Hoffmann, the Hungarian consul in Berlin, is seen as a channel of communication for North American ideas on sterilization to German racial hygienists. Hoffmann was active in the *Berliner Gesellschaft für Rassenhygiene* during the First World War. Marius Turda has commendably reconstructed Hoffmann's career, which includes a set of delineated eugenic programs and an active role in Hungarian racial politics. Hoffmann was secretary of the Berlin society from 1916 until July 1917, at which point he was recalled to assist the new Hungarian welfare ministry.[37] The Berlin group, moreover, took a close interest in the development of population policy and infant welfare in Austria, Bohemia and Hungary. But this is not surprising considering that the *Mitteilungen der Berliner Gesellschaft für Rassenhygiene* contain much information on Hungarian developments in the field of population policy.

I would suggest that the following factors may be seen as helpful for shaping patriotically-mobilized academics:

1. The sense of a defensive reaction—that German identity and the Germanic races (however defined) were under threat of extermination.
2. A desire for racial rejuvenation (*Erneuerung*)—namely restoring a primal vigor to the German race—as an antidote to racial emasculation.
3. The search for an alternative form of modernity.
4. The aim of eliminating a parasitic species. Here biological and bacteriological notions of total extermination were wholly transposed to human races. Although the process of extinction was known, actually realizing extermination through the wholesale elimination of a pathogenic species was a fundamentally novel idea in the 1930s.

## A Hygienic *Mitteleuropa*?

The histories of "racial hygiene" and *Mitteleuropa* are not conventionally associated with each other, and the central progenitors of the terms, like Ploetz and Naumann moved in different spheres. Naumann's *Mitteleuropa* was not initially eugenic or racial, despite his defining German interest in Central Europe while projecting a sense of superior German moral and psychological values.[38] Translated into a German economically-dominated Central Europe, Naumann included Austria-Hungary in referring to the

37 Marius Turda, "The Biology of War: Eugenics in Hungary, 1914–1918," *Austrian History Yearbook* 40, 1 (2009), 238–264.

38 Friedrich Naumann, *Mitteleuropa* (Berlin: Georg Reimer, 1916). Translated as *Central Europe* (New York: Knopf, 1917).

"South Slavs." He only excluded Greece, Serbia and Romania on the basis that they were still unproven nations which might, in time, join the German/Austro-Hungarian federation.[39] He recognized the Germanic virtues of the Transylvanian Saxons alongside similar areas of German settlement that would strengthen this Germanic confederation.[40]

But over the course of World War I the concept of *Mitteleuropa* acquired a more racial meaning. Naumann asserted that it was a duty to have more children. Thus, his inclusive territorial scheme ran in tandem with a need to promote population increase.[41] Ploetz, for his part, framed his internationalism in terms of developing a Germanic racial network. The Germanic race [*Germanentum*] predated—and was far more extensive than—the modern political entity that was Germany. If we go back to the early Racial Hygiene Society, we find a scattering of members from Central and Southeastern Europe. Ideally, this network would extend from German emigrant groups in North and South America (Ploetz had Argentinian links through his wife, Anita) to the [*volksdeutsche*] in the East.

Yet there were also a number of Southeastern European offshoots. Tudor Georgescu has researched the longstanding links between the racial hygienist Ploetz and Transylvanian Saxons in interwar Greater Romania.[42] We also find a stray Greek physician, Georgios N. Papanikolaou (1893–1962) as a member of the Racial Hygiene Society, first listed as a zoologist in Munich in the 3 March 1910 membership list and later as coming from Euboea.[43]

The Austrian hygienist Heinrich Reichel is a good example of a consistent Ploetz contact. Through Max von Gruber, Ploetz developed substantive links in hygiene and public health.[44] Ploetz and Thurnwald had decided views on the future of Central Europe. They hoped for a "mitteleuropäisches Staatenbund" [Central European federation] with ties to Switzerland, the Netherlands, Belgium and Scandinavia forming a larger political entity. More specifically, they wanted an Austria separated from Hungary as part of a Greater German Reich.

39 Naumann, *Mitteleuropa*, 2–3.
40 Naumann, *Mitteleuropa*, 100–102.
41 Naumann, *Mitteleuropa*, 186–187.
42 See Tudor Georgescu's chapter in this volume.
43 He joined between 10 December 1909 and 11 February 1910.
44 Weindling, *Health, Race*, 142–143.

World War I brought ideas of *Mitteleuropa* together with eugenics and race. This is visible in the *Vaterlandspartei* [Fatherland party] as well Ludendorff's plans for a Greater German Reich with vassal states providing Germany with security and resources in the East. German public health policies imposed delousing and sanitary policies in the *Generalgouvernement* and the Baltics. Racial hygienists like Fritz Lenz acquired important experience while undertaking anti-typhus [*Fleckfieber*] delousing measures in the East. Thus he observed the hygienic conditions of racial types in occupied Poland and the Ukraine. The control of infectious disease remained a theme in East European eugenics long after it fused with racial thinking. Was one race more prone to infection than another? These were not just academic debates; Austria-Hungary was faced with wartime sanitary challenges in the Balkans and consequently, Austrian planning for Southeastern Europe included special authorities for public health. In the case of health administration, the eugenicist and expert on occupational health Ignaz Kaup was appointed. He was succeeded by the anatomist Julius Tandler, whose focus was on social conditions in Vienna.

By the interwar period, we can see the formation of a dense network of Central European eugenics societies. What is interesting is the frequency with which the term "Racial Hygiene" is used by these new organizations. This definitely suggests a German orientation. But by his time clear differences had emerged in Germany between ideas for national regeneration through welfare state agencies and an ultra-nationalism advanced by the likes of the publisher Julius Lehmann and Ploetz. We thus find such key political figures as Michael Hainisch in Austria and Count Pál Teleki (1879–1941) in Hungary sharing interests in rural geography and ethnic groups as well as a population policy agenda.

The more Germanic ethos of the Society after 1910 and especially its goal to support racial hygiene "unter Personen mit deutscher Muttersprache" (among persons of German mother tongue) and the requirement that all members had to belong to the "white race" attracted Lehmann. Fritz Lenz contributed to Lehmann's bellicose journal *Osteuropäische Zukunft* [East European future] in 1916/7, discussing the racial value of Eastern settlement. Between 1917 and 1938, Lehmann's pugnacious *völkisch* journal, *Deutschlands Erneuerung* [Germany's renewal], contained between 1917 and 1938 at least 75 articles on racial hygiene, many of which were also issued as pamphlets. Contributors in these area include academic racial hygienists—for example, Lenz and von Gruber initially and later the geneticists Otmar von Verschuer (1896–1969) and Heinrich Schade (1876–1935)—as well as Nordic propagandists like Houston Stewart Chamberlain (1855–1927), Hans F. K. Günther (1891–1968), Walter Darré (1895–1953) and Lothar Tirala (1886–1974).[45] Lehmann also launched the journal *Volk und Rasse* [Nation and race] and, from 1926, the racially minded *Deutsche Gesellschaft für Blutgruppenforschung* [German Society for Blood Group Research] which was critical of any idea of an inclusive welfare state. Yet we still understand all too little about how academics and populists interacted. How, for instance, did Günther gain the support of conservatives within the anthropological establishment while arousing left-liberal animosities for his populist outline of Nordic racialism? This intermediate Germanic, populist form of racial knowledge has yet to find adequate historical appraisal. Although it appealed to and inspired many within the emergent Nazi movement, Günther maintained a degree of autonomy and distinct ultimate aims. While acknowledging that the above are well-known features of the German eugenics movement, there are many issues that still have to be documented and clarified concerning the international impact of the German Racial Hygiene Society.

In sharp contrast to Naumann, Hitler's *Mein Kampf*, (first published in 1925) offered an alternative view on the break up of the Austro-Hungarian Empire and the rise of nationalism. He charted the rise of nationalism after the failed Germanization policies of Emperor Joseph II (1741–1790) and noted each blow against the Dual Monarchy with delight. While racists praised the virtues of ancient Athens and Sparta, Hitler derided the Hellenic style of the Austrian parliament. For him democracy was "a plague." Hitler diagnosed a life-and-death struggle between Germans and Slavs and denounced Imperial Austria-Hungary as biased towards the latter. He set out to purge *Mitteleuropa* of its strong Southeastern European orientation.[46] We find beside his ardent anti-Semitism, anti-Bolshevism and desire to expand the German *Lebensraum* a sense of eugenics: "that the most

45 Paul Weindling, "The Medical Publisher J. F. Lehmann and Racial Hygiene," in Sigrid Stöckel, ed., *Die 'rechte Nation und ihr Verleger'. Politik und Popularisierung im J. F. Lehmanns Verlag 1890-1979* (Berlin: Lehmanns Media, 2002), 159–170.

46 Adolf Hitler, *Mein Kampf* (New York: Reynal and Hitchcock, 1941), 92–96.

beautiful bodies find one another and thus help in giving the nation a new beauty." This was an aesthetic, physical form of eugenics. He denounced biological inferiors and "ingenious weaklings," as well as those with sexually-transmitted diseases.[47]

## Jewish Eugenics

German racial hygienists also grappled with problems like Jewish racial identity and psychological attributes, raising the issue of similarities between the assimilated German Jews wit the *Ostjuden* (Eastern Jews) and oriental Sephardic types. Jewish communities were distributed across Central and Southeastern Europe, with sizable communities in Romania (with 984,213 Jews in the 1930 census) and Bulgaria, Yugoslavia and Greece (with about 50,000 Jews living in Salonica alone at the eve of World War II). The ethnic and anthropological composition of Sephardic and Ashkenazi ethnic groups was complex. We might see the Ashkenazi/Sephardic dichotomy as representative of the polarity between Central and Southeastern Europe—notwithstanding the inter-marriages between the two ethnic groupings in Judaism.[48]

The (non-Jewish) professor of anthropology Felix von Luschan was castigated as a "philosemite" for his critique of *fin de siècle* anti-Semitism. The issue revolved around Semite origins, with *Judentum* standing in contrast to *Deutschtum*. Eugenicists showed similar caution regarding Arthur de Gobineau's (1816–1882) linguistic criteria for the so-called Aryan race. Ploetz, in turn, recruited Jewish medical scientists and public health experts, including Max Neisser from Breslau and the statistician Wilhelm Weinberg from Stuttgart. While Ploetz kept his nascent Society clear of anti-Semitic organizations, he remained ambivalent about the involvement of assimilated Jewish members. He did not want a Jewish partner supporting the *Archiv für Rassen- und Gesellschaftsbiologie* [Journal for racial and social biology] and noted with concern both the Society's Jewish or half-Jewish members as well as those who were open anti-Semites. While it is clear that the

47 Hitler, *Mein Kampf*, 614.
48 M.J. Gutmann, "Die sephardischen Juden," *Archiv für Rassen- und Gesellschaftsbiologie* 17, 3 (1925–26), 324–325.

inner core of racial hygienists, notably Ploetz and Lenz, were Nordic racial idealists, they were cautious in articulating such sentiments, at least until the patriotic fervor of World War I brought about an intensification of ideas for *Lebensraum* and the promoting of German racial health.[49]

In 1895 Ploetz prophesied that democracy and science meant that anti-Semitism would become extinct. While Jews were deemed a *Kulturrasse* (civilized race), on a par with the German, other races were ranked as primitive. Eugenicists thus condemned the nomadic "Gypsies" as degenerate. But the trend toward degeneration was diagnosed also in the midst of advanced industrial society, especially as linked to urbanization. The eugenicists stigmatized social types like single mothers and juvenile delinquents as degenerates threatening the racial health of the nation. Racial experts correlated the incidence of diseases and disabilities with idiocy, feeble-mindedness and a range of asocial behavior. Estimates of the numbers of degenerate individuals in Germany considered ripe for sterilization ran to as high as one third of the population.

The rise of nationalist fervor prior to World War I only increased the links between eugenics and the *völkisch* movement; indeed Lehmann produced the catalogue for the Dresden Hygiene Exhibition of 1911. He took a leading role in racializing eugenics by sponsoring the anthropologist Hans F. K. Günther and more generally the journal *Volk und Rasse* beginning in the mid-1920s. By way of contrast, non-racist forms of eugenics also penetrated deep into the Weimar welfare state, with measures to curb the propagation of the asocial, the feeble-minded and the spread of the so-called *Rassengifte*—racial poisons, notably tuberculosis, venereal diseases and alcoholism. Race and welfare were final fused in 1933; once racial hygiene was "gleichgeschaltet" (the Nazi term for subordination to the Party and state), the rapid Nazification of the German welfare state was facilitated.

Viewed internationally in terms of their social recruitment, there were striking similarities between all the eugenics movements in interwar Europe. These were generally led by professional elites, often public health officers or demographers, but also administrators and lawyers. There were common concerns with countering physical and psychological degenera-

49 Weindling, *Health, Race*, 301–304, 307–315.

tion, particularly on the basis of re-directing the welfare state away from universalist toward selective social measures. Eugenicists were unrepentantly meritocratic, supporting the professionalisation of social administration. What eugenicists offered were schemes like national hereditary surveys and databanks, mechanisms to segregate deviants and undesirables and maps of the distribution of presumed racial types.

Racism in its myriad forms was a ubiquitous international phenomenon with a far wider, stronger and more populist appeal than the expert and elite ideology of eugenics. Yet German eugenicists nevertheless adopted a number of distinctive ideological traits. *Völkisch* racism was populist in its appeal to the masses, deploying evocative and deeply rooted cultural symbols of blood and rebirth.[50] *Völkisch* racists and eugenicists shared the aim of regenerating Germany by way of recovering a Teutonic primitive racial vigor. But their methods and categories differed sharply. We have, in effect, two forms of regenerative ideology: one scientistic, the other ultranationalist and anti-Semitic. If we accept that eugenics and populist racism were distinct, this allows one to examine their fusion *after* 1933 and to gain greater insight into why Nazi racial policy was to take such lethal forms.

While Jews were certainly of interest to eugenicists, they were not fixated on the so called Jewish problem. A comparison of the views of Ploetz and the more public-health oriented psychiatrist Schallmayer, who argued for the development of a corporate national racial service (*Rassedienst*) and the purification of the hereditary elements in the population (*Volkseugenik*) by means of health passports is telling. The latter insisted that nations were conglomerates of races and his writings show no clear hierarchy between races. Apart from noting that European Jews had low birth-rates, he paid no special attention to the Jewish race.[51]

Unlike Schallmayer, Ploetz considered Jews to be a gifted *Kulturrasse.* Although not overtly discriminatory, this had the potential for perceiving German and Jewish races as locked in a Social Darwinist struggle for survival. One can debate whether Ploetz steered the *Deutsche Gesellschaft für Rassenhygiene* away from *völkisch* groupings for tactical reasons or out of genuine conviction. Among the latter were the *Gobineau-Vereinigung,* the

---

50 George Mosse, *The Crisis of German Ideology* (New York: Grosset and Dunlap, 1964).

51 Wilhelm Schallmayer, *Vererbung und Auslese*, 4 ed. (Jena: Gustav Fischer, 1920).

*Mittgart-Bund*, the *Alldeutscher Verband* and a range of other organizations variously deemed to have provided Hitler many of his racial ideas. Above all, Ploetz wanted acceptance of racial hygiene as a science. He saw this as a branch of public health (or what in Germany was the academic discipline of "hygiene") stressing the hereditary factors in a population. At this level a race was merely a breeding group without distinctive psychological or hereditary biological characteristics. Ploetz proclaimed the *Gesellschaft für Rassenhygiene* to be an elite breeding group, encouraging the admission of wives and children and of students. What was important was the academic and professional status of its members. Here the Racial Hygiene Society offered a model for community-based eugenic schemes.

In the 1920s Jews withdrew progressively from the Racial Hygiene Society. Liberal and left-leaning eugenicists gravitated to the *Bund für Volksaufartung* [League for national regeneration], and Nordic racists went to the journal *Volk und Rasse*. The geneticist Herman Poll is a representative example. Publications by the *Archiv für Rassen- und Gesellschaftsbiologie* show an increasing focus on the problem supposedly posed by *Ostjuden* and in the Jews' more general life expectancy and health. In 1923 a paper by Arthur Czellitzer (1871–1942) appeared, while still others continued to maintain a Jewish presence in racial hygiene. A paper on Jewish health was published in the next year by Moses Julius Gutmann,[52] who had earned his doctorate on "Über den heutigen Stand der Rassen- und Krankheitsfragen der Juden" [On the current state of questions of race and health among Jews] from the University of Munich in 1920.[53] Gutmann supplied reviews on Jewish health and statistics, on deaths among Jews in Petrograd and published these in the *Blätter für Demographie, Statistik und Wirtschaftskunde der Juden* [Papers for demography, statistics and economic science], which first appeared in 1923. An active Zionist, Gutmann emigrated to Palestine in November 1936. The mid-1920s saw the publication of further papers in the *Archiv für Rassen- und Gesellschaftsbiologie*, but thereafter articles diminished; critical reviews by Jewish authors had ceased entirely by 1932.[54]

52 Moses Julius Gutmann, in *Archiv für Rassen- und Gesellschaftsbiologie* 16 (1924), 456–457.

53 Moses Julius Gutmann, *Über den heutigen Stand der Rasse- und Krankheitsfrage der Juden. Eine Feststellung und neue Wege zu ihrer Lösung* (Munich: Müller & Steinicke, 1911).

54 Hans Ullmann, "Zur Frage der Vitalität und Morbität der jüdischen Bevölkerung" *Archiv für Rassen- und Gesellschaftsbiologie* 18 (1925): 1–54, and S. Weissenberg, "Zur Sozialbiologie und Sozialhygiene der Juden," *Archiv für Rassen- und Gesellschaftsbiologie* 19 (1926): 402–406.

As Jews withdrew, the character of eugenic anti-Semitism changed. From the later 1920s it became much more medicalized. The Nazi daily newspaper *Völkischer Beobachter* [The people's observer] for example began to print articles by doctors and geneticists.[55] When the *Deutsche Gesellschaft für Blutgruppenforschung* was founded in Vienna in 1926, Jewish academics were only selectively admitted, as Otto Reche (1879–1966), its founder, had pronounced Nordic *völkisch* sympathies. The mathematician Felix Bernstein (1878–1956) and serologist Fritz Schiff (1889–1940) were among those excluded, although Karl Landsteiner (1878–1956), a discoverer of blood groups, and a mathematician and engineer Siegmund Wellisch (1864–1938) were taken up as members.[56]

## Scientific Networks

The racial anthropologists Eugen Fischer, Otmar von Verschuer and Otto Reche developed even more nationalist Central and Southeastern European contacts. Reche worked closely with the Austrian sero-anthropologist Michael Hesch (born in Romania) while Fischer collaborated with the anthropologist Wolfgang Abel (1905–1997), who was to take an important role in German racial surveys for the Nazis in an occupied Russia. A number of Central and Southeastern European researchers worked at the KWI in Berlin:[57]

| Name | Country of Origin | Year |
|---|---|---|
| Ladislau Apor | Hungary | 1942 |
| Lajos Csík | Hungary | 1940 |
| Carl Role Czapnik | Poland | 1942 |
| Miklós Fehér | Hungary | 1940 |
| Nicolaus Ilkow [i.e. Nikolay Ilkov] | Bulgaria | 1942 |

55 Detlef Mühlberger, *Hitler's Voice. The Völkischer Beobachter 1920–1933*, vol. 1 (London: Peter Lang, 2003), 62.

56 Pauline Mazumdar, "Blood and Soil: The Serology of the Aryan Racial State," *Bulletin for the History of Medicine* 64 (1990): 187–219; and Weindling, *Health, Race*, 464–469.

57 See Hans-Walter Schmuhl, *The Kaiser Wilhelm Institute for Anthropology, Human Heredity, and Eugenics, 1927–1945. Crossing Boundaries* (Dordrecht: Springer, 2008).

| | | |
|---|---|---|
| Franjo Ivaniček | Croatia | 1941 |
| Mihály Malán | Hungary | 1936 |
| Božo Škerlj | Yugoslavia | 1932 |
| Anton Steif | Hungary | 1941 |
| Marius Sulică | Romania | 1942 |
| Witold Sylwanowicz | Poland | 1937 |

Ernst Rüdin had a similar pattern of visitors at the German Institute for Psychiatry in Munich.

Politically, different ends were being served. First, sustaining such ethnic German communities undermined the despised Versailles settlement. It proved the existence of a Greater Germany beyond contemporary state borders. The role of the Volga and Ukrainian ethnic German groups took a significant turn in German–Soviet relations during the interwar period. The advocate of geo-epidemiology Heinrich Zeiss (1888–1949) studied chronic diseases among ethnic groups in the East. What happened in terms of medical and eugenic politics in the "lands in-between" is only just beginning to be clarified. Here, we need to add an anthropological and medical dimension to the *völkische Wissenschaften* and to examine other forms of contact, such as Lamarckian socialist populists or the autonomous and varied forms of Southeastern European eugenics as these interacted with the German racial hygiene.

If we further examine hygiene and the control of diseases, a variety of frameworks in Central and Eastern Europe prove to have existed: the interstate Committee for Prevention of Diseases, which was linked to the Red Cross, the Rockefeller Foundation, several networks of the *Hamburger Tropeninstitut* [Hamburg Tropical Institute], and individual hygienists like Rodenwaldt in Heidelberg, Zeiss in Berlin and Mrugowsky of the Waffen-SS Hygiene Institute. For the most part, these frameworks were not eugenic, raising issues of demarcation between hygiene, social hygiene and racial hygiene. Correspondingly, there is a need to reconstruct how anthropology, demography and race remained focal points of eugenic attention in Southeastern Europe. Such frameworks could work with both center-periphery models and one of multiple cultural centers that collectively shape the practices of eugenics and race. This suggests how, in specific contexts, ethnic and racial factors could shape other aspects of preventive medicine, particularly regarding the control of infectious diseases. The overall implications are, simply, whether or not developments on the peripheries allow one to reconceptualize race as a central idea in twentieth-century German history.

What is interesting is the attitude to ethnicity and disease, and especially the role of the German ethnic groups. These represent some consistent, longer term concerns in racial hygiene and *völkisch* circles. Here, we find a range of German scientific networks measuring and evaluating ethnic Germans in the East. These form an essential feature of—the still only very partially documented—project of racial repatriation found in the *Heim ins Reich* [Home into the Reich] repatriation scheme initiated in 1938. Indeed, the *Generalplan Ost* [General plan for the East], while a highly aggressive policy, can be seen as the culmination of efforts to strengthen German racial identity in transplanting prime racial stock to the peripheries of the expanding Reich. To date, such medical and anthropological dimensions of Nazi racial policies have only been marginally investigated.

This chapter has created space for reconstructing networks of health, hygiene and eugenics in Southeastern Europe.[58] German racial hygiene was a forceful presence: as a model, as a collaborative enterprise and as expansionist policy. What is emerging is a fuller understanding of health, welfare and population issues and how these intersected with larger state health and welfare organizations. Finally, the place of race and hygiene in Nazi genocide may be better understood, both in terms of its dynamics and responses to it among eugenic and racial experts in Southeastern Europe as well as the diverse positions held in public health and eugenics circles regarding Jews and other ethnic and social groups.

---

58 Here I would cite the admirable studies by Margit Berner and Maria Teschler-Nichola. See M. Teschler-Nicola, M. Berner, Die anthropologische Abteilung des Naturhistorischen Museums in der NS-Zeit; Berichte und Dokumentation von Forschungs- und Sammlungsaktivitäten 1938–1945 [unpublished report, 1998)], 333–358; and Margit Berner "Judentypologisierungen in der Anthropologie am Beispiel der Bestände des Naturhistorischen Museums, Wien" *Zeitgeschichte* 32 (2005): 111–116.

# PART II

# HYGIENE AND HEALTH POLITICS

# ORIENTALIZING DISEASE

## AUSTRO-HUNGARIAN POLICIES OF 'RACE', GENDER, AND HYGIENE IN BOSNIA AND HERZEGOVINA, 1874–1914

*Brigitte Fuchs*

The Austro-Hungarian cavalry captain Alexander Spaits remarked in his 1907 book on the history of Bosnia and Herzegovina that, as compared to Croatians and Serbs, Bosnians had proven themselves spineless, since the middle ages its elites had been induced to convert to Islam. As Spaits explained it, only those lacking sufficient financial means remained with the discordant Christian churches.[1] Bosnia's population in 1910 numbered approximately 1.5 million people (consisting of approximately 35% Muslims, 33% Serbian Orthodox and 20% Roman Catholics)[2] and was defined by Spaits as a homogenous "Slav nation," despite the fact that it was split into competing national groups of Slavic 'Turks' (Muslims), Serbs (Orthodox) and Croatians (Roman Catholics). Notwithstanding differing religious beliefs, the formation and popularization of a homogenous "Bos-

1 Alexander Spaits, *Der Weg zum Berliner Kongress. Historische Entwicklung Bosniens und der Herzegowina bis zur Occupation* (Vienna: Stern, 1907), 27; all translations from the original are my own unless otherwise cited.

2 For census data and data on confession, see *Ortschafts- und Bevölkerungs-Statistik von Bosnien und Hercegovina. Statistika miesta i pucanstva Bosne i Hercegovina* (Sarajevo: Landesdruckerei, 1880); *Ortschafts- und Bevölkerungs-Statistik von Bosnien und der Hercegovina nach dem Volkszählungs-Ergebnisse 1885* (Sarajevo: Landesdruckerei, 1886); Heinold Helczmanowszki, "Die Bevölkerung Österreich-Ungarns," in *Geschichte und Ergebnisse der zentralen amtlichen Statistik in Österreich 1829–1979*, ed. by Österreichisches Statistisches Zentralamt (Vienna: Österreichsche Staatsdruckerei, 1979), 369–402. It must be noted that in 1895 8,213 "Israelites" were also counted, the largest majority being native Sephardic Jews who primarily inhabited the larger cities, particularly Sarajevo. The presence of some thousand "Protestants" was due to the immigration of mainly Hungarian, German-speaking "colonists" into Bosnia and Herzegovina. See Hans Maier, *Die deutschen Siedlungen in Bosnien* (Stuttgart: Ausland und Heimat-Verlag, 1924).

niak" nation was a crucial concern for the Austro-Hungarian authorities in the former Ottoman province "Bosnia and Herzegovina."

Spaits described the historical fact of the conversion to Islam by one section of the "nation" using phrases such as "spinelessness," "weakness" and "degeneration" which, as will be demonstrated, also shaped Austro-Hungarian medical discourses on Bosnia. As a result of skilled diplomacy, Austria-Hungary was able to implement its expansion plans for the Balkans by occupying Bosnia and Herzegovina in 1878/9, first as a territory that it administrated but whose sovereign was still formally the Ottoman Sultan, then eventually annexing it in 1908. These colonial acts were legitimized as Austria-Hungary's "civilizing mission" [*Kulturmission*] in a former Ottoman province which was without doubt both socially and economically one of the least developed parts of Europe. However, the Austro-Hungarian occupiers' description of Bosnia as "primitive" was primarily due to the lack of a public health and hygiene system. In the course of the occupation, Austro-Hungarian health officers had already noted a state of general "degeneration" in the Bosnian population, urgently demanding the implementation of a modern public health regime in order to combat the poor state of health.[3]

The introduction of Austro-Hungarian public health and hygiene policies in Bosnia could be considered as examples of what scholars—following Foucault—have termed modern Western biopolitics.[4] As Balibar and others have demonstrated, biopolitics gained momentum after the 1880s, characterized increasingly by state intervention into the "female" private sphere.[5] In this context, women as child-bearers became specific subjects of biopolitical discourses and practices, as Western governments moved

3 See Dr Spanner, "Die sanitären Verhältnisse der Hauptstadt Mostar seit der Occupation der Herzegowina bis Ende des Jahres 1879," *Der Militärärzt* 14 (1880): 161–163 and 169–171; Dr Schorr, "Aus meiner Praxis im Sandschak Novi-Bazar," *Der Militärarzt* 15 (1881): 90–93; Dr Ulmer, "Von den sanitären Verhältnissen der Truppe im Okkupationsgebiete," *Der Militärarzt* 18 (1884): 121–124, 127–129 and 131–133. Whereas Austro-Hungarian health officers judged the "sanitary condition" of Bosnia and Herzegovina negatively, the official journal of the Austrian supreme public health administration (at the Ministry of the Interior) reported that generally, the Bosnians were of extraordinary good health. See "Die sanitären Verhältnisse in Bosnien und der Hercegovina," *Das österreichische Sanitätswesen. Organ für die Publicationen des k.k. Obersten Sanitätsrates* 1 (1889): 71.

4 Michel Foucault, *"Society Must Be Defended." Lectures at the Collège de France, 1975–76* (London: Penguin Press, 2003), 239–264.

5 Etienne Balibar, "The Nation Form: History and Ideology," in Etienne Balibar and Immanuel Wallerstein, *Race, Nation, Class. Ambiguous Identities* (London: Verso, 1991), 86–106.

to "improve" the quantity and quality of their populations, both in central and in more marginal and colonial settings.[6] Such measures and discourses of public health and welfare were not only gendered, but also aimed at specific population groups. Therefore, the questions must be raised about how precisely certain social classes, ethnicities or religious communities were affected, with the final consequence being social segregation along medically defined lines.[7]

Initially, the focus here will be on Austro-Hungarian anthropological and medical discourses on the "nature" and physical quality of the Bosnian population. Emphasis will be placed on a description of the establishment of a public health regime, in addition to discourses on combating endemic and epidemic diseases developed against a background of contemporary Orientalist and Balkanizing discourses[8] on degeneration, in particular, the "syphilitic" degeneration of the Bosnians. Austro-Hungarian campaigns to combat syphilis and to foster hygiene will also be investigated, especially the involvement of female physicians, who not only medicated Bosnian Muslim women but contributed more widely to changing attitudes towards gender differences. Finally, the focus will shift to the emancipation campaign of Austro-Hungarian female physicians which, as will be demonstrated, was based heavily on Orientalist stereotyping and was largely in an attempt to raise feminist awareness in the capitals of the Empire, rather than in Bosnia.

## Narratives on Bosnian 'Bones' and 'Bosniak Nationality'

In the course of the nineteenth century, the Porte gradually lost grip over Bosnia, and its efforts to modernize the Ottoman state were met with great hostility from Bosnian Muslim land holders fearing a loss of privileges, culminating in a rebellions against Ottoman rule. An 1875 insurrection of Christian small tenants against their Muslim landlords spread from Her-

[6] Seth Koven and Sonya Michel, eds., *Mothers of a New World. Maternalist Politics and the Origins of Welfare States* (New York: Routledge, 1993); and Philippa Levine, ed., *Gender and Empire* (New York: Oxford University Press, 2004).

[7] Alison Bashford, *Imperial Hygiene: A Critical History of Colonialism, Nationalism and Public Health* (Basingstoke: Palgrave Macmillan, 2004).

[8] Edward Said, *Orientalism* (London: Penguin Books, 1979) and Maria Todorova, *Imagining the Balkans* (New York–Oxford: Oxford University Press, 1997).

zegovina and sparked a peasant uprising, which eventually involved several Balkan states as well as the military intervention of Russia in the so called Eastern Crisis (1875–1878). The developments on the battlefields and the ensuing brokering of the Great Powers on the political future of the Balkan Peninsula at the 1878 Treaty of Berlin finally forced the Ottomans to cede administration of the country to Austria-Hungary. Although the Dual Monarchy never seriously considered either curtailing landlord privileges or implementing an urgently needed land reform, Muslim land holders organized an insurrection against the occupier in southern Herzegovina, which was successfully put down by Austro-Hungarian forces in 1882. The Habsburg Monarchy nevertheless managed to quickly establish some stability, allowing Herzegovina Muslim refugees who had emigrated after their armed resistance had failed to return. The Monarchy later embarked on a number of administrative reforms aimed at modernization, intending to transform Bosnia and Herzegovina into a "model colony."[9]

In 1882, Vienna granted responsibility for Bosnia and Herzegovina, to the joint Austro-Hungarian Ministry of Finance. The Ministers Bénjamin Kállay (1882–1903) and István Burián (1903–1912), served as de facto governors of Bosnia, whose integration into the Habsburg Monarchy as a "future province" was intended from the very outset. As a consequence, the "imperialist" discourse on Austria-Hungary's "civilizing mission" in Bosnia and Herzegovina was closely intertwined with the goal of constructing a universalist narrative for "Bosniak nationality," one dovetailing with larger Austro-Hungarian plans for regional integration. Austro-Hungarian modernization policies, at least in the Western part of the empire, focused on a multi-confessional identity set against irredentist, ethnic and national identities respectively. Consequently, in the specific case of Bosnia and Herzegovina the cultural policies of a multi-confessional "Bosniak" identity were intended to snuff out rising Southern Slav nationalism and aimed particularly at winning the loyalty of the Muslim elites.[10]

---

9 Barbara Jelavich, *Modern Austria. Empire and Republic 1800–1986* (Cambridge: Cambridge University Press., 1987); Noel Malcolm, *Bosnia: A Short History* (New York: New York University Press, 1994); and Ferdinand Hauptmann, *Die österreichisch-ungarische Herrschaft in Bosnien und der Hercegovina 1878–1918. Wirtschaftspolitik und Wirtschaftsentwicklung* (Graz: Institut für Geschichte der Univ. Graz, 1983).

10 For a detailed account, see Robin Okey, *Taming Balkan Nationalism: The Habsburg 'Civilizing Mission' in Bosnia 1878–1914* (Oxford: Oxford University Press, 2007).

In the 1880s much political writing dealt with the Muslim population in occupied Bosnia and Herzegovina. The common attitude expressed in such writing was—as demonstrated by an anonymous author in the typically racist language of the time—that Bosnia was a country "without culture", its "Asian" population being viewed simply as raw "material" from which the Austro-Hungarian authorities had to manufacture "Europeans."[11] While many self-styled experts agreed that it was the "Turks" who could limit or prevent "modernization," the official line was to express the conviction that Bosnian Muslims were ready to modernize, meaning that they were willing to accept Bosnia as a future Habsburg province. Pointing to the willingness of some Muslim officials to accept "progress" (i.e. Austro-Hungarian rule) in administration and political practice, great optimism was expressed that Bosnians in general and Muslims in particular would also readily accept a European lifestyle. The bourgeois, Western universalist form of gendering and family structure was considered the highest form of human social evolution, and particular emphasis was placed on the transformation of "Oriental" into "Western" gender relations.[12] Specifically, the most important means for achieving "Europeanization" was the implementation of a non-confessional system of schooling designed to prevail over confessional, ethnic education, which was doomed to failure.[13] The same is true for the Austro-Hungarian modernist meta-narrative on Bosnia and Herzegovina, placing former Ottoman subjects among the Monarchy's many nationalities.

In the nineteenth century, universalist discourses on nation and race, class and gender, was firmly founded in pseudo-scientific, pseudo-empirical, medical and anthropological narratives. In the last third of the eighteenth century, enlightened science had established gender difference on the basis of anatomical disparities between "male" and "female" pelvises, the "broadness" of the latter considered "proof" of women's exclusive aptitude to bear children. From the 1840s onwards, nations and races became defined increasingly in terms of anatomical differences between certain skeletal characteristics, especially regarding the cranium. In order to constitute these nations as historical kinship-groups, continuities—or discontinuities—between prehistoric and present populations were constructed

[11] *Bosniens Gegenwart und nächste Zukunft* (Leipzig: F.A. Brockhaus, 1886), IV.

[12] Okey, *Taming Balkan Nationalism*, 101–2.

[13] For a detailed account, see Okey, *Taming Balkan Nationalism*.

by comparing ancient and contemporary human bones. As a consequence, national, ethnic and gender differences became inscribed into both Western narratives and specific national European histories.

From the very beginning, the Austro-Hungarian authorities promoted a homogenous "Bosniak" identity for the multi-confessional population by accentuating their common Slavic origin. In order to give this "Bosniak" identity a "scientific" basis, Austro-Hungarian authorities consulted with the most eminent Austrian representatives in the various anthropological disciplines. This is above all noteworthy because most of the relevant petitions launched up to that point on the part of the Vienna Anthropological Society—which either aimed at the academic institutionalization of anthropology or at ethnographic and anthropological research into Austro-Hungarian nationalities—were rejected on the grounds that the authorities correctly assumed that anthropological measuring of differences would encourage rather than reduce nationalist aspirations. Consequently, much ethnographic and anthropological research was conducted in Austria itself by amateurs, the results of which did in fact fuel competing nationalisms. Bosnia, however, would become an exception.

An "Ethnographic Commission" [*Ethnographische Commission*]—established in 1884 by the "Imperial Academy of Sciences in Vienna" [*Kaiserliche Akademie der Wissenschaften in Wien*] and presided over by Friedrich Salomon Krauss (1859–1938)[14]—recorded Bosnian "monuments of the Slav language," such as local costumes and farmhouses.[15] An archaeological commission presided over by Moritz Hoernes (1852–1917), later Professor for Prehistoric Archaeology at the University of Vienna, documented the country's prehistoric and Roman sites. Roman sites served to place Bosnian and Herzegovinian prehistory and its human remnants in a Western tradition and were elaborated into a myth of the contemporary Bosnians' common origin with the population of the Austrian crown

14 Krauss also collected extensive ethnographic materials in the Balkans, which Freud used to develop his theory of sexuality. Thus, "southern Slav" popular culture came to represent the "primitive" state of mind of (Austro-Hungarian) bourgeois society, see Raymond L. Burt, *Friedrich Salomo Krauss (1859–1938). Selbstzeugnisse und Materialien zur Biobibliographie des Volkskundlers, Literatur- und Sexualforschers mit einem Nachlassverzeichnis* (Mitteilungen des Instituts für Gegenwartskunde 3) (Vienna: Österreichische Akademie der Wissenschaften, 1990).

15 Friedrich Salomon Krauss, *Bericht* (Vienna: author's edition, 1885). See Christian Marchetti. "Scientists with Guns. On the Ethnographic Explorations of the Balkans by Austrian-Hungarian Scientists before and during World War I," *Ab Imperio* 1 (2007): 170–173.

[lands]. As Hoernes argued, Bosnians as well as Albanians, Herzegovinians and Serbs belonged to the "Dinaric race", characterized by an extraordinary body height and brachycephaly, and was deemed to constitute the main stock of the population in the Austrian Alps.[16]

This attempt to create common origins for both Balkan and Alpine populations was further supported by Austro-Hungarian anthropological studies of the contemporary population. The first large-scale study was based on anthropometrical data from 3,803 individuals from all districts of Bosnia and Herzegovina. It was conducted by Augustin Weisbach (1837–1914), who had measured thousands of Austro-Hungarian individuals of all nationalities since the 1850s in order to develop a hierarchical system of "white races" (based on George Buffon's polygenic model of human descent).[17] However, in the case of Bosnia and Herzegovina, Weisbach largely refused to draw unequivocal conclusions concerning "racial" differences. In 1895 he stated that the homogeneous ethnic character of Bosnia's population, supported by the above-average body height, was also found among other Southern Slavs. Yet he still considered the agrarian "Greek oriental" to be the most "pure" race, especially in comparison with the more urban Muslims, seen by Weisbach as the most racially mixed "element" of Bosnia's population (thus leaving space for an ascription of different "racial" value to each of the different confessions).[18] A more explicit interpretation was presented by Leopold Glueck (1845–1907), director of the provincial hospital in Sarajevo, who corrected Weisbach on the basis of Hoernes' findings. Based on anthropometrical data from a larger sample than that measured by Weisbach, Glueck pointed to the incidental character of many anthropometrical correlations. In turn, he assumed that the "explicitly brachycephalic" population was of "Illyrian" origin—just like the population of the Alps. Glueck thus concluded that "the population of

16 Moritz Hoernes, *Natur- und Urgeschichte des Menschen*, vol. I (Vienna–Leipzig: Hartleben, 1909), 349–51. The creation of this myth mirrors the invention of another myth about the common "Etrurian" origins of populations both in the (Western) Alps and in Italy in order to legitimize Habsburg rule in northern Italy. See Brigitte Fuchs, *Rasse, Volk, Geschlecht. Anthropologische Diskurse in Österreich 1800–1960* (Frankfurt: Campus-Verlag, 2003), 160–4.

17 The relative rank of each "race" was determined by the degree of "miscegenation" with other races, the more "mixed" races ranking lower than the "purer" ones on a scale of nine "white races," the lowest of which Weisbach identified as the (Ashkenazi) Jews. See Fuchs, *Rasse, Volk, Geschlecht*, 134–137.

18 Augustin Weisbach, "Die Bosnier," *Mitteilungen der Anthropologischen Gesellschaft in Wien* 25 (1895): 236.

Bosnia and Herzegovina had long been an ethnically as well as historically homogenous people."[19]

Glueck also found Bosnian men to be four millimeters shorter than did Weisbach. Yet with an average recorded height of 1722 mm, this was still extremely similar to that of the Norwegians.[20] This was one of the determinant factors contributing to the stereotypical ascription of great "manly beauty" to Southern Slav males. However, this stereotype was applied to Bosnians only in their modified, Orientalist form: most Austro-Hungarian travelers and health officers found the obvious "healthy" appearance and "good looks" of male Bosnians in marked contrast to their ascribed "actual" frailty and general weakness.[21] Bosnians even came to exemplify the inaccuracy of contemporary anti-urban ideology, which had long dominated Austrian "German" nationalism. Such a view, based mainly on medical examinations of Bosnian individuals after 1900, was utilized by Austro-Hungarian health officers who tended to describe "degeneration" in Bosnia and Herzegovina with terms like "neurasthenia,"[22] "hysteria,"[23] and various "nervous diseases."[24] Numerous Austro-Hungarian health officers stated that Bosnia's rural but "degenerate" population, indeed, "proved" that "degeneration" was not characteristic of urban areas. The Bosnian discourse on mental diseases, moreover, superseded other Orientalist discourses focused on combating contagious and infectious diseases, particularly "syphilis"[25] as a "national epidemic" [*Volksseuche*], followed by the rapid establishment of a regime of public health and hygiene by the Austro-Hungarian authorities.

19 Leopold Glueck, "Physische Beschaffenheit der Bevölkerung Bosniens," in: *Die österreichisch-ungarische Monarchie in Wort und Bild. Bosnien und Hercegovina* (Vienna: k.k. Hof- und Staatsdruckerei, 1901), 284.

20 Glueck, "Physische Beschaffenheit der Bevölkerung," 279.

21 See Emil Mattauschek, "Einiges über die Degeneration des bosnisch-herzegowinischen Volkes," *Jahrbücher für Psychiatrie und Neurologie* 49 (1909): 138–9; Peter Stachel, "Der koloniale Blick auf Bosnien-Herzegowina in der ethnographischen Populärliteratur der Habsburgermonarchie," in Johannes Feichtinger, Ursula Prutsch, and Moritz Csáky, eds., *Habsburg Postcolonial. Machtstrukturen und kollektives Gedächtnis* (Innsbruck: Studien-Verlag, 2003), 259–76.

22 Geza Kobler, "Die Neurasthenie bei den Landbewohnern (mit besonderer Berücksichtigung auf die bosnische bäuerliche Bevölkerung)," *Wiener Medizinische Wochenschrift* 53 (1903): 1238–40.

23 Mattauschek, "Einiges über die Degeneration," 134–148.

24 See Mark Bermann, "Ein Beitrag zur Kasuistik der Lehre von der Cerebrospinalsklerose. Unilaterales Intentionszittern," *Wiener Medizinische Wochenschrift* 54 (1904): 933–936 and 1006–1009; Josef Löwenthal, "Über einen Fall von hysterischem Mutismus," *Wiener Medizinische Wochenschrift* 56 (1906): 864–866.

25 Syphilis does not just refer to the specific disease but to a socially constructed discourse which bore a particular threat to the welfare of nations or races, thus providing a pretext for the application of coercive measures vis-à-vis individuals or populations as a whole.

## The Establishment of the Austro-Hungarian Public Health Regime

During the occupation of Bosnia and Herzegovina in 1878/9, the Austro-Hungarian military authorities stated that cases of typhoid, dysentery and malaria were occurring far more than normal among their forces.[26] The situation was considered even more serious in the predominantly Muslim south of Bosnia and in Herzegovina where, in 1882, a Muslim insurrection had to be put down.[27] By the early 1880s, Austro-Hungarian military health officers complained repeatedly about the impassability of the mountainous regions, its rough climate, the lack of potable water, the lack of fresh food (responsible for an epidemic of scurvy in 1879)[28] and the lack of adequate accommodation and hospitals to care for the numerous servicemen suffering from epidemic or endemic diseases.[29] What Austro-Hungarian military health officers characterized as Bosnia's miserable sanitary conditions did not come as a surprise, since they were already convinced of having been sent to a country that was described as "cut off from civilization for centuries."[30] From the Austro-Hungarian perspective, "civilization" was primarily identified with the implementation of a modern "sanitary policy." From the early eighteenth century onwards, an almost impenetrable territorial "sanitary cordon" had been established along the Habsburg Military Frontier [*Militärgrenze*] in order to prevent the spread of contagious diseases from Ottoman lands.[31] As Sheldon Watts has remarked, Western policies of maritime and continental sanitary cordons vis-à-vis the Ottoman Empire are closely

26 Paul Myrdacz, *Sanitäts-Geschichte und Statistik der Occupation Bosniens und der Hercegovina im Jahr 1878* (Vienna-Leipzig: Urban & Schwarzenberg, 1881), 263–270.

27 Paul Myrdacz, *Sanitäts-Geschichte der Bekämpfung des Aufstandes in der Hercegovina, Süd-Bosnien und Süd-Dalmatien im Jahre 1882* (Vienna: Seidel, 1885), 158–159.

28 Adolf Zemanek, "Der Skorbut und seine militärhygienische Bedeutung. Mit specieller Berücksichtigung der unter den k.k. Occupationstruppen in Bosnien und der Herzegowina aufgetretenen Epidemie," *Der Militärarzt* 13 (1879): 169–172, 180–183 and 189–191.

29 Myrdacz, *Sanitäts-Geschichte und Statistik der Occupation*, 89–90; see also Spanner. "Die sanitären Verhältnisse der Hauptstadt Mostar," 160, 169–170.

30 Mattauschek, "Einiges über die Degeneration," 137.

31 Erna Lesky, "Die östereichische Pestfront an der k.k. Militärgrenze," *Saeculum* vol. 8 (1957): 82–105; Gunther E. Rothenberg, "The Austrian Sanitary Cordon and the Control of the Bubonic Plague, 1710–1871," *Journal of the History of Medicine* 28 (1973): 15–32; Daniel Panzac, *La peste dans l'Empire Ottoman 1700–1850* (Leuven: Peeters, 1985) and *Quarantaines et Lazarets. L'Europe et la peste d'Orient (XVIIe–XXe siècles)* (Aix-en-Provence: Édisud, 1987).

associated with the redefinition of the boundaries between "East" and "West" and between "the primitive" and "the civilized."[32] Once the Porte had introduced sanitary police to its dominions in the 1830s—joining Western efforts to develop an international plague control—the sanitary cordon was essentially reduced after the Crimean War and abolished in the course of a gradual deregulation of the Military Frontier, which was fully completed as late as in 1882. In this context, the cultural anthropologist André Gingrich brings a specific historic Austro-Hungarian "Frontier Orientalism" into play, whom Gringrich considers the central structure of the Habsburg imperial meta-narrative; the latter, as a consequence of the occupation of Bosnia, changed from purely negative to more ambivalent stereotyping.[33] Whereas Bosnian Muslims had been otherwise depicted since the 1880s as loyal subjects of the Habsburgs, we can also observe an Orientalist discourse that established a close relation between Islam, Ottoman Rule and the dissemination of epidemic and infectious diseases in the region.[34] Stereotypes about Muslim "fatalism" and "corruption" were applied continuously in order to explain an assumed Ottoman unwillingness or incapability to implement measures to improve public health and hygiene.

Orientalist prejudice was certainly of some importance when, in the autumn of 1878 the Austro-Hungarian military authorities claimed emergency power, introducing the "sanitary police", which were empowered to control epidemic and endemic diseases in the occupied territories.[35] In February 1879 the "Austrian Public Health Law" [*Reichssanitätsgesetz*] dating from 1870 was enacted in Bosnia and Herzegovina.[36] The *Reichssanitätsgesetz* stipulated, first, that health departments had to be set up at the administrative levels of provinces and districts [*Bezirke*]; secondly, that small municipalities had to join "sanitary municipalities" [*Sanitätsgemein-*

32 Sheldon Watts, *Epidemics and History. Disease, Power and Imperialism* (New Haven–London: Yale University Press 1997), XI–XVI. See also Božidar Jezernik, *Wild Europe. The Balkans in the Gaze of Western Travellers* (London: Saqi Books, 2004).

33 André Gingrich. "Frontier Myth of Orientalism; The Muslim World in Public and Popular Cultures of Central Europe," in Bojan Baskar and Borut Brumen, eds., *Mediterranean Ethnological Summer School*, vol. 2 (Ljubljana: Inštitut za multikulturne raziskave, 1998), 99–127.

34 See Philipp Sarasin, *'Anthrax'. Bioterror als Phantasma* (Frankfurt: Suhrkamp, 2004).

35 Landesregierung für Bosnien und die Hercegovina, ed., *Das Sanitätswesen in Bosnien und der Hercegovina 1878–1901* (Sarajevo: Landesdruckerei, 1903), 4–5.

36 Landesregierung für Bosnien und die Hercegovina, ed., *Sammlung der für Bosnien und die Hercegovina erlassenen Gesetze, Verordnungen und Normalweisungen I.* (Sarajevo: Landesdruckerei, 1905), 89–90.

*den*] and were obliged to pay public health officers; thirdly, that medical practice was restricted to physicians holding an academic degree *Doktor der gesamten Heilkunde* [*MD, Doctor medicinae universale*]) from an Austrian university; and finally mandated the introduction of health and population statistics which, in Austria-Hungary, in principle distinguished people according to their religion.

Medical reports reveal that during the 1880s and 1890s, public health legislation was only reluctantly implemented in most Austrian provinces, particularly at the administrative level in smaller municipalities. Yet in Bosnia and Herzegovina, largely as a result of quasi-colonial rule, there was an opportunity to implement an exemplary system of public health and hygiene. In August 1879, a legal provision was enacted for Bosnia and Herzegovina that obliged all midwives, dentists, surgeons and veterinarians to register with their local public health departments. Furthermore, those health professionals, whose training had not been officially approved by the Austrian authorities, were prohibited to practice.[37] At the same time, all Turkish physicians, traditional healers and midwives were declared illegal, the latter because they had not obtained the training required according to an Austrian Decree from 1853.[38]

Folk medicine, which was the topic of systematic and intense ethnographic research by Austro-Hungarian health officers, was soon replaced by a modern system of public health and hygiene.[39] In 1879 just one hospital existed, namely the Ottoman military hospital in Sarajevo. By 1901, the Austro-Hungarian administration had developed a health infrastructure which included the "provincial hospital" in Sarajevo with 325 beds, nine uniform "district hospitals" with twenty-four beds each, and twenty-four infirmaries of different sizes that were funded by larger municipalities.[40] Additionally, four small private hospitals had been established and the former Ottoman military hospital had been adapted to serve as a public san-

37 Landesregierung, *Sammlung der für Bosnien und die Hercegovina erlassenen Gesetze*, 110–1.

38 Josef Daimer, *Handbuch der österreichischen Sanitäts-Gesetze und Verordnungen für Behörden und Gemeinden, Districts- und Gemeindeärzte, Aerzte, Thierärzte, Apotheker, Verwaltungen von Heil- und Humanitätsanstalten etc. etc.*, part I (Leipzig–Vienna: Franz Deuticke, 1896), 398.

39 See Leopold Glueck, "Skizzen aus der Volksmedizin und dem medicinischen Aberglauben in Bosnien und der Hercegovina," *Wissenschaftliche Mittheilungen aus Bosnien und der Hercegovina* II (1894): 392– 454; and Leopold Glueck, *Medicinska narodna terminologija u Bosni i Hercegovini* (Sarajevo: Zemaljska štamparija 1898).

40 Landesregierung, *Das Sanitätswesen in Bosnien*, 201, 212.

atorium. In 1901, there was one hospital bed available per 1,569 people in Bosnia and Herzegovina, far below the Cisleithanian average of 630.7 people per hospital bed. However, similar conditions were found in Galicia (1,559.5 people per hospital bed) and in Bukovina (1576.1 persons per hospital bed).[41] The number of physicians—which in 1881 was limited to the sixteen health officers of the Austro-Hungarian military forces—had increased to 142 in 1909, twenty-one of whom were "private physicians."[42] One of the head physicians from Sarajevo's provincial hospital, Géza Georg Kobler (1864–1935), proudly remarked in 1910 that "public health in Bosnia and Herzegovina was approaching the ideal" of an entirely "nationalized" public health system. As Kobler pointed out, this meant that the "efficient combating of epidemic and endemic diseases" in Bosnia was more advanced than anywhere else in the Monarchy.[43]

In fact, the occurrence of infectious diseases in general and of small pox in particular, had decreased rapidly shortly following the first large-scale compulsory vaccination campaign in the late 1880s.[44] Between 1897 and 1901 the most frequently reported infectious diseases were malaria (40.29%), measles (18.43%), pertussis (whooping cough) (9.05%), diphtheria (7.23%) and influenza (7.22%).[45] Mortality rates due to infectious diseases accounted for 4.4% of total deaths, with the most lethal diseases being diphtheria (25.58%), measles (18.90%), pertussis (12.08%), scarlet fever (10.5%), small pox (9.29%), typhoid (8.17%) and malaria (6.40%).[46] Total mortality due to infectious diseases between 1893 and 1897 averaged 205 per 100,000 people which, again, was far above the Cisleithanian average of 138 per 100,000, yet still below comparable death rates in Galicia (253 per 100,000) and Bukovina (241 per 100,000).[47] The Austro-Hungarian author-

41 Anton Ullmann, "Die Krankenanstalten Österreichs diesseits der Leitha in den Jahren 1848 bis 1896," in *"Gesundheitspflege." Oesterreichs Wohlfahrtseinrichtungen 1848–1898* vol. III (Vienna: Perles, 1900), 252.

42 Geza Kobler, "Über das Vorkommen und die Bekämpfung der Lepra in Bosnien und Hercegovina," *Wiener Medizinische Wochenschrift* 60 (1910): 158.

43 Kobler, "Über das Vorkommen und die Bekämpfung der Lepra," 158.

44 Landesregierung, *Das Sanitätswesen in Bosnien*, 98.

45 The category of "infectious diseases" included smallpox, varicella, pertussis, scarlet fever, diphtheria, typhus, typhoid, recurrent fever, dysentery, malaria, influenza, mumps, anthrax and cholera.

46 Landesregierung, *Das Sanitätswesen in Bosnien*, 131.

47 As calculated by the Landesregierung, *Das Sanitätswesen in Bosnien*, 131 and Josef Daimer, "Die Erkrankungs- und Sterblichkeitsverhältnisse in Österreich im Allgemeinen und mit Rücksicht auf übertragbare Krankheiten und deren Bekämpfung insbesondere," in *"Gesundheitspflege." Oesterreichs Wohlfahrtseinrichtungen 1848-1898*, vol. III (Vienna: Perles, 1900), 45.

ities championed public health and public hospitals as icons of modernization, a process far less effective in Habsburg stewardship of the economy, agriculture and the development of the infrastructure.[48] This rapid establishment of public health and hygiene, developed against a background of discourses on the "syphilitic degeneration" of Bosnia, embedded in broader contemporary debates on syphilis as the main threat to "modern nations" (or/and races), needing to be controlled and combated as a matter of urgency.[49]

## 'Syphilis' as a Bosnian Epidemic

During the Habsburg occupation of Bosnia a particular cause of concern to Austro-Hungarian health officers was the infection of servicemen with venereal diseases.[50] These health officers expressed their conviction that, in Bosnia, a "well acclimatized syphilis" represented a significant threat to the military, since "Christian women affected by luetic diseases were rarely, while the Turkish women not at all" ready, to consult physicians in order to have themselves cured.[51] In 1879, Austro-Hungarian military authorities—who generally presumed a compelling correlation between syphilis and prostitution—introduced registration and compulsory medical inspection of all prostitutes in the occupied territories, representing one of the first such Austro-Hungarian public health measures.[52] Although morbidity within the armed forces, due to venereal diseases, had not increased over 1878/9 when compared with previous years, military authori-

48 But it might be noted that the authorities even thought of transforming "public health" during economic boom periods. In the 1890s the authorities began to consider the touristic development of Bosnia and Herzegovina, particularly by advertising Bosnian mineral waters and sanitariums in Austrian medical journals. See, for example, Insp. Pojmann, "Klimatologisches und Touristisches aus Bosnien und der Hercegovina mit besonderer Berücksichtigung des Bades Ildiže," *Wiener Medizinische. Wochenschrift* 59 (1900): 1341–1344, 1395–1397 and 1433–1435.

49 See, for example, Claude Quétel, *History of Syphilis,* trans. by Judith Braddock and Brian Pike (Baltimore: Johns Hopkins University Press, 1990); Alain Corbain, *Wunde Sinne. Über die Begierde, den Schrecken und die Ordnung der Zeit im 19. Jahrhundert* (Stuttgart: Klett-Cotta, 1973), 125–150; and Christine Zschiegner, *Die Syphilis in Österreich und ihre sozialen Folgen in der zweiten Hälfte des 19. und im frühen 20. Jahrhundert* (Innsbruck: unpubl. phil. Diss., 1996).

50 At that point, venereal diseases in general began to be commonly referred to as syphilis; even in medical statistics, the category of syphilis often also included the much more frequent gonorrhea.

51 Dr Ulmer, "Die sanitären Verhältnisse der Truppen im Okkupationsgebiete," *Der Militärarzt* 18 (1884): 123.

52 Myrdacz, *Sanitäts-Geschichte und Statistik der Occupation,* 275–6.

ties in Bosnia considered "syphilis" the greatest threat to their men.[53] Such a conclusion had been reached due to military authorities believing that syphilis in Bosnia and Herzegovina was a "national epidemic." Again, the assumption that syphilis was widespread in Ottoman countries was partly due to Orientalist stereotypes of Muslim "moral decay" and "polygamy", the latter—incorrectly—seen as a general institution within Muslim societies. Obviously, such prejudices were fostered by popular depictions of "the harem" and Oriental "slave markets."[54] However, such views were also further evident in contemporary medical literature, for example in the works of Eduard Reich (1836–1919), who, in 1858, had attempted to establish "Hygiene" as a separate academic discipline, or of Franz Pruner (1808–1882), who had been a professional physician in Egypt during the 1830s and 1840s.[55]

The fact that syphilis —or to be more exact "endemic syphilis"— was also widespread in all neighboring Austrian and Hungarian provinces received far less publicity. Endemic syphilis, (non-sexually transmitted syphilis), had been introduced into many agrarian regions of Europe in the late eighteenth century, including Istria, northern Italy, Inner and Lower Carniola, Slavonia, and Dalmatia.[56] By about 1790, local physicians reported a "new disease" spreading quickly and which came to be known as *škrljevo*—a term derived from the presumed place of its first occurrence, the village of Škerlj near Rijeka (Fiume).[57] Because of its obviously contagious character, the government in Vienna reacted with a large-scale public health campaign, the responsibility for which was given to a commission headed by the physician Andreas Joseph (Freiherr von) Stifft (1760–1836). With the aid of the Austrian military, local clergy and physicians carried out compulsory visits to entire populations of affected areas. All those found to be infected were compulsorily medicated with mercury and sulfur preparations and

53 Myrdacz, *Sanitäts-Geschichte und Statistik der Occupation*, 275.

54 See, for example, Meyda Yeğenoğlu, *Colonial Fantasies. Towards a Feminist Reading of Orientalism* (Cambridge: Cambridge University Press, 1998). Contemporary travel writers such as Arthur Evans, for example, had to point explicitly at the falseness of this idea. See Arthur J. Evans, *Through Bosnia and the Herzegovina on Foot during the Insurrections August and September 1873* (London: Longman, Green & Co, 1876), 191–196.

55 Eduard Reich, *Lehrbuch der allgemeinen Aetiologie und Hygieine* (Erlangen: Enke, 1858), 451–452; and Franz Pruner. *Die Krankheiten des Orients vom Standpunkte der vergleichenden Nosologie* (Erlangen: Palm & Enke, 1847).

56 Etienne Lancereaux, *Traitè Historique et pratique de la syphilis* (Paris: Bailliere, 1873), 38–39.

57 German spelling also "škerljewo" ("Morbus Škerljewo"), Italian spelling also "scherlievo" and "scharlievo"; other contemporary terms were "Grobnigger Krankheit (Übel)" and "mal di fiume."

then sent to hospitals established ad hoc.[58] In 1859 *škrljevo* was declared "eradicated" in Carniola, Istria and the Kingdom of Croatia. Nevertheless, the campaign continued in Dalmatia and Galicia well into the 1880s. Furthermore, the Austro-Hungarian supreme sanitary administration decided to "eradicate syphilis" in all provinces where recently introduced health statistics reported that over 10% of people treated in hospitals suffered from such venereal diseases.[59]

In the early nineteenth century, government-authorized experts in Vienna had identified the *škrljevo* disease either as "syphilis" or as some variation combined with scabies, scurvy or leprosy. Throughout the century, debates over whether or not "endemic syphilis" and "proper syphilis" had to be differentiated remained unsolved. Venereologists in Vienna and Prague who founded their reputation on expertise derived from the campaign against *škrljevo* tended to avoid this question. Many of them pursued the theory that syphilis was probably a transformation of leprosy, as they assumed at the time that both diseases were sexually transmitted and hereditary.[60] Other venereologists suggested that the specific characteristics of *škrljevo* resulted from the recent introduction of "syphilis" into rural populations. Not even the 1905 identification of the causing agent of (sexually transmitted) syphilis, the bacterium *Treponema pallidum pallidum* (a spirochaete bacterium) was able to resolve this controversy, for at that point there was simply no way to differentiate between various forms of treponematosis.[61]

58 See Friedrich W. Lorinser, "Ueber die Skerlievo-Krankheit im österreichischen Küstenlande," *Wiener Medizinische Wochenschrift* 15 (1865): 1689; Franjo Gruber, "Škrljevo Disease – Two Centuries of History," *International Journal of STD & AIDS* 13 (2000): 209; and Zvonka Zupanič Slavec, "Morbus Škerljevo – An Unknown Disease among Slovenians in the First Half of the Nineteenth Century," *Wiener Klinische Wochenschrift* 108 (1996): 766–767.

59 Hugo Zechmeister, "Über die endemische Syphilis in Dalmatien und im westlichen Kroatien," *Das Österreichische Sanitätswesen* vol. 15 (supplement) (1903): 151; "Vorkehrungen gegen Volkskrankheiten in Österreich. – III. Vorkehrungen gegen Syphilis und venerische Krankheiten in Galizien," *Das österreichische Sanitätswesen* vol. 29 (1907): 293. See Brigitte Fuchs, "Zur Geschichte und Statistik der venerischen Erkrankungen in den Ländern der österreichischen Krone (1815 bis 1914)," in Tom Buchner et. al., eds., *Demographie – Arbeit – Migration – Wissenschaftsgeschichte* (Munich: Oldenbourg, 2008), 433–459.

60 See Hermann Zeissl, *Lehrbuch der Syphilis und der nut dieser verwandten örtlichen venerischen Krankheiten* vol. II (Erlangen: Enke, 1872), 7–8; for a short overview of nineteenth century Austrian venereology (which does not take into consideration "endemic syphilis"). See Karl Holubar, "29 March 1841 – 29 September 1869: Two Tales in One City," in Karl Holubar, Cathrin Schmitt, and Klaus Wolff, eds., *Challenge Dermatology. Vienna 1841–1992*, (Vienna: Verlag der Österreichischen Akademie der Wissenschaften, 1993), 33–61.

61 Endemic forms of syphilis—the recent forms being *yaws* (West Africa) caused by *Treponema pertenue*, *bejel* (Middle East, Near East) caused by *Treponema endemicum* and *pinta* (South-America) caused by *Trepeonema carateum*—occur as an effect of poor social and hygienic conditions. The treponemes are transmitted by direct contact, most commonly among children. The primary stage of the disease is characterized by

Once health officers in Bosnia had established "syphilis" as a "social disease," the authorities in Vienna reacted quickly. Maximilian Zeissl (1853–1925), who had been involved in combating *škrljevo* in Carniola, Istria and Croatia, claimed to have been informed by an officer from the Ministry of the Interior in 1885 that "in the occupied territories an endemic disease is prevailing that is referred to as *frenjak* and which presumably is identical with *škrljevo*."[62] Zeissl, who had already been commissioned to examine the occurrence of cases of *škrljevo* in Dalmatia, made his way to Bosnia, where Austro-Hungarian health officers in different areas enabled him to visit cases of *frenjak*. Zeissl asserted that *frenjak* and *škrljevo* were identical, which he concluded from the prevalence of "primary" and "tertiary" symptoms of syphilis. However, according to the doctrine held by the universities of Vienna and Prague, Zeissl interpreted the sudden occurrence of "chancres" (ulcers) as cases of "hereditary (congenital) syphilis," explaining the frequency of the occurrence of gummata through insufficient medication, poor social conditions and the "indolence" of Bosnia's population.[63] He concluded that *frenjak* and *škrljevo* were nothing but "partly … acquired, partly … hereditary syphilis."[64] Once this expert had defined the prevalence of endemic syphilis in Bosnia and Herzegovina, a campaign to "eradicate syphilis" similar to that carried out previously in neighboring Habsburg provinces was inevitable.

In 1889 the Joint Ministry of Finance commissioned the first professor of Dermatology and Venereology at the University of Vienna, Isidor Neumann (1837–1906), to conduct "exhaustive studies on the occurrence and spread of syphilis and other skin diseases" in Bosnia "in order to advise the authorities on how to combat it in administrative ways."[65] By 1890 all of the districts' populations had been tested for "syphilis" by Austro-Hungarian health officers. According to Neumann, Austro-Hungarian authori-

---

a cutaneous lesion at the site of infection, which in the nineteenth century was identified with syphilitic exanthema. The secondary stage, during which the treponemes are disseminated, is without pathological evidence. The third stage, occurring after fifteen to twenty years, is characterized by multiple cutaneous lesions (particularly of throat and mouth) together with the decomposition of bones or cartilage (gummata), corresponding to some of the possible long-term effects of "proper syphilis." Endemic syphilis occurs as an effect of poor social and hygienic conditions.

62 Maximilian Zeissl, "Ueber 'Škerljevo'. Ein Reisebericht," *Archiv für Dermatologie und Syphilis* 19 (1887): 299.

63 *Medicinisch-chirurgische Rundschau* 18 (1887): 663.

64 Zeissl, "Ueber 'Škerljevo'," 313.

65 Landesregierung, *Das Sanitätswesen in Bosnien*, 137.

ties were "now informed thoroughly about provenience, forms and dispersal" of "syphilis."[66] Although the authorities later denied the use of coercion on local populations, this obviously was the case—at least initially. Clearly, none of the individuals visited by the *škrljevo* expert Maximilian Zeissl had been asked for their consent.[67] During subsequent campaigns for eradicating "syphilis," the appearance of military health personnel, together with armed soldiers, was probably sufficient to make most people "agree" to medical examinations.

Neumann traveled to Bosnia in 1890, where he diagnosed 336 individuals with "syphilis" and noted the prevalence of both its manifest and its latent forms.[68] In Herzegovina, he examined 845 individuals believed to suffer from syphilis of whom he found 389 people to be actually infected.[69] Neumann subsequently calculated the exact percentage of Bosnia and Herzegovina's syphilitics to be 0.026% of the total population. Such total rates, of course, were not comparable with rates of other Austrian or Hungarian provinces, simply because these latter statistics were confined to people who voluntarily submitted to treatment in public hospitals.[70] The category of venereal diseases also included the more frequent cases of gonorrhea. However, these accounted for approximately 7.6% of all diseases according to Cisleithanian health statistics from 1896, thus ranking them higher than any other diseases (including tuberculosis) treated in 1896 in Cisleithanian hospitals. Provided that venereologists were right in their estimate that only about 5% of all those who suffered from a venereal disease (including gonorrhea) were treated in a public hospital, the total percentage of people suffering from a venereal disease in Cisleithania would have been about 0.074% in 1896.[71] Since gonorrhea was estimated to account for at least two thirds of all people suffering from venereal diseases, even under the (false) presumption that *frenjak* and "syphilis" were identical diseases, the total syphilis rates for Bosnia and Herzegovina (0.026/1000) and Cisleithania

66 Isidor Neumann, *Syphilis* (Vienna: Hölder, 1896), 32.

67 Zeissl, "Ueber 'Škerljevo'," 300–310.

68 Neumann, *Syphilis*, 33.

69 Neumann, *Syphilis*, 33–34.

70 On the problem of the statistics of "venereal diseases" in Austria, see Zschiegner, *Die Syphilis in Österreich*, 17–25.

71 Calculated from figures given by Ullmann, "Die Krankenanstalten Österreichs," 253 and Zschiegner, *Die Syphilis in Österreich*, 24.

(0.024/1000) would have been similar. While Austrian statisticians were proud of Austria's comparatively moderate syphilis rates, combating syphilis in the occupied territories was still considered an urgent priority, which ensured that Habsburg public health funds continued to flow abundantly into the occupied territories.

Neumann had previously stated that "so-called endemic syphilis" was due to the populations' "poor conditions."[72] He now recommended another area-wide medical examination of the population in order to investigate its "sanitary and hygienic conditions" more systematically. Following the example of earlier Austrian syphilis campaigns, he instructed the authorities not only to search for cases of syphilis, but the spread of lupus, scabies, favus and leprosy all over Bosnia and Herzegovina as well.[73] He advised the authorities to establish hospitals in all districts as well as hospitals in areas that were particularly "contaminated" with syphilis. Moreover, Neumann suggested that systematic medical examinations should be combined with a rigorous campaign promoting public hygiene, for which he recommended the employment of female physicians, especially for the female Muslim population. These area-wide examinations were carried out in the 1890s, resulting in the discovery that syphilis was endemic only in the districts of Sarajevo and Banja Luka, where provisional hospitals [*Barrackenhospitäler*] were set up in order to isolate those infected.[74] The Austro-Hungarian authorities obviously assumed a priori that syphilis was more widespread among the Muslim population than among people of other faiths (notably Christians).[75] The Orientalist nature of this assumption is indicated by the difficulty of validating it by statistical

72 Statement of Isidor Neumann in "Verhandlungen Der Wiener Dermatologischen Gesellschaft," *Archiv für. Dermatologie und Syphilis* 33 (1895): 431.

73 Since the 1880s British medical experts, in particular, considered "leprosy" (in India) as a potential pandemic threat for the West, see Watts, *Epidemics and History*, 40–41. As a consequence, Austrian medical journals comprehensively covered "leprosy," based on cases in Bosnia and Dalmatia, see Geza Kobler, "Über das Vorkommen und die Bekämpfung der Lepra in Bosnien und Hercegovina," *Wiener Medizinische Wochenschrift* 60 (1910): 121–124, 129–131, and 137–139.

74 Ferdinand Schmid, *Bosnien und die Herzegowina unter der Verwaltung Österreich-Ungarns* (Leipzig: Veit, 1914), 279.

75 "Die sanitären Verhältnisse in Bosnien und der Hercegovina," *Das österreichische Sanitätswesen. Organ für die Publicationen des k.k. Obersten Sanitätsrates* 1 (1889): 71. See also Leopold Glueck, *Mittheilungen. aus der Abtheilung. für Syphilis und Hautkranke des bosnisch-hercegovinischen Landesspitales in Sarajevo* (Vienna: Josef Šafář, 1898), 86.

data.[76] In 1889, however, Leopold Glueck found another Orientalist explanation for the spread of venereal disease in the occupied territories. Against all the evidence indicating the movement of syphilis from the northern, Habsburg regions to Bosnia, he assumed that syphilis had been introduced to Bosnia and Herzegovina from the south by Ottoman forces in the early nineteenth century.[77] This theory was soon adopted as an official explanation for the spread of syphilis in Austria, Hungary and other countries which had been, or still were, under Ottoman rule.[78]

However, the assumption of an Ottoman origin of *frenjak* might also have served to establish "endemic syphilis" as a specific phenomenon, different from *škrljevo,* which had been classified as "proper syphilis" by an earlier generation of experts. With respect to *frenjak,* Neumann had already remarked that infection was evidently due to poor hygienic conditions. Leopold Glueck, head of the Department of Dermatology and Venereology at the provincial hospital in Sarajevo, observed the continued prevalence of "syphilis of the mouth and pharynx."[79] He concluded that medical staff in Bosnia and Herzegovina were mainly confronted with "secondary syphilis", a term he used to describe the non-sexually transmitted character of *frenjak.* As he claimed, *frenjak* was mainly transmitted by family members sharing commodities like towels or crockery.[80] Yet so-called "secondary syphilis"[81] seemed to be a disease whose spread could be easily combated by observing hygienic practices which, according to Western gender values, were an agenda directed at women. Bosnian and Herzegovinian women, particularly women of Muslim faith, therefore became the

76 We found only one document in this respect: For the years 1897 to 1800, Glueck reported from the Department of Venereology and Dermatology of the provincial hospital in Sarajevo that more than one third of all patients were syphilitic, of whom 45.83% were Muslims, 37.7% Roman Catholics and 32.14% Orthodox, cf. Leopold Glueck, "Bericht der Abtheilung für Syphilis- und Hautkranke 1897–1900," in *Jahrbuch des bosnisch-hercegovinischen Landesspitales in Sarajewo für 1897, 1898, 1899 und 1900* (Vienna: Josef Šafář, 1901), 371.

77 Leopold Glueck, "Ueber das Alter, den Ursprung und die Benennung der Syphilis in Bosnien und der Hercegovina," *Archiv für Dermatologie und Syphilis* 21 (1889): 348.

78 See Wladimir St. Maneff, *Über die endemische Syphilis in Bulgarien* (Leipzig: phil. Diss., 1934), 5.

79 Leopold Glueck, Geza Kobler, "Zur Kenntnis der Kehlkopfsyphilis," in Ernest Finger et al., eds., *Beiträge zur Dermatologie und Syphilis. Festschrift, gewidmet Herrn Hofrath Dr. I. Neumann zu seinem fünfundzwanzigjährigen Professoren-Jubiläum* (Vienna: Franz Deuticke, 1900), 173–283.

80 Leopold Glueck, "Die volkstümliche Behandlung der Syphilis in Bosnien und der Herzegowina," *Wiener Medizinische Wochenschrift* 40 (1890): 300–1, 350.

81 It was not until the 1920s that academic medicine acknowledged "endemic syphilis" as a disease distinct from "syphilis", a finding which owed much to Leopold Glueck and particularly his son, Aleksander Glueck (1884–1925).

target of "education" carried out primarily by Western female physicians, who pursued their own specific "female"—and feminist—"civilizing mission" vis-à-vis non-Western women.

## Campaigning for Hygiene: Austro-Hungarian 'Lady Doctors for Bosnia'

According to a 1903 public health report on Bosnia and Herzegovina, the Austro-Hungarian authorities had decided to employ female health officers because "they are free to visit the harems and therefore to gain insight into the living conditions of the residents in order to act upon the suppression of unhealthy customs and on the implementation of hygienic conditions in the families."[82] This program referred to arguments made by British feminists since the 1870s, who argued that because Indian and "Oriental" women were "imprisoned" in *zenana* or *harems,* British professional women should be employed by the state to educate "passive Indian (Oriental)" women in both personal hygiene and the use of public health.[83] It is likely that these claims were influenced by an ongoing project to educate Christian Bosnian girls and women, initiated by Adelina Paulina Irby (1831–1912), the famous British traveler.[84] Irby, who in 1872 was joined by Priscilla Johnston, had set up a girls' school in 1871 intended to better the status of women in a society which, in the nineteenth century, was commonly described as "uncivilized" and "patriarchal" by Westerners. On the occasion of the Serb uprising against Ottoman rule in 1875, Irby and Johnston not only raised funds to support Bosnian refugees crossing the border into Croatia, they also evacuated their five brightest pupils from Sarajevo to a girls' school in Prague — supported by Czech feminists who had also

82 Landesregierung, *Das Sanitätswesen in Bosnien,* 17.

83 Antoinette Burton, "Contesting the Zenana. The Mission to Make 'Lady Doctors for India', 1874–1885," *Journal of British Studies* 35 (1996): 369.

84 Dorothy Anderson, "Two Women Travellers in the Balkans in the 1860s: Georgina Muir Mackenzie, Adeline Paulina Irby," *Proceedings of the BRLSI (Bath Roval Literay and Scientific Institution)* 8 (2004), http://www.brlsi.org/proceed04/lunch200311.htm (Last accessed 3 October 2008). See Omer Hadžiselimović, "Two Victorian Ladies and Bosnian Realities, 1862–1875. G.H. MacKenzie and A.P. Irby," in John B. Allcock and Antonia Young, eds., *Black Lambs and Grey Falcons. Women Travellers in the Balkans* (New York: Berghahn Books, 2000), 4.

made the right to medical education and practice of medicine by women a key focus of the Czech women's movement.[85] Although Austro-Hungarian authorities had been reluctant to tolerate Irby and Johnston in Sarajevo in 1878—both being reputed partisans of Serb nationalism—their project to educate and modernize Bosnian women was nonetheless adopted. In particular, de facto governor Benjamin Kállay made "lady doctors for Bosnia and Herzegovina" his personal concern. Isidor Neumann recommended that a position be created for a female public health officer [*Amtsärztin*] (at a fixed remuneration) in Donja Tuzla (today's Tuzla) in 1891. The future "lady doctor" (with the military rank of captain) was required to speak the "Slav language" [sic!], in addition to holding the required medical degree, which meant necessarily that it be obtained in Switzerland because in Austria-Hungary women were at that point neither permitted to take up medical studies nor to pursue a medical profession.[86] Since Austrian medical pressure groups had exhibited great animosity toward the accreditation of female physicians, Habsburg authorities needed to justify this extraordinary measure by pointing to the need of Muslim women for female physicians, given that "religious shame" prevented them from consulting male doctors.

Kállay, who personally chose the female public health officers, first appointed the Czech physician Anna Bayerová (1854–1924), who had finished her medical studies in Zurich and had practiced in Bern. Bayerová, who was reputed to be the "first Czech female physician" (in the Czech women's movement), assumed her duties in 1892. As she reported, Kállay himself instructed her to "enlighten" Bosnian-Herzegovinian women, particularly in regard to hygiene.[87] Bayerová, however, avoided unduly exercising her office. Officially on account of health reasons, she abandoned her position in Tuzla after a few months. In March 1893 her vacant position in Tuzla was filled by a Russian citizen, Teodora (Theodora) Krajewska, née Kosmowska (1854–1935), who was given an Austrian citizenship

---

85 Karen J. Freeze, "Medical Education for Women in Austria: A Study in the Politics of the Czech Women's Movement in the 1890s" in Sharon L. Wolchik and Alfred G. Meyer, eds., *Women, State, and Party in Eastern Europe* (Durham, N.C.: Duke University Press, 1985), 54.

86 "Weibliche Aerzte in Bosnien (Nach amtlichen Quellen)," *Wiener Medizinische Wochenschrift* 36 (1886): 1349–51.

87 Lenka Vytlacilova, "Dr. Anna Bayerová (1853–1924). Second Czech Female Physician," http://www.pinn.net/~sunshine/czech/byerova.html (Last accessed 3 October 2008).

for the purpose. Krajewska had studied medicine in Geneva, was former president of the *Société des étudiants polonaise* [Association of Polish Students] and had also worked as an assistant at Geneva University.[88] From February 1893, a newly-created position in Mostar was staffed by another Czech physician, Bohuslava Kecková (Keck) (1854–1911), who had also completed her medical studies in Zurich, but—unable to pursue her profession in Austria-Hungary—practiced as a midwife in Prague.[89] In 1899 additional positions for female health officers were created in Sarajevo and Banja Luka, these towns having been classified as areas with "endemic syphilis." Krajewska was transferred to Sarajevo and the position in Tuzla was taken up by the native Polish physician, Jadviga (Hedwig) Olszewska (1855–1932).[90] For the position in Banja Luka, Gisela Kuhn, née Rosenfeld (1867–1943), who had finished her medical studies in Zurich in 1898, was appointed. After marrying her commanding officer, Ladislaus Januszwewki (1847–1916), she was forced to leave her position and began to practice privately in Banja Luka in 1903.[91] Her position went to Rosa Einhorn (Einhorn-Bloch) (1874–?), who in 1902 was transferred to a newly created post in Travnik.[92]

The female public health officers had a broad field of duties to cover. They were tasked with "educating" Bosnian and Herzegovinian women, to hold clinics at outpatient wards in the district hospitals (within the scope of regular service hours) and to "approach" and visit native women. The latter task was deemed to be of special importance since health statistics indicated that in Bosnia and Herzegovina the percentage of women turning to public health facilities was well below the Austrian average of one third of all patients. Between 1895 and 1900 between 21% and 30% of women in Bosnia and Herzegovina saw a physician, tendency rising. Health authorities were particularly content to find that the total percentage of Muslim

88 Zbigniew Danielak, "Krajewska z Kosmowskich Teodora," in *Polski Słownik Biograficzny* XV. (Warsaw: Zakład Narodowy imienia Ossolińskich Wyd. Polskiej Akad. Nauk. 1970), 101–103.

89 Lenka Vytlacilova, "Bohuslava Keckova (March 18, 1854), First Female, University-Trained Czech Physician," http://www.pinn.net/~sunshine/czech/medicine.html (Last accessed 3 October 2008).

90 See Omer C. Ibrahimagić et. al., "Liječnici u tuzlanskom kraju u devetnaestom stoljeću," *Medicinski archiv* 60, no 6, supl. 2 (2006): 119–20.

91 See Reinhold Aigner, "Die Grazer Ärztinnen aus der Zeit der Monarchie," *Zeitschrift des Historischen Vereines für Steiermark* 70 (1979): 69–2; Beate Ziegeler, *Weibliche Ärzte und Krankenkassen. Anfänge ärztlicher Berufstätigkeit von Frauen in Berlin 1893–1935* (Weinheim: Deutscher Studien-Verlag, 1995), 78–81.

92 Landesregierung, *Das Sanitätswesen in Bosnien*, 338.

women turning to public hospitals had risen from approximately 10% in 1895 to about 14% in 1900.[93] They ascribed this success, above all, to the activities of the female health officers, who were obliged to treat all patients asking for aid free of charge. In fact, one third of the 19,000 patients treated by female health officers between 1893 and 1901 were children, and only two percent were men.[94] While Islamic practice had originally justified the employment of female physicians who, consequently, were based in predominantly Muslim districts, female health officers did not treat Muslim women alone. Between 1892 and 1900, the percentage of female Muslim patients fluctuated between 50% and 64% in Mostar and between 36% and 74% in Tuzla. In Sarajevo, a total of 70% (1899–1901) of patients were Muslim women while in Banja Luka the figure only reached 46.2% (1900).[95]

Austro-Hungarian female health officers participated in the systematic medical examination of the entire Bosnian and Herzegovinian population as well as in vaccination campaigns that were combined with efforts to popularize modern obstetrics, hygiene and the utilization of public health. As Krajewska noted, the hygienic education of Bosnian women was primarily designed to inform on the dangers of infection, particularly syphilis, and offered ways in which to prevent their spread.[96] As the reports of female health officers reveal, their secondary concern was to popularize modern obstetrics, infant care and child rearing. One official argument for employing female physicians had been the Muslim women's need for modern obstetrics which, according to the public health report of 1903, had merely led them to consult "superstitious charlatans who do much mischief."[97] As female health officers indeed reported, they were only called upon to assist in complicated deliveries—after relatives and native midwives had failed—or when it was already "too late."[98] In order to improve this situation, female health officers were employed to "influence their ways of living and of child

93 Landesregierung, *Das Sanitätswesen in Bosnien*, 400–401 (chart no. 74).

94 Landesregierung, *Das Sanitätswesen in Bosnien*, 398–399.

95 Landesregierung, *Das Sanitätswesen in Bosnien*, 402 (chart no XXXVI).

96 Theodora Krajewska, "Expérience d'une Femme Médicin à Dolnja Tuzla (Bosnia)," in Rosalie Schoenflies et al., ed., *Der Internationale Kongress für Frauenwerke und Frauenbestrebungen in Berlin, 10. bis 26. Sept. 1896* (Berlin: Hermann Walther, 1897), 188.

97 Landesregierung, *Das Sanitätswesen in Bosnien*, 17.

98 See Theodora Krajewska, "Jahresbericht der Amtsärztin Dr. T. Krajewska in D.-Tuzla für das Jahr 1897," *Wiener Klinische Rundschau* 12 (1898): 567.

education."[99] These officers were also taught to collect ethnographic data on Muslim customs, which they did, drawing more or less heavily upon Orientalist stereotypes. Krajewska, for example, did not appear to exhibit much sympathy for her Muslim clientele. She stated that the "idle, passive way of living" for married women meant that staying in "locked and unaired rooms" was as unfavorable for health and hygiene as the custom of veiling. Moreover, Krajewska criticize them for their "false ideas" on "hygiene in childbed," as well as over their "nourishment of women in childbed and of infants," for breast-feeding children over one year, for their assumed reluctance or inability to apply measures of modern infant care (such as feeding their babies with sterilized cow's milk on a regular basis) and for their assumed inability to "educate their children."[100]

At an 1896 international feminist congress in Berlin, Krajewska expressed her conviction that Bosnians and Herzegovinians were unable to articulate their needs on their own. She thus defended the colonial prerogative to "import civilization and progress from outside" and, as she put it, impose it "from above."[101] Moreover, Krajewska made it very clear that educating "Bosnian women", and in particular Muslim women without political rights or social status, constituted the very core of the "civilizing mission" for Western vis-à-vis "Oriental" women—ultimately aimed at transforming gender and family relations according to a Western model.

This Orientalizing discourse culminated in a campaign to combat osteomalacia as a specific disease afflicting Bosnian and Herzegovinian Muslim women. In 1900, Krajewska published a detailed article on "endemic osteomalacia" in the district of Tuzla.[102] Osteomalacia, the adult equivalent of rickets in the nineteenth and early twentieth century, was considered a disease of unknown aetiology.[103] Since at least the late eighteenth century, casual observation had suggested that osteomalacia was a purely

99 Krajewska, "Expérience d'une Femme Médicin," 188.

100 See Krajewska. "Jahresbericht 1897," 566–7, 581–2; Theodora Krajewska, "Jahresbericht der Amtsärztin Dr. T. Krajewska in Sarajewo für das Jahr 1902," *Wiener Medizinische Wochenschrift* 53 (1903): 1778–82.

101 Krajewska, "Expérience d'une Femme Médicin," 185.

102 Theodora Krajewska, "Osteomalacie in Bosnien (Kreis Dolnja Tuzla)," *Wiener Medizinische Wochenschrift* 50 (1900): 1785–1788, 1824–1828, 1893–1896, 1930–1935, 1982–1986, 2022–2024, 2074–2078, and 2134–2138.

103 Adolf Strümpell, *Lehrbuch der speciellen Pathologie und Therapie der inneren Krankheiten*, vol. 2 (Leipzig: Vogel, 1922), 55 and 165.

female disorder, resulting in the deformation and narrowing of the pelvis, making childbirth an increased risk. In the 1870s, osteomalacia became increasingly described as an endemic disease occurring especially in mining regions like Flanders, the Rhineland and Westphalia. Male osteomalacia was also observed at this time, but its different aetiology was taken for granted. Osteomalacia itself referred to both the "abnormal flexibility" and the "abnormal brittleness" of the bones; in other words, the term osteomalacia was used to describe various bone disorders such as osteopenia and osteoporosis.[104] Osteomalacia had become a heavily debated disease in medical journals by this time. It was mainly classified as a female sexual disorder, first treated by German gynecologist Hermann Fehling (1847–1925) in 1879 by means of "female castration."[105] Also under discussion was the supposedly endemic nature of osteomalacia—which Krajewska had established purely on the basis of fifty cases—affecting Muslim women living in the mountainous regions of the district of Tuzla.[106] Krajewska mentioned the "damp climate," "lack of sun," poverty, malnutrition and "Muslim customs" as causes for the endemic character of the disease. With regard to the latter, she singled out early marriage in particular, "sometimes as soon as at the age of 12–13 years, i.e. even before menstruation has appeared," too much time spent indoor and the custom of veiling in public in addition to excessive sexual demands on he part of Muslim men, overly frequently numerous pregnancies and extended lactation periods.[107]

A peculiarity of "endemic osteomalacia" in Bosnia, so Krajewska was convinced, was that the disease did not lead to a distortion of the pelvis, generally considered a characteristic of osteomalacia. Instead, she identified the cause of this difference by explaining that Muslim women "turned their upper body in a way that they moved liked quadrupeds," meaning that pressure was taken off the pelvis, but affected the "spine, sacrum and coccyx."[108] Krajewska therefore treated her patients with cod liver oil and

104 Markus Wernly, *Die Osteomalazie* (Stuttgart: Thieme, 1952), 2.

105 Jutta Bönninger, *Die Osteomalazie als Indikation für bilaterale Oophorektomie im späten 19. und frühen 20 Jahrhundert.* Inaugural-Dissertation (University of Erlangen-Nürnberg, 1980) and Chandak Sengoopta, "The Modern Ovary Constructions, Meanings, Uses," *History of Science* 38 (2000): 435–7.

106 Krajewska reported elsewhere on numerous cases of osteomalacia in Sarajevo and its neighboring villages. See Krajewska, "Jahresbericht 1902," 1781.

107 Krajewska, "Osteomalacie in Bosnien," 1786 and 1824.

108 Krajewska, "Osteomalacie in Bosnien," 1824.

phosphorus, usually resulting in an alleviation of suffering. She also discussed "female castration," the treatment earlier recommended by Fehling, who had assumed a relation between "trophic skeletal disorders and the ovaries." Krajewska was skeptical of Fehling's recommendation, although she agreed with him "in principle." However, she denied his identification of the "pathological function of the sexual organs" with the "function of the ovaries." She identified "gravidity" and "other functions of the sexual sphere (sexual intercourse, menstruation, lactation)" as causes of osteomalacia.[109]

When Krajewska bluntly characterized her female Muslim clientele as animals, she drew heavily on an Orientalist topos previously applied by many nineteenth century female travel writers. The fact that only women travelers were able to report on "harem" and "Oriental" women ensured their characterization of "Oriental" women as unprincipled, indolent and "spineless."[110] Krajewska, indeed, continually emphasized that only Muslim women, and never Christian women, suffered from osteomalacia. Thus, Krajewska—as with so many of her contemporaries—resorted to a specific female Orientalism that used these "uncivilized" woman "without rights" to underline the necessity and urgency of female emancipation at home.[111] When debating the lot of "Oriental woman," feminists were able to make claims that could not have been otherwise voiced. Krajewska's report on osteomalacia may also be read as a plea for birth control, contemporaneously claimed by Neo-Malthusians, feminists and eugenicists, ensuring their close cooperation. Neo-Malthusians, who formed an international movement after 1880, saw social reforms as based in eugenic improvements in the working classes and by restricting family size. They objected particularly to the physical and psychological degradation of women as forced by faith and tradition to bear a great number of children.[112] In fact, female health officers clearly took certain birth control measures that were not

109 Krajewska, "Osteomalacie in Bosnien," 1827.

110 See, for example, Gabriele Habinger, *Frauen reisen in die Fremde. Diskurse und Repräsentationen von reisenden Europäerinnen im 19. und beginnenden 20. Jahrhundert* (Vienna: Promedia, 2006), 262–77 and Christina von Braun and Bettina Mathes, *Verschleierte Wirklichkeit. Die Frau, der Islam und der Westen* (Berlin: Aufbau-Verlag, 2007), 210–29.

111 See Reina Lewis, *Gendering Orientalism: Race, Femininity and Representation* (London: Routledge, 1996); and Yeğenoğlu, *Colonial Fantasies.*

112 See Ann T. Allen, *Feminism and Motherhood in Western Europe 1890–1970. The Maternal Dilemma* (Basingstoke: Palgrave Macmillan, 2005).

objectionable to authorities in Bosnia. In 1903, for example, the authorities expressed their satisfaction that infant mortality in Bosnia and Herzegovina did not exceed the Cistleithanian average of 25%.[113] Their concern for infant mortality (like elsewhere in the Dual Monarchy) was limited since the birth surplus, which in Bosnia was 1.30% per annum over the period 1886 to 1901 and thus exceeded the birth surplus of any province of the Monarchy, remained considerable.[114] Krajewska and Kecková were thus free to claim that they often advised their patients to have pessaries fitted, as they put it, to "buttress their uterus." Kecková, however, noted that "Muslim women hardly ever, and Christian women only reluctantly [...] decided to take this step."[115] For Krajewska in particular, this failure on the part of (Muslim) women to take the initiative and limit the danger of bearing and breast-feeding resulted in their suffering from osteomalacia, that is, literal spinelessness.

This Neo-Malthusian "scientific" discourse was, without doubt, also aimed in part at women in metropolitan centers like Vienna, as underlined by the fact that the first female health officer, Anna Bayerová, shortly after leaving Bosnia, participated in a feminist campaign organized by the Czech feminist organization "Minerva," which was subsequently joined by the Austrian Social Democrats. As the activists argued, not only "religious," but also "natural" "female shame" should be acknowledged as a reason to further empower female physicians.[116] In 1900, as a result of this campaign, women were finally admitted to medical studies at Austrian universities. Therefore it may be said that a feminism based on the imperial notion of "female Orientalism" in Austria (as anywhere), proved to be one of the most effective strategies to get feminist demands accepted, even if Bosnia remained the only province where female health officers were employed.

113 Landesregierung, *Das Sanitätswesen in Bosnien*, 55.

114 Heinold Helczmanovszki, "Die Bevölkerung Österreich-Ungarns," in *Geschichte und Ergebnisse der zentralen amtlichen Statistik in Österreich*, 379 and 381.

115 Bohuslava Keck, "Jahrebericht der Amtsärztin Dr. Bohuslava Keck über ihre Thätigkeit in Mostar im Jahre 1897," *Wiener Klinische Rundschau* 12 (1898): 502.

116 Martina Gamper, *Die Ärztin gehört für die Frau. Niedergelassene Ärztinnen und Ärztinnen im Sozialwesen in Wien 1900–1938* (Vienna: unpublished phil. master's thesis, 2001), 26.

## Conclusion

The Austro-Hungarian imperial discourse on its "civilizing mission" was particularly closely associated with the establishment of a system of public health and hygiene in Bosnia and Herzegovina. These medical discourses and practices may be considered to be part of a modernizing biopolitics aimed at transforming Bosnia's population into a modern, homogenous "Bosniak" nation. To achieve this goal, special awareness was given to revising traditional gender relations, characterized as "patriarchal" according to the Western bourgeois model of sex difference. However, such cultural politics of "Bosniak identity" coincided with medical discourses and practices which, though aiming at the whole Bosnian population, denounced Bosnia's Ottoman history and the local Muslim populations in particular, the latter specifically being considered representatives of a stereotypical Oriental "culture," which was both antagonistic to modern hygiene and Western-style "modernity" in general.

The conscious denunciation of "Ottomans" became particularly apparent in Austro-Hungarian medical debates over syphilis. Evidently, endemic syphilis in Bosnia gradually spread southward from northern Italy and Carniola. Yet Austro-Hungarian authorities insisted on syphilis being an "Ottoman plague" which, like scabies, favus and leprosy, was initially considered to be more widespread among Muslim than Christian populations. The Austro-Hungarian effort to "eradicate syphilis," however, soon became standard practice across the empire.[117]

The employment of female health officers, whether physicians or not, also became a standard in hygiene campaigns principally addressed at rural women, who were defined as inferior to Western (bourgeois) women in terms of ethnicity, religion or social class. Their education was a main concern for reformers and feminists of the imperialist era, who more or less successfully campaigned for female emancipation at home by pointing to their modernizing achievements for "inferior" women abroad. The cam-

[117] Due to the efforts of Austro-Hungarian health officers in Bosnia, syphilis was acknowledged as a sexually transmitted, endemic disease in the early 1920s. Methods for combating the disease, i.e. by means of more or less compulsory examinations, did not change either in the interwar years, when Serbia continued to combat endemic syphilis, or after World War II, when the Yugoslav government had to do the same.

paign of "female physicians for women" in the German and Czech parts of the Empire is an outstanding example of the way colonialism advanced female emancipation. This feminist discourse on a particularly female "civilizing mission" in Bosnia reinforced Austro-Hungarian claims about a country which, though never subject to colonialism in socio-economic terms, was the object of a characteristically colonial discourse.

# TYPHUS, TURKS, AND ROMA

## HYGIENE AND ETHNIC DIFFERENCE IN BULGARIA, 1912–1944

*Christian Promitzer*

In the spring of 1928, a local typhus epidemic broke out in Sofia—two people, among them a fiscal officer, died. At that time, typhus was one of the most dangerous infectious diseases, feared for its high mortality rate, with doctors putting their lives at risk in suppressing such epidemics. The appearance of this lethal disease (in the capital city of Bulgaria and, furthermore, among the urban middle class) was a matter for discussion for the Bulgarian parliament, which agreed unanimously on the issue. Iliya Yanulov (1880–1962), professor of jurisprudence and deputy of the Social Democrat party, used the incident to criticize Sofia's sanitary services. He complained that the Roma could walk unmolested through the city in their search for garbage. Georgi Danailov (1872–1939), professor of political economy and member of the ruling Democratic Alliance, took up Yanulov's criticism and proclaimed that in the interest of public health, Sofia's Roma population should have been removed from the city long ago.[1]

Two years later, the "Central Directorate of Public Health" (CDPH) [*Glavna direktsiya na narodnoto zdrave*], the central health authority in Bulgaria, published an official report on the general health situation in the country. Here one could read that the highest rates of typhus, although small in number, were to be found in the southwest Bulgarian border districts of Petrich and Kyustendil as well as in the northern Bulgarian district of Ruse. The latter was renowned for its the size of its Turkish population.

1 *Stenografski Dnevnitsi na 22. obiknoveno sabranie* 1, 91 (1928): 1813.

The conclusion of this assessment was rather dry: "The central sanitary administration calls special attention to the delousing of the populations of Turks and Roma of these districts as careful investigations have shown that these groups are the cause for typhus."[2]

It would be problematic to simply interpret such warnings within the narrow framework of medicine. In the 1920s Bulgaria was haunted by various warnings together with both real and perceived threats. This wary state of Bulgarian public and state institutions had its roots in the unfavorable outcome of World War I. The defeat had frustrated aspirations for a Greater Bulgaria: the peace treaty with its reparation payments and postwar inflation had shaken the state's already weak economy and ultimately, resulted in political instability as witnessed in, for example, the *coup d'état* that toppled the government of the left-wing Bulgarian Agrarian Union in 1923; an unsuccessful Communist uprising in the same year; or a severe assault on the elite of the country in 1925, also organized by the Communists, contributing to a feeling of uncertainty among the middle classes and to an introspective mood among Bulgarian intellectuals.[3] The possible danger of typhus emanating from the Turks and Roma—two ethnic groups who were already subject to suspicion because of their perceived failure to integrate into the Bulgarian nation-state—only aggravated an already tense situation. These two ethnic groups were connected with the despised Ottoman past, (the so-called "Turkish Yoke"), which Bulgarian intellectuals held responsible for the Orientalization of their country, and its state of relative backwardness, which became particularly apparent during the interwar period.[4]

The ascription of particular diseases to certain ethnic groups seems to be the resumption of an old tune that had already made the Jews responsible for the Black Death in 1348–49 and which, in the nineteenth century, was recreated by the representatives of classical epidemiology. As Peter Baldwin has pointed out in his work on prevention politics in the modern era: "Each nation had its favorite epidemiological whipping boy: Poles and Galicians for the Germans, Russians for the Swedes, Irish for the English, Spanish for

---

2 *Arhiv na Glavnata direktsiyata na narodnoto zdrave* 1 (1932): 75.

3 See Maria Todorova, "The Course and Discourses of Bulgarian Nationalism," in Peter Sugar, ed., *Eastern European Nationalism in the Twentieth Century* (Washington: American University Press, 1995), 85–88.

4 Mary Neuburger, *The Orient Within. Muslim Minorities and the Negotiation of Nationhood in Modern Bulgaria* (Ithaca: Cornell University Press, 2004).

the French, Catholics for Protestants, while (to paraphrase Tom Lehrer) everybody feared the Orient."[5] It is evident that in the nineteenth century the importance of this phenomenon has to be considered within the context of nation-building. Ethnic groups and nations are created by ascription and self-ascription of certain cultural features—and by stressing possible differences with respect to other groups. With respect to the latter, one can assume that if medical language is used to expose a certain group as the focus of diseases, two possible directions can be taken: either their threat derives from an alleged lack of hygiene, which is still interpreted as expression of certain essentialized *cultural* features, or they pose a threat to the majority population because of their imagined *biological* difference. In his seminal work on typhus in Eastern Europe, Paul Weindling has shown how German medicine assumed the existence of a correlation between typhus and Jews and how the biologization of this connection played a part in planning the Holocaust.[6] But where is the line between the "explicit racist belief in biological differences" on the one hand, and the "essentialization of cultural features"—which proceeds from a *Leitkultur*, from a "leading culture", and according to Etienne Balibar can be described as "racism without races"[7]—on the other? Is this line detectable on the basis of the possible ideological uses of hygiene and health? One can assume that a "softer," assimilative option—standing for overcoming cultural differences under the sign of a *Leitkultur*—expresses itself in the intention to teach a certain group to adopt the hygienic norms of the majority population and abandon its own traditional customs and forms of folk hygiene, ensuring, within a preexisting hierarchy of powers and values, both the health of the minority group and of the majority. The alternative, that is, the hygienic containment of "dangerous" minorities, in order to protect the majority, stands for the hard-core option of biological racism. In both situations, hygiene appears as a complex of norms and as a discipline that can be interpreted either in the sense of Norbert Elias' "civilizing process" or Michel Foucault's "disci-

5 Peter Baldwin, *Contagion and the State in Europe, 1830–1930* (Cambridge: Cambridge University Press, 1999), 552.

6 Paul J. Weindling, *Epidemics and Genocide in Eastern Europe, 1890–1945* (Oxford: Oxford University Press, 2000).

7 Etienne Balibar, "Is There a Neo-Racism?" in Etienne Balibar and Immanuel Wallerstein, *Race, Nation, Class: Ambiguous Identities* (London: Verso, 1991), 17–28.

plinary society." In either case, it stands for the use of certain techniques for controlling people.[8] Examining the case of Turks, Roma and other Muslim minorities in Bulgaria offers the possibility to explore the above mentioned argument. Here, in a Southeastern European state significantly influenced by Central European medicine, typhus was connected strongly with Muslims, similar to the role played by the Jews in Nazi Germany.

The Bulgarian example may be significant in another respect, shedding light upon the connection between medicine and the concept of Orientalism implicit in Peter Baldwin's quotation. In the nineteenth century the plague made way for Asiatic cholera, which became one of the main epidemiological threats for the European continent in that time. The Mediterranean countries and the Austrian Empire, (whose border with the Ottoman Empire stretched across the Balkan Peninsula), had a long record of epidemic defense and were in the forefront of the fight against diseases coming from the Orient.[9] In the late nineteenth and early twentieth centuries, many young Bulgarians appear to have learnt about these threat scenarios as students of medicine in the Habsburg Empire, namely in Vienna, Prague, Graz and Innsbruck.[10] After their return home, the Muslim minorities of their country formed the most vivid legacy of the Ottoman Empire for them.

Before the Eastern Crisis, around 1.5 million Muslims lived within the territory of Bulgaria; making up roughly 40% of the population. During the fighting or as a consequence thereof, 260,000 died, and until 1889 a further half a million Muslims left the cities and the countryside, seeking refuge behind the new borders of the Ottoman Empire.[11] Other sources refer to 350,000 Turkish emigrants from Bulgaria in the years 1878–1912.[12] When Bulgaria annexed further Ottoman territories in the First Balkan War of 1912–1913, massacres and expulsions were once again on the agenda.[13]

8 While Peter Baldwin supports Elias' view of the civilizing process, Philipp Sarasin stands for a Foucauldian approach in the history of medicine. See Baldwin, *Contagion and the State*, 540–541 and Philipp Sarasin, *Reizbare Maschinen. Eine Geschichte des Körpers 1765–1914* (Frankfurt a/M: Suhrkamp, 2001), 452–465.

9 Baldwin, *Contagion and the State*, 211–236; Erna Lesky, "Die österreichische Pestfront an der k.k. Militärgrenze," *Saeculum* 8 (1957): 82–104.

10 Ivan Balkanski, "Lekarskiyat kadar na Bulgariya," *Arhiv na Glavnata direktsiyata na narodnoto zdrave* 1 (1932): 438–439.

11 Figures according to Justin McCarthy, *Death and Exile. The Ethnic Cleansing of Ottoman Muslims, 1821–1922* (Princeton: The Darwin Press, 1995), 88–91.

12 Ali Eminov, *Turkish and other Muslim Minorities of Bulgaria* (London: Hurst & Company, 1997), 78–79.

13 McCarthy, *Death and Exile*, 135–177.

These incidents and further emigration notwithstanding, the share of Muslims among the whole population of about six million was still about 15% in 1926. Among them, Turks formed the principal minority with about 580,000 members. They tended to live in ethnically homogenous villages in the northeastern and southeastern parts of Bulgaria. The Roma were the second largest minority group in Bulgaria with about 130,000 members, who were in large part also Muslim, with many of them identifying themselves as Turks.[14] Two thirds of Roma lived dispersed across the country, the rest in (often but not always) separate *mahalas* [town quarters].[15] The Pomaks were the third largest minority group with approximately 100,000 members and inhabiting the Rhodope mountain range in the southern part of Bulgaria. Since the Pomaks spoke Bulgarian, it was attempted to convert them (by force) to the Orthodox Church.[16] In 1878 the urban Muslim upper class had already left the country. The Muslims who remained were in their overwhelming majority a rural population. Illiteracy among this group was about 80% in 1926, which was partly the consequence of the Bulgarian state's restrictive policy vis-à-vis Turkish private schools.[17] Their agrarian way of life together with their religious beliefs ensured that they remained secluded and patriarchal. From the perspective of Bulgarian city dwellers, Muslims were considered resistant both to modern life and to the rules of hygiene.[18] The unhygienic conditions under which these minorities lived were indeed worse than those of the Bulgarian urban middle class. This

14 Figures according to Georgi T. Danailov, *Izsledvaniya varhu demografiyata na Balgariya* (Sofia: Darzhavna Pechatnitsa, 1931), 353.

15 Actually, religious affiliation among the Roma is rather difficult to estimate since they use it as a strategy of survival, adopting the religion of their respective neighbors. See Eminov, *Turkish and other Muslim Minorities*, 114; on the history and religious practice of the Roma in Bulgaria, see Elena Marushiakova and Vesselin Popov, *Gypsies (Roma) in Bulgaria* (Frankfurt: Peter Lang, 1997), 27–34 and 122–129.

16 Velichko Georgiev and Stajko Trifonov, *Pokrastvaneto na Balgarite mohamedani 1912–1913. Dokumenti* (Sofia: Akademichno Izdatelstvo 1995); Neuburger, *The Orient Within*, 92–95, 123–124 and 148–149; Evangelos Karagiannis, *Flexibilität und Definitionsvielfalt pomakischer Marginalität* (Wiesbaden: Harrassowitz, 2005) (Balkanologische Veröffentlichungen #42), 77–87; Ulf Brunnbauer, "Ökologie, Gesellschaft und Kultur in den Rhodopen: christliche und muslimische Haushalte im späten 19. und frühen 20. Jahrhundert," in Klaus Steinke and Christian Voss, eds., *The Pomaks in Greece and Bulgaria. A model case for borderland minorities in the Balkans* (Munich: Verlag Otto Sagner, 2007), 15–47.

17 See Mila Mancheva, "Image and Policy: the Case of Turks and Pomaks in Inter-war Bulgaria, 1918–44 (with special reference to education)," *Islam and Christian–Muslim Relations* 12, 3 (2001): 366–367.

18 For a contemporary, explicitly Bulgarian perspective, see Danailov, *Izsledvaniya varhu demografiyata na Balgariya*, 385–403; for more details cf. R. J. Crampton, "The Turks in Bulgaria, 1878–1944," *International Journal of Turkish Studies* 4, 2 (1989): 43–78.

circumstance alone explains the abundant occurrence of lice, which were the carriers of typhus, and consequently the imputed endemic character of typhus among these groups. But we should keep in mind that these conditions say as much about political power relations, social inequality and lack of access to basic hygiene and health care as about cultural difference.

In the following, this chapter will give a survey on the first epidemics of typhus during the Balkan Wars and World War I and then show how, in the interwar period, medicine and hygiene served the power relations between the Bulgarian nation-state and its Muslim minorities. The chapter will also explore how this medical discourse manifested itself in administrative orders and medical practice, namely in a massive delousing campaign and the quarantine of ethnic minorities starting in the mid-1930s to reach its peak in 1943–44. Finally, this chapter will discuss the phantasm of "impurity" with which marginalized groups were burdened within the Bulgarian nation-state and whether this phantasm was an echo of Western views about different regimes of "cleanliness" in European "core zones" on the one hand and in the Balkans on the other.

## Typhus as War Epidemic (1912–1918)

> After the stage of incubation, which [...] lasts about 14–21 days, there is a feeling of sickness: [ ... ] gradual loss of appetite, malaise, avoidance of any mental and physical exertion, general fatigue; pains [...] in the neck, in the muscles of the extremities, particularly in the legs and joints; nerve pain in the upper jaw cavities, catarrh, eventually nose bleeding, drowsiness, restless and interrupted sleep, frequent appearance of chills and constipation. The body temperature [...] goes up to 40° and more. The disturbances of the nervous system, which have already appeared, become very intense. There is a deep apathy, [ ... ] and soon an abnormally deep sleep, accompanied by delirium. If awakened, the sick are reluctant to answer and are often dizzy. Between the 3rd and 5th days after the temperature has increased, the exanthem appears in the shape of small spots the size of millet grains up to lentil grains, and sometimes even up to maize grains. [...] The exanthem appears first on the back [ ... ] in the shape of tiny, reddish spots, which spread [ ... ] out from the back to the lumbar spine, buttocks and the forward parts of the breast, and finally to the upper and lower extremities. When the spots have fully developed the

> skin assumes a smudgy, leaden-bluish pigmented color. In malignant cases the spots usually do not change color until death occurs.[19]

This description of the initial symptoms of typhus was presented in May 1916 in Warsaw to the German medical elite by the Bulgarian Sanitary Officer Dr. Bogdan Chavov (1869–1936) at an extraordinary convention of the *German Congress of Internal Medicine* [*Außerordentliche Tagung des Deutschen Kongresses für Innere Medizin*] devoted to war epidemics. Chavov was one of the first Bulgarian physicians to gain experience with typhus during the Balkan Wars. His performance at the congress owed less to what he had to say than to the necessary integration of army doctors from the other Central Powers. Although for the German army the main combat zone where this disease was relevant was the Eastern Front,[20] in the Balkans the Serbian typhus pandemic that had lasted from late 1914 until mid-1915 and affected 200,000 persons, of which 60,000 died, was still quite present.[21]

Typhus made its first major appearance in Bulgaria during the First Balkan War (1912–1913). Public health authorities at that time believed that this disease had been previously unknown in the country.[22] However, a close examination of earlier sources indicates that in the late nineteenth century typhus was not unknown in Lovech, (a town in north-central Bulgaria). Allegedly, typhus had been brought to the town from Svishtov, then one of the largest Bulgarian Danube ports.[23] Other authors writing in the 1940s believed that typhus had been endemic in Bulgaria at least since the Eastern Crisis of 1875–1878, when the country was ravaged by Ottoman, Russian and Romanian troops.[24]

19 See Wilhelm His and Wilhelm Weintraud, eds., *Verhandlungen der ausserordentlichen Tagung des Deutschen Kongresses für Innere Medizin in Warschau am 1. und 2. Mai 1916. Kriegsseuchen und Kriegskrankheiten* (Wiesbaden: J. F. Bergmann, 1916), 184–185.

20 For the most thorough survey on German efforts to combat typhus and lice see Weindling, *Epidemics and Genocide*, 73–108.

21 Richard Pearson, et al., eds., *Typhus Fever with Particular Reference to the Serbian Epidemic* (Cambridge, Mass.: Harvard University Press, 1920); and William Hunter, "The Serbian Epidemics of Typhus and Relapsing Fever in 1915: Their Origin, Course and Preventive Measures employed for their Arrest," *Proceedings of the Royal Society of Medicine* 13, 2 (1919): 29–158.

22 Stefan Vatev, *Opasnite zarazitelni bolesti* (s. l. s.d. [1915]), 1.

23 Petar Orahovac, "Sanitarnata chast v Lovchanskiy okrag II," *Meditsina* 1, 3 (1894): 8.

24 Ivan Golosmanov, "Voyni i epidemii," *Balgarski meditsinski pregled zar zarazni bolesti, mikrobiologiya i parazitologiya* 4 (1–2) (1940): 9–10, 121; and Lyubomir Tsvetkov, "Petnist tif (sashtnost, razprostranenie, inaparentni formi i borba s bolestta)," *Godishnik na Sofiyskiya universitet. Meditsinski fakultet* 21 (1941–1942): 447.

But it is true that typhus only became widely known to the Bulgarian public at the beginning of 1913, when the first cases were documented among Ottoman prisoners of war who had brought the infection from Asia Minor and Arabia. Bulgarian civilians who had come into contact with the Ottoman army also contracted the disease. Furthermore, Serbian detachments aiding the Bulgarian army on the Thracian front spread the disease. A few hundred cases occurred, but they were by far outnumbered by cholera victims, costing the lives of at least a thousand Bulgarian soldiers.[25]

A regimental doctor, Minko Balevski (1876–?), provided the first comprehensive report detailing the outbreak of an epidemic of typhus shortly after the Second Balkan War, in the summer of 1913. This epidemic occurred among 500 Ottoman prisoners of war sent to the barracks of the Bulgarian town of Sevlievo. It also affected 187 Bulgarian artillerymen who had recently returned from the front. The infected were isolated, all other POWs and soldiers being lodged in bivouacs. The barracks were disinfected with insecticide, carbolic acid, green liquid soap and fuel oil, while the barrack's yard was washed with boiled milk. Since the bathhouse was out of order, the soldiers were quartered in tents without washing facilities or having changed their underwear.[26] This induced the regimental doctor to challenge the recent discovery of the French bacteriologist Charles Nicolle (1866–1936) that lice were the carrier of typhus: "They [the soldiers] went there with their lice and with their nits; nothing would have prevented the latter from playing their infectious role, if they have such at all, but we observed, however, that the disease quickly stopped spreading."[27] The regimental doctor mentioned further that he had chased away the wife of a sergeant, who had snuck to her sick husband in his tent; the woman fell ill ten days later. In contrast to subsequent observations about the incredible mobility of lice, the doctor deemed 1–2 minutes of close contact too short for lice to have crept over to her. "Therefore with

---

25 See Petar Orahovac, "Sanitarnata sluzhba v Sofiya prez voynite v 1912–1913 g.," *Sbornik na Balgarskata akademiya na naukite* 5 (klon prirodo-matematichen 2) (1915): 23–28; Vatev, *Opasnite zarazitelni bolesti*, 1–2; Golosmanov, "Voyni i epidemii", 121;

26 Minko Balevski, "Varhu petnistya tif v Sevlievskiya garnizon," *Balgarski lekar* 8, 1 (1914): 20–21.

27 Balevski, "Varhu petnistya tif," 22.

respect to lice," he came to the improbable conclusion, "they hardly play a more important role in the spread of the disease than they do in case of other infectious diseases."[28] Due to the low degree of lethality (4.8%),[29] one cannot even be totally sure in retrospect that typhus was really the cause of this epidemic.

Typhus had definitely reached Bulgaria with the outbreak of World War I. In late 1914, infected Macedonian POWs who had served in the Serbian army were transferred from Austria-Hungary to Bulgaria. Disinfection was conducted haphazardly. Consequently, the disease spread among the medical personnel and then to the civilian population. By the end of January 1915 about 1,400 cases of typhus had been recorded. Another source of infection were Serbian deserters and refugees who had sought asylum in Bulgaria.[30] The *Directorate for the Maintenance of Common Health* [*Direktsiya za opazvane na obshtestveno zdrave*], as the central Bulgarian health authority was known at that time, met with harsh criticism due to its alleged carelessness in dealing with the epidemics. Not only were the public alarmed by articles in Bulgarian dailies, also the Bulgarian Doctors' Union offered scathing criticism.[31]

In March 1915 a *Law on the fight against Epidemics: Typhus, Cholera, and the Plague* [*Zakon za borba protiv epidemite: petnist tif, holera i chuma*] was passed.[32] In the implementation provisions the louse was officially recognized as the carrier of typhus. For the delousing of body and hair, the application of a lotion that contained Creolin and the fungicidal agent Itraconazole (*Itasol*) was suggested. For the disinfection of clothes the provision anticipated the use of disinfecting machines or simply hot flatirons.[33] The implementation of the law, however, came too late to have an effect on the typhus epidemic in the winter of 1914/1915.

28 Balevski, "Varhu petnistya tif," 22.

29 Balevski, "Varhu petnistya tif," 24.

30 Wasil Mollow, "Beitrag zur Therapie des Flecktyphus," *Wiener Medizinische Wochenschrift* 65, 23 (1915): 885; Konstantin Karadzhov and Ivan Teodorov, "Epidemiyata petnist tif v Ruse prez zimata 1918–1919 godina," *Letopisi na Lekarskiya Sajuz* 13, 2 (1920): 5–6; Golosmanov, "Voyni i epidemii", 121–122.

31 "Hronika,"*Letopisi na Lekarskiya Sajuz* 12, 1–2 (1915): 99–100; "Hronika,"*Balgarski lekar* 8, 2 (1915): 103–105, 108, 112–115; and "Sofiyskiyat klon na Lekarskiya Sajuz i borbata s epidemite ot petnist tif, holera i pr.," *Balgarski lekar* 8, 3–4 (1915): 188–191.

32 "Zakon za borba protiv epidemite: petnist tif, holera i chuma,"*Letopisi na Lekarskiya Sajuz* 12, 1–2 (1915): 87; and "Hronika," *Letopisi na Lekarskiya Sajuz* 12, 3–4 (1915): 177.

33 *Nastavleniya i pravila za predpazvane ot zarazitelni bolesti* (Sofia: Darzhavna Pechatnitsa, 1915), 17, 19–20.

When Bulgaria joined the Central Powers and attacked Serbia in autumn 1915, almost all physicians were drafted into the army. In view of the seriously weakened public health system the already mentioned Social Democrat Iliya Yanulov demanded the "public health mobilization of the population."[34] On his initiative, a *Central Hygiene Council* (CHC) [*Tsentralen Higienicheski Savet*] was established to preside over 2,500 local hygiene councils in villages and towns throughout Bulgaria. These were manned by the few remaining local physicians, mayors, local representatives and teachers. Using brochures, the members of the councils instructed the population in preventative measures and controlled the street, and public and private facilities. In an appeal to the population, the Central Hygiene Council, literally declared war on the lice and haughtily announced their impending death.[35]

Yet the ongoing war created favorable conditions for this "disease of misery and uncleanliness."[36] Due to the lack of manpower, harvests were bad and food shortages became a permanent plague, bringing the population of the towns to the brink of famine. This was the result of massive requisitioning as well as, of the abuse of the food purchasing privilege that the Austrians and Germans enjoyed in Bulgaria.[37] When in early 1917 the cases of typhus more than tripled within a month, the CHC called in vain on the government and the "Central Committee for Economic and Social Welfare," insisting that without a sufficient food supply an "irreparable situation," i.e. the outbreak of a pandemic, could not be excluded.[38] The local hygiene councils could do nothing more than organize provisional disinfection programs, the promised disinfection machines being unavailable until September 1917.[39]

The situation was also critical on the front. While in action Bulgarian soldiers had already been in contact with infected Serbian soldiers, and after the occupation of the Serbian part of Macedonia, the Bulgarian army feared infec-

34 Iliya Yanulov, "Razvitie na socialnoto zakonodatelstvo v Balgariya," *Godishnik na Sofiyskiya universitet. Yuridicheski fakultet* 33, 6 (1939): 120.

35 Russev, *Petnistiya tif*, 15.

36 *Nastavleniya i pravila za predpazvane ot zarazitelni bolesti*, 17.

37 R. J. Crampton, *Bulgaria* (Oxford: Oxford University Press, 2007), 212–214.

38 Yanulov, "Razvitie," 119.

39 Yanulov, "Razvitie," 117.

tion from the local population.[40] However, the Bulgarian army was in a better situation than the civilian population with respect to medical treatment. In late 1915 the headquarters of the army published instructions for combating typhus and relapsing fever, with detailed directions for delousing and disinfection and, what is more, an extensive description of typhoid, typhus and relapsing fever in order to diagnose the patient's condition.[41] It appears that this strategy was influenced by German advisers. In the late autumn of 1915 the German army sent a group of hygienists to the newly established Macedonian front against the forces of the Western *Entente* that had disembarked in Salonica. Among the German doctors were Ludolf Brauer (1865–1951), who was well-known for his expertise in the study of tuberculosis, the bacteriologist Wilhelm Kolle (1868–1935) and Friedrich Fülleborn (1866–1933), an expert in tropical medicine.[42] It is also important to mention the German hygienist Peter Mühlens (1874–1943) who, after having been personally enlisted by the Bulgarian King Ferdinand, held the post of chief hygienist in the Second Bulgarian Army between 1915 and 1918.[43] In the town of Strumica the Bulgarian hygienists Toshko Petrov (1872–1942) and Ivan Kesyakov (1871–1965) became familiar with the clinical picture of typhus under Brauer's guidance. Petrov served as Sanitary Inspector at the headquarters of the Bulgarian army. In his reports he emphasized the progress of the organization of the sanitary service since the Balkan Wars. But he also noted the inexperience of the Bulgarian doctors in the application of disinfection machines. Furthermore, he downplayed the role of the German hygienists in combating typhus. According to their respective accounts both Petrov and Mühlens claimed for themselves the leading role in the introduction of mass delousing operations among soldiers and civilians, thereby preventing the outbreak of larger typhus epidemics.[44] (See fig. 1)

40 Golosmanov, "Voyni i epidemii," 10.

41 Ministerstvo na Voynata, Voenno-sanitarna chast, *Nastavlenie za borba s petnistiya i vazvratniya tif* (Sofia: Darzhavna Pechatnitsa, 1915).

42 Tsvetkov, "Petnist tif," 447.

43 Peter Mühlens, "Kriegshygienische Erinnerungen, "*Archiv für Schiffs- und Tropenhygiene* 43, 12 (1939): 531; Mühlen's connections to Bulgaria were rekindled prior to and during the early years of World War II, when he advised the Bulgarian Army on ways to combat malaria. See Stefan Wulf, *Das Hamburger Tropeninstitut 1919 bis 1945. Auswärtige Kulturpolitik und Kolonialrevisionismus nach Versailles* (Berlin: Dietrich Reimer Verlag, 1994), 110–116.

44 Toshko Petrov, "Iz istoriyata na zdravnata sluzhba v armiyata v nas prez svetovnata voyna 1915–1919," *Balgarski higienen pregled* 11, 1 (1941): 7–18 and 11, 2 (1942): 109–146; and Mühlens, "Kriegshygienische Erinnerungen," 534–535.

Fig. 1. Delousing in the communications zone (Simitli, Southwestern Bulgaria). Source: Peter Mühlens, "Kriegshygienische Erinnerungen," *Archiv für Schiffs- und Tropenhygiene* 43, 12 (1939): 531–561.

Between 1914 and 1918 official statistics recorded 12,148 cases of typhus, of which 1,445 were lethal. These numbers peaked in 1917 with 6,697 cases and 764 deaths from typhus.[45] The 4,089 cases (and 554 fatalities) of 1918 counted by Toshko Petrov in the district of Drama, appear to have been excluded from the official statistics, possibly because after the war, Drama was reabsorbed into Greece.[46] If we are to believe Iliya Yanulov, there were even 61,144 cases and 5,111 deaths among the civilian population of Bulgaria for the years 1916–1918. However, these high numbers have never been discussed by other Bulgarian authors.[47] Officially, it was recognized that in the Bulgarian army there were 2,145 cases of typhus (of which 247 were fatal). Once again, Yanulov offered different estimates, cit-

45 Petar E. Verbev, *Petnistiya tif i vashkata* (Sofia: Pridvorna pechatnitsa, Sofia 1935), 18.

46 Toshko Petrov, "Po borbata s petnistiya tif v dramskata voenna inspekcionna oblast prez svetovnata voyna 1917–1918," *Balgarski higienen pregled* 11, 3 (1941): 209.

47 Ilya Yanulov, "Socialna politika na Balgariya prez vreme na voynata ot 1915–1918 god," *Spisanie na balgarskata akademiya na naukite i izkustvata* 62, 1 (1941): 115.

ing approximately 4,000 cases of which 10% were fatal.[48] We can assume that the official data only showed the lower threshold because not each case of typhus was diagnosed as such.

The end of the war brought the return of soldiers and a continuous influx of refugees from the lost territories of Macedonia and the Aegean coast. In late 1918 a typhus epidemic in Ruse (which had been thought to have been contained already in October) again flared up and lasted until the middle of 1919. Because of its persistence, experts believed that typhus had already become endemic in the city and indeed the whole district of Ruse.[49] Undeniably, 5,671 cases of typhus (of which 726 were fatal) were registered in 1919 throughout the whole of Bulgaria—the second highest figure recorded to that date.[50] Typhus was especially proliferate in places where people lived in crowded conditions, suffered from malnutrition and had no washing facilities, i.e. in the POW and refugee camps, prisons and in extremely poor urban quarters, particularly Roma residential areas.[51] A noticeable decrease was observed in 1921. This was also the last year in which the Central Hygienic Council was active. In 1924 only 197 cases were registered, which was the lowest figure of the interwar period.[52]

## Muslims and Typhus: The Establishment of a Discourse

The first conflation of typhus and lice with the Muslim population occurred during World War I. It appears that this connection was first made in the southern Dobrudja, which Bulgaria had ceded to Romania after the Second Balkan War. In the course of the successful 1916 campaign, southern Dobrudja once again came under Bulgarian control. Toshko Petrov arrived there in early 1917. He noticed that typhus was common among Russian POWs in the town of Dobrich and could also be found in the Romani, Turkish and Tartar *mahalas*.[53] Discipline and control followed.

48 Golosmanov, "Voyni i epidemii," 10; Yanulov, "Socialna politika na Bulgariya," 115.

49 Karadzhov and Teodorov, "Epidemiyata petnist tif v Ruse," 6–7.

50 Verbev, *Petnistiya tif*, 18.

51 Yanulov, "Socialna politika na Balgariya," 192.

52 Yanulov, "Razvitie", 120; Verbev, *Petnistiya tif*, 18.

53 Toshko Petrov, "Iz istoriyata na zdravnata sluzhba v armiyata v nas prez svetovnata voyna 1915–1918," *Balgarski higienen pregled* 10, 4 (1940): 252–253, 261–262.

Petrov asked the mayor of the town to inform the population—in particular the Turks and Roma—that they had to rid themselves of parasites by boiling their clothes and bed linens. A commission registered the diseased in the town and threatened to return after eight days to examine the citizens. If lice-infested individuals were found their whole families would be sterilized by the military authorities in the disinfection station. After the deadline elapsed, the commission ascertained that the citizens had complied and were without parasites, with one exception: "Three Roma women of salacious character were escorted to the disinfection station and cleansed there. This serves as a good example for the unclean Turks, who adhere to their Mohammedan morale; it will compel them to change their notion that 'a man without lice cannot live.'"[54]

The notion of cultural difference became the leitmotif of the typhus campaign of 1917 in Dobrudja. Petrov held the view that typhus was endemic in the numerous Turkish villages, from whence the disease spread into the Turkish quarters of the towns in the region.[55] In the Turkish villages around the town of Silistra the plan was that should an individual be diagnosed with typhus, the family of the diseased person as well as their neighbors would be deloused. However, if typhus was discovered among Roma, the whole group would be deloused. During these actions the shaving of women became a bone of contention between the health officers and the Muslim population. Petrov conceded that this procedure, although absolutely justified from a medical point of view, was an affront to the moral and religious feelings of the Turks. But, "in view of the current cultural level and the curious understanding of the Turkish population with respect to elementary cleanliness and hygiene, such drastic measures are necessary. And only in this way can adequate results in the fight with this disease be achieved."[56] The only concession Petrov was willing to make was that rubbing the skin of Turkish women with a Creolin lotion and the shaving of their heads could be done by other household members.[57]

54 Petrov, "Iz istoriyata," 262.
55 Petrov, "Iz istoriyata," 252, 262, 270.
56 Petrov, "Iz istoriyata," 273.
57 Petrov, "Iz istoriyata," 276.

It should also be mentioned that in Shumen—a district bordering the southern Dobrudja with a similarly large Turkish population—the same equation of typhus with Muslims took place. In a popular book on "Infection and Disease" published in 1924, the district physician Petar D. Skorchev (1882–?) wrote:

> In the district of Shumen, typhus from its first occurrence (1916) to its disappearance was an almost exclusively Turkish disease. Among the Bulgarian population it never became epidemic. The reason for this selective spreading among others must be found in the fact that a multitude of relatives and neighbors gather around a diseased Turk; women in particular sit together like anchovies in the small room, where they spend the whole day and eat from a common table. Their wide, ballooned pants are pressed together in closest contact and it comes to an intensive exchange of abundantly cultivated lice.[58]

The same doctor observed how the diseased inhabitants of the Romani *mahala* of Shumen used quixotic methods to disguise ill health, either by smoking a pipe or by chewing bread in order to hide the dry and rust-colored tongue:

> This adamant endeavor at concealment and simulation has to be ascribed to the harsh measures of isolation—all who come down with typhus (all of them are poor Turks and Roma) are sent to the hospital (doubtlessly to better conditions than at home), while their closest family members are confined to their houses for 21 days and the *mahala* is blocked by a military cordon.[59]

How can we explain these observations? Muslim groups nourished a deep distrust vis-à-vis the Christian Bulgarian state and its institutions (which were planning to obliterate the Ottoman legacy). Instead they clung to their traditions so that one can even speak of parallel worlds. As Mary Neuburger in her essential work on Bulgarian Muslims has argued: "In many cases, Muslim tradition was often retained or reinvented not as an indication of some sort of essential backwardness, but rather as a response to the dislocation that modernity generated."[60] We should bear in mind that the reli-

58 Petar D. Skorchev, *Zaraza i bolest* (Shumen: Pechatnitsa "Spas Popov", 1924), 127–128.

59 Skorchev, *Zaraza*, 231–232.

60 Neuburger, *The Orient Within*, 12.

gious convictions that restrained the Muslims from making use of medical care reflected a state of "folk Islam" as it had been preserved since the Eastern Crisis. Most urban Muslims who could have been supporters of a modern approach towards medicine had left the country during and in the wake of the establishment of the Bulgarian nation-state. Consequently, a typical sentence in the sanitary reports of the district and county physicians from the beginning of the twentieth century was: "In general the [Bulgarian] population in the county [...] is sufficiently accustomed to medical care and make use of it adequately. A difficulty is the fight against infectious diseases among the Turkish population, who for religious reasons reject medical help."[61] The Bulgarian contempt for this medical ignorance was packaged in instructive anecdotes. When in 1908 a localized typhoid epidemic broke out in the ethnically mixed village of Kostandovo in the central Rhodope Mountains, the Pomak population concealed their dead. The Bulgarian population of the village called a sanitary detachment for help. A female doctor—a remarkable exception to the rule and possibly a gesture to the sensibilities of Pomak women—commanded the detachment. During the examination of a Pomak house the doctor realized that she was being presented with a healthy person lying in the bed of a diseased individual concealed in the wardrobe. She predicted that the healthy person would fall ill and die, while the diseased one would recover. The accidental verification of her prediction impressed the population, so much so that they brought forward their dead and diseased.[62] But even in the late 1920s, male Pomaks would only call on the local surgeon [*feldscher*], when they needed his consent to bury a corpse. The common Bulgarian image of the relationship between Muslims and health care reads as follows:

> They are placid when faced with death; that they do not look for medical help does not bother their conscience because they consider disease a divine affair in which they have no right to interfere. "What is foreordained for man will happen," and every worry is useless—that is the philosophy of the Mohammedan religion. They hold the view that a sick person, if he is not cured and dies,

61 B. T. Aramov, "Otchet za sanitarnoto sastoyanie na balchikskata okoliya prez 1902 g.," *Letopisi na lekarskiya sajuz* 1, 8–9 (1903): 429.

62 N. V. Troyanov, "Okolo edna epidemiya," *Narodno zdrave* 3, 6 (1929): 1.

> honors Allah. This led to the idea that he commits a great sin if he looks for medical help, and in the case of his death his soul will go to hell instead of paradise.[63]

In an exceptional piece of reasoning, Velichko Georgiev (1872–1924), district physician in the county of Peshtera (in the foothills of the western Rhodope Mountains), associated the Muslim refusal of medical care with their rejection of material enrichment, which was an attribute of modernity and typical for the Bulgarian population:

> Although he lives in a miserable dwelling, the Mohammedan has a satiated mind and he is morally satisfied. Even though he sometimes considers himself dejected, he strives for more distant spiritual pleasures like eternal life, moving to the Holy Land and considers himself master of other peoples. In most cases he uses all that he earns for food and keeps his body clean. The intelligent Turks have a contempt for medicine, which interferes in the natural way of life, as it strengthens the weak and invalid elements of society, mocking natural selection, natural evolution and as such causes degeneration in society, rather than improving it physically. Their arguments are based on strong facts: "If medicine is able to achieve something," they say, "then why do people, who do not make use of it, not vanish from the globe?"[64]

These arguments, which Georgiev imputed to Bulgarian Muslims, were apparently influenced by the critique of Western civilization and by the idealization of rural life typical for the "Bulgarian Agrarian Union," to which Georgiev tended politically and for whom, when in power (1918–1923), he would act as head of the "Directorate of Public Health." Georgiev further mentioned that surveys of school children and the medical examination of conscripts in the district had shown that the Muslim population was in a better physical state than the Bulgarian one—with one significant exception: "The weakest, scrawniest and most sickly elements of the county are the Roma. These unlucky people live in absolute misery and indeed represent the pariah in our society."[65] But in general the distrust of the Muslim population was met with disinterest by public health authori-

63 N. V. Trayanov, "Zdravnoto nevezhestvo vsred pomashkoto naselenie," *Narodno zdrave* 3, 3 (1929): 23.

64 Velichko Georgiev, "Otchet za santiarnoto sastoyanie na Peshterskata okoliya prez 1902 g.," *Letopisi na lekarskiya sajuz* 1, 8–9 (1903): 399–400.

65 Georgiev, "Otchet za santiarnoto," 400.

ties, which had, furthermore, to grapple with the problem that, due to the lack of a native medical faculty (which was to be established in Sofia only in 1918), there were not enough physicians in the country to establish an extensive medical care network. Furthermore, it should be acknowledged that it was not very attractive for the young physicians of bourgeois families to serve in distant rural settlements, not to mention villages with a Muslim population.[66] The observations, quoted above, without doubt contributed to an image of Muslims as unclean and pathogenic agents of infectious diseases. However, there certainly existed positive opinions about Muslim hygiene, mainly concentrated on the role of bathhouses, fountains and wells, as well as on the religious instruction to wash before entering or worshipping in a mosque.[67] But these voices were in the minority.

## A New Health Policy toward Muslim Minorities

The nexus between Muslims and typhus was established during World War I. With the retreat of the disease it apparently lost significance until the mid-1920s. How was it that this nexus was revived in the second half of the 1920s, forming the backbone of a new health care policy toward Muslim minorities? We only know for sure that in 1927 the "Directorate of Public Health"—in order to ascertain the spread of typhus and other infectious diseases—introduced the serological Weil-Felix reaction and other bacteriological tests for detecting typhus, which theoretically should have been in use since at least 1922.[68] In 1927 six anti-epidemic detachments with full-time staffs were also founded with each detachment responsible for a specific territory in Bulgaria.[69] One of the areas of activity was Deli Orman, (a hilly region in northeastern Bulgaria) with a compact Turkish popula-

66 P. K. P, "Zashto lekarite byagat ot selata?," *Balgarski lekar* 8, 1 (1914): 5–16 and 8, 2 (1914): 72–82.

67 P. Kiryakov, "Hayrat," *Narodno zdrave* 1, 8 (1927): 1; and Radi Genov, "Higienata na tyaloto u myuslyumanite," *Narodno zdrave* 1, 11 (1928): 83–84.

68 "Okrazhno № 13471, ot 4. X. 1927," *Prituka kam Izvestiya na Direktsiya na narodnoto zdrave* 12, 15 (1927): 229–232.

69 "Okrazhno № 4852, ot 11. IV 1927," *Prituka kam Izvestiya na Direktsiya na narodnoto zdrave* 12, 7 (1927): 90–91; Hristofor Mihaylov, "Usilvane borbata na demokratichnata obshtestvenost za narodno zdraveopazvane, zdravno zakonodatelstvo i progresvini napravleniya v meditsinata prez perioda na fashistkoto upravlenie u nas (1924–1944 )," in Vera Pavlova, ed., *Istoriya na meditsinata v Balgaria* (Sofia: Meditsina i Fizkultura 1980), 194.

tion. A report from one of the detachment's leaders was published in the "Gazette of the Directorate of Public Health" [*Izvestiya na Direktsiyata na narodnoto zdrave*] and presented as best practice. The article described the fight against a typhus epidemic in the village of Mirhanlu (renamed Trem today), a Turkish village of roughly 900 inhabitants situated north of the district town of Shumen where in the first months of 1928, 47 people fell ill and 13 deaths were recorded. The procedure for the containment of the disease ran as follows: isolation of the village; isolation and disinfection of the ill individuals in the school building; quarantine and triple disinfection of the infected houses and their inhabitants—who often were so poor that they only owned one set of clothes—investigation of further cases by visiting the houses of relatives and neighbors who also had to be disinfected; inspection of the neighboring villages; disinfection of lice-infested persons; disinfection of the mosque and, finally, the successful discovery of the original source of the epidemic.[70] The head of the detachment, Georgi Dementev (1877–?), remarked: "The population, who were already quite frightened by the enormous dimension of this misfortune, received our arrival with appropriate confidence and with a rare preparedness to meet all demands."[71] Already aware of the delicacy of shaving the heads of unmarried women, he tried to let wash their hair with paraffin and a lotion of vinegar and mercury chloride. Since this procedure proved unsuccessful, he ordered his detachment to march out in full regalia, armed with scissors and hair clippers, together with a police escort, in order to shave the first group of lice-infested virgins. The following morning, the fathers of the girls next on the list brought them to the commander of the detachment. They were neatly combed and perfectly clean: "The whole night the scared mothers had cleansed the heads of their children under the lamp and pulled out the nits with their own fingernails."[72] The dubious lesson that Bulgarian physicians were intended to learn from this example was that it was not always necessary to use force; instead the mere threat of it could compel an obstinate Muslim population to comply with the requirements of modern hygiene.

70 Georgi Dementev, "Petnist tif v s. Mirhanlu," *Izvestiya na Glavnata direktsiya na narodnoto zdrave* 13, 9 (1928): 354–365.

71 Dementev, "Petnist tif," 355.

72 Dementev, "Petnist tif," 364.

At the end of the 1920s it is possible to observe three different developments in Bulgaria, which would ultimately fuse into a common health policy towards Muslims. We first observe that the official policy of the right-wing government of the "Democratic Alliance" [*Demokraticheski Sgovor*] became restrictive toward the Turkish minority as Turkish schools were closed, allegedly for financial reasons. Kemalist organizations were observed, less because they wanted to persuade the Turkish population to adopt a modern way of life but because of their attempts to strengthen a Turkish national conscience. At the same time, attacks by small fascist groups against Turkish persons, houses and mosques became commonplace.[73]

The second development is connected with the discussion of typhus. In the 1920s, 385,809 persons in Bulgaria suffered from seriously infectious diseases of which 3,644, or less than one percent, suffered from typhus. However, deaths from typhus accounted for 12% of all fatalities.[74] Some experts considered medical statistics from the countryside as untrustworthy; believing instead that the number of typhus cases was increasing.[75] This same supposition formed the basis for the parliamentary discussion of 1928 with which this chapter was introduced.

The final development resulted in health legislation, namely the "Law for Public Health," which became effective in March 1929. It prescribed a novel health policy: curative medicine, which was costly and unaffordable in times of fiscal and economic crisis, faded into the background, while prevention became a stock phrase. For the successful fulfillment of this program the "Central Directorate of Public Health" was granted a broad mandate.[76] Although not explicitly mentioned in the law, it was the Bulgarian nation that was protected by measures of prevention. This unspoken principle put Muslims minorities at disadvantage. Until then, the health authorities had combated typhus among Muslims both for curative reasons and with the intention of protecting the Bulgarian majority population. At this point the CDPH appears to have raised the question of in how far Mus-

---

73 Bilâl N. Şimşir, *The Turks of Bulgaria (1878–1985)* (London: K. Rustem & Brother, 1988), 76–78, 105–111; and Ibrahim Jalumov, *Istoriya na turskata obshtnost v Balgaria* (Sofia: IMIR, 2002), 154–158 and 248–250.

74 *Arhiv na Glavnata direktsiyata na narodnoto zdrave* 1 (1932): 66–77, 72–73.

75 "Ot kakvo umira naroda. Petnist tif," *Narodno zdrave* 1, 16 (1928): 123.

76 *Darzhaven vestnik* 277, March 9, 1929). See Rumen Daskalov, *Balgarskoto obshtestvo 1878–1939*, vol. 2 (Sofia: IK "Gutenberg", 2005), 69–70

lims should be treated other than as possible hosts of parasites endangering ethnic Bulgarians' health. It is arguable whether this anti-Muslim and anti-Turkish stance was simply fueled by a rejection of Islam, which was held responsible for the lack of willingness to accept modern hygienic norms, or whether it was also a reaction to an awakening nationalism among the Turkish minority.

One of the first campaigns of the CDPH was the provision of hygiene books to peddlers and heads of nomadic families in order to control both their wandering and their state of health.[77] This action was clearly aimed at the Roma population, However, in the long run it proved to be impracticable. The second measure was directed at school children, each morning school heads and teachers had to examine their pupils, to check for lice.[78] In spite of these initiatives, in spring 1931 a local typhus epidemic broke out in the central Bulgarian town of Zlatitsa. It was extensively studied by Sanitary Inspector Petar Verbev (1891–1977). Verbev, a former fellow of the Rockefeller Foundation and specialist in epidemiology from the John Hopkins Bloomberg School of Hygiene and Public Health in Baltimore,[79] became head of the "Department for Infectious Diseases" of the CDPH. In this role he designed and coordinated the anti-epidemic activities within the Bulgarian public health sector in the following years. According to Verbev, the epidemic in Zlatitsa had started among the local Roma families before it infected the inhabitants of a neighboring Bulgarian family; a child from that family subsequently infected his Bulgarian schoolmates (see fig. 2). In total, 45 people fell ill, three of whom died. Consequently, the town was sealed, inns and coffee houses were closed and access to religious meeting-places was restricted. Infected households and their inhabitants were disinfected, along with those neighbors with whom they had previous contact. The heads of all lice-infested individuals were shaved. Verbev further recommended the creation of a separate quarter for the Roma population—erected from building materials supplied by the municipality—because "their present housing among the Bulgarian population is dangerous and the supervision of their cleanliness proves to be more dif-

77 "№ 29238-III, ot 19. XII. 1929," *Izvestiya na Glavnata direktsiya na narodnoto zdrave* 15, 58 (1930): 699–700.

78 "№ 17209-III, ot 25. VIII. t. g.," *Izvestiya na Glavnata direktsiya na narodnoto zdrave* 15, 67 (1930): 894–895.

79 Bulgarian Central State Archive – *Tsentralen Durzhaven Archiv* (TsDA) (f-372k, op-1, ae- 586, l-123)

80 Petar Verbev, "Petnist tif v gr. Zlatitsa," *Arhiv na Glavnata direktsiyata na narodnoto zdrave* 1 (1932): 399.

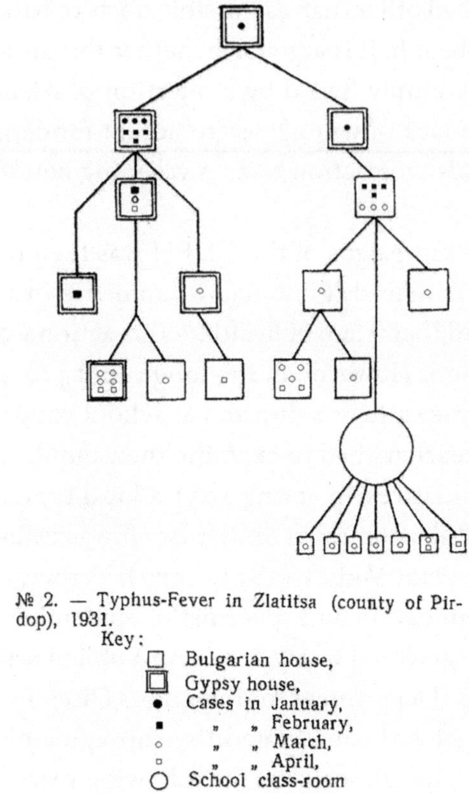


Fig. 2. Source: Petar Verbev, "Petnist tif v gr. Zlatitsa," *Arhiv na Glavnata direktsiyata na narodnoto zdrave* 1 (1932): 390–399.

ficult."[80] The Zlatitsa epidemic led the CDPH to rashly issue circulars addressed to the various exponents of the health care system in Bulgaria, reprimanding district physicians for failing to ascertain infections in time, either because of negligence or wrong diagnoses. Furthermore it was alleged that these doctors had neglected "the important epidemiological fact" that the first cases of typhus had been observed among migrants and Roma, from where they were transmitted to the majority population. "It is a shame for our country that it is still one of the foci of typhus in Europe. It is the duty of the Bulgarian physicians to annihilate this focus."[81] Hospi-

81 "№ 11465-III, ot 12. VI. 1931," *Izvestiya na Glavnata direktsiya na narodnoto zdrave* 16, 79–80 (1931): 1202–1203.

tal heads also came under fire from the CDPH, which became aware of the hospitalization of wrongly diagnosed Roma. They lacked the rash typical for typhus, although they otherwise showed signs of a typical typhus infection. Therefore it was ordered that "it is obligatory to take a blood sample for serological examination of every sick Roma, Turk or migrant, who is hospitalized and in a state of stupor."[82] Finally, the CDPH alerted the Ministry of Education that in two ethnically mixed villages the louse infestation of the pupils of the Turkish schools was at 50–80% considerably higher than those from Bulgarian schools at 14–20%. The conclusion was that if school directors and teachers had followed the rules more vigilantly i.e. if the children had been subject to the appropriate screening, the dissemination of typhus in these villages would have been prevented. This assumption led the CDPH to demand that the fight against louse-infestation be enforced among school children.[83] In the following years the CDPH reiterated these demands, which demonstrates the difficulty in applying them as well as their general ineffectiveness.

In spring 1934 the CDPH made it clear that "a genuine war has to be started against lice." Razgrad, in the region of Deli Orman, was the first district to be subject to an action carried out by anti-epidemic detachments. The Turkish and Roma population, in fear of seclusion, quarantine, delousing and shaving, fled to neighboring villages. To avoid such extreme reactions the CDPH ordered the sanitary detachments to act more quickly and more prudently in future, suggesting briefing the heads of the neighboring villages about the pest that could come over them if they accepted such fugitives. In any case the detachments could resort to the support of the police.[84] A constant bone of contention throughout the campaign was the institute of "*[açık] oda*"; the provision of public guest rooms for visitors and religious people, of which at least one existed per village in the house of a wealthy and prestigious villager. Because they were perceived as "genuine hotbeds for typhus" the detachments were expected to close them down, but only if there was no serious resistance by the population. Other initia-

82 "№ 11466-III, ot 12. VI. 1931," 1205–1206.

83 "№ 11467-III, ot 12. VI. 1931," 1203–1205.

84 "№ 8640-II, ot 10. V. 1934," *Izvestiya na Glavnata direktsiya na narodnoto zdrave* 19, 28–129 (1934): 2237–2238.

tives included engaging the help of the local religious leader, the *hodja,* in order to combat the Turkish superstition that killing a louse was a sin, as well as, providing the poor (who had limited or no underwear) with discarded underwear from the hospitals.[85]

As of July 1934 the war against lice was announced in the neighboring district of Shumen, lasting until the end of the year (see fig. 3). Here the mortality rate during the last years had been the highest among all Bulgarian districts and in 1934 it was four times higher than the national average.[86] In this district not only villages where typhus had been endemic but also neighboring and distant villages were submitted to systematic delousing with disinfection machines, autoclaves with sulphur fumes as well as ordinary barrels and kettles. Most of the Turkish guest rooms were closed and abolished, with all flats cleaned and lime-washed. In order to prevent the concealment of clothes in other villages, the examination of the family members was undertaken with the help of communal registers in each village. In total, 97,127 people had to undergo delousing.[87] We can assume that these campaigns were facilitated by the successful coup d'etat of the authoritarian *Zveno* [*Link*] group together with the Military League [*Voenen sajuz*] on 19 May 1934, which had led to the abolishment of the parliament. In 1936 another 40,000 persons were deloused, with the focus on the Turkish areas of eastern Bulgaria.[88]

The focus on the systematic delousing of the Muslim population could no longer be justified as a simple public health measure. On the contrary, the general public needed to be informed about the pathogenic role of the Muslims, as illustrated in a popular brochure on vermin: "The most genuine custodian of typhus are the Roma, for whom the lice are like livestock; then come the Mohammedans, who consider lice a sacred animal and do not kill them; and finally, those lice infested families, who believe that the earth must shake before they clean themselves of their lice."[89] In a widely distrib-

85 "№ 8640-II, ot 10. V. 1934," 2238–2240; on the "açık oda." See Stoyan Genev, "Konfesionalni varianti na narodnata kultura," *Balgarska ethnologiya* 2, 4 (1991): 9.

86 Petar Verbev, "Razprostranenie na ostrite zarazni bolesti v Balgariya prez 14-lyatieto 1921–1934 g.,"' *Izvestiya na Glavnata direktsiya na narodnoto zdrave* 24, 184–188 (1939): 4495.

87 Dimitar Verbev, "Borba s vashlivostta v Shumenska oblast," *Izvestiya na Glavnata direktsiya na narodnoto zdrave* 20, 140 (1935): 2578.

88 TsDA (f-372k, op-1, ae- 2200, l-514).

89 Zahariev, *Borba s domashnite paraziti,* 15.

Fig. 3. Idealized illustration of delousing in the Turkish village of Vrani-kon (district of Shumen) in 1934. Source: *Izvestiya na Glavnata direktsiya na narodnoto zdrave* 30, 140 (1935): 2578.

uted public health pamphlet published in 1939 under the auspices of the CDPH, the public was also informed of the role played by nomadic Roma in the dissemination of typhus. In order to win the fight against typhus, the CDPH equipped most of the sanitary services in the country with new disinfection machines. Although morbidity was already relatively low, greater efforts were needed in order "to annihilate the disease totally."[90] This was an indirect concession, firstly, that the epidemical situation had not drastically worsened throughout the whole interwar period and, second, that preventive public health measures were promoted by a discourse in which the Muslim way of life and "cleanliness" were diametrically opposed. A doctoral thesis by Bulgarian student (published in Leipzig in 1939) focusing on typhus in Bulgaria, discussed this issue in a tone reminiscent of what could be described as "medical Orientalism":

90 *Narodno Zdrave* (Sofia: Druzhestvo na narodno zdrave pri Glavnata direktsiyata na narodnoto zdrave, 1939), 81–82.

> These groups, Turks and gypsies, are the origin and the reason for the dissemination of typhus in Bulgaria, while the admittedly barely hygienic lives and dwellings of the true Bulgarian population is of secondary importance [...] Here I do not want to fail to mention that in spite of this lack of hygiene the Bulgarian peasant is very meticulous with respect to lice. Like any other European he also feels aversion against lice and eradicates them wherever he is able to do so, while the gypsies and the Turk do not feel bothered at all by this vermin.[91]

This argument has its origins in the fact that Bulgarian intellectuals in general, and physicians in particular, felt that within the European milieu they were treated as part of the dreaded Orient. They thus felt obligated to demonstrate the Europeanness of the Bulgarian population, who were ready to learn the principles of personal hygiene. Bulgarian doctors wanted to see their nation on the superior side of what Paul Weindling succinctly described as the "metaphorical divide between the advanced sanitary conditions of Western Europe and a pathogenic and primitive East."[92] Consequently, the Muslims who still clung to their traditional and unhygienic way of life reminiscent of the Ottoman period had to stay on the other side of the figurative divide; for the Bulgarians they became their Oriental "epidemiological whipping boy."

## 'Total Eradication'—Typhus Becomes War Epidemic Again

But not all Muslims were deemed to be equal. In Bulgarian eyes their treatment depended on their respective ethnic background. Thus, the Slavic-speaking Pomaks became subject to assimilation to include them as *Bulgaro-Mohammedani* [Bulgarian-Muslims] within a national modernization project. This project was undertaken by the so-called *Rodina* [Homeland] movement in the late 1930s and early 1940s with the support of the government. This group supported the wearing of Western clothes, the discarding of Muslim names together with conversion to Orthodox Christianity.[93]

91 Angel Ignatoff, *Der Flecktyphus in Bulgarien* (Leipzig: Alexander Edelmann, 1939), 24.

92 Weindling, *Epidemics and Genocide*, xv.

93 On this issue in detail, see Mancheva, "Image and Policy: the Case of Turks and Pomaks," 359–264; Neuburger, *The Orient Within*, 46–48, 52–54, 95–100, 124–126, 147–153; Alexander Velinov, *Religiöse Identität im Zeitalter des Nationalismus. Die Pomakenfrage in Bulgarien* (phil. diss.: University of Cologne, 2001), 116–132.

Hygiene measures had to be handled with care, even when lice infestation reached up to 90% in many Pomak villages in the Rhodope Mountains, as illustrated by an article in the medical press.[94] Instead of mass delousing by the anti-epidemic detachments of the CDPH, one author suggested measures of preventive risk management—the control and delousing of the population should be performed by the community members themselves while the activities of the CDPH should be restricted to "broad and inexorable cultural and education work, the implementation of healthy customs, cultivation of a clear awareness and of a consciousness for cleanliness as well as the development of material and moral conditions for a bearable human life in the villages."[95]

Such voices indeed found resonance within the public health administration. In early December 1939, for instance, the regional Director for Public Health in Plovdiv wrote to the Ministry of the Interior and the CDPH, full of indignation about the despotism of anti-epidemic detachments. In some Pomak villages in the Rhodope Mountains "a unit consisting of a surgeon, a midwife and an attendant rush in upon a settlement for delousing, in other villages they often burn the household goods making up the dowry and in some houses they unclothe young maidens. Early in the morning before dawn the women of whole villages flee into the forests and only return late in the evening."[96] The regional director had justified reasons to ask for a pause in the delousing campaign. He argued that delousing could be risky in winter since people often waited naked during the procedure because they had no other clothing and that—and this was probably the principal reason—interested persons could use the campaign to willfully create a panic in order to influence the result of the forthcoming parliamentary elections. The personal regime, (which had followed the luckless Zveno government), had granted controlled elections on the basis of a single-winner voting system. And indeed, the Minister stopped the campaign "for the moment."[97]

This example shows that in the case of the Pomaks the medical discourse largely followed the mainstream Bulgarian national policy toward

94 Vl. Ivanov, "Borba s parazitite," *Zdrave i zhivot* 8, 31 (1938): 3.
95 Ivanov, "Borba," 3–4.
96 TsDA (f-372k, op-1, ae-2197, l-1).
97 TsDA (f-372k, op-1, ae-2197, l-1).

this group. In the case of the Turks the relationship was more complicated. The Bulgarian government was unable to foster a modernization movement among the Turks similar to the Rodina movement that in the long run would integrate them into the Bulgarian nation. Existing modernist tendencies among the Turkish minority were influenced by Kemalist nationalism, which was oriented towards Turkey, and therefore a red flag to the Bulgarian government. The authorities thus sought refuge in cementing the existing power relations, by supporting the "Old Turks" both in the Islamic hierarchy and in the population. This group was considered religiously conservative and "inert," and therefore, impervious to modern Turkish nationalism.[98]

But this attempt to support the Turkish population in maintaining their traditional way of life was not shared by the public health institutions. The CDPH only attached importance to the fact that the "Old Turks" were not susceptible to sanitizing measures. The best solution was to "contain" them within their traditional environment. But before containment could take place, the danger of epidemics had to be eliminated. This meant nothing less than a crusade for the "total eradication of lice," by way of compulsory delousing conducted by anti-epidemic detachments with police assistance. This plan was put into motion for the first time in early 1940.[99] Careful provisions had to be made in order to override the existing policy toward the conservative Turkish population. First, the CDPH asked the Ministry of Foreign and Religious Affairs to inform the Islamic hierarchy of the impending campaign and to ask it for its support in order to calm the population. Such support was necessary because the application of force could not be excluded in case the population would not bow to disinfection measures, in particular with respect to shearing the hair of women.[100] The Head Mufti of Sofia acceded, subsequently asking the regional muftis to inform the Muslim population that they should comply because they were "obliged to fulfill the commandments of the Sharia in which cleanliness is one of the conditions of faith and that Muslims must not allow various parasites like the louse to settle on their body."[101] With respect to shaving, the Head Mufti

98 See Şimşir, *The Turks of Bulgaria*, 51–52; Neuburger, *The Orient Within*, 44–46, 50–52, 95–100, 124–125.

99 "№ 14874 ot 26. VII. 1940 g.," *Izvestiya na Glavnata direktsiya na narodnoto zdrave* 25, 209 (1940): 420.

100 "№ 14874 ot 26. VII. 1940 g.," 420–421.

101 "№ 14874 ot 26. VII. 1940 g.," 421–422.

gave a rather complicated and not completely consistent explanation: "If a woman approves the clipping of her hair she becomes a sinner and accursed. She cannot do so even on command of her husband. [ ... ] He who by force clips the hair of a woman so that she is without hair is guilty and must pay her a blood tax." However, in the event of necessity, a prohibited course of action might become a permitted one. This was the case if the cleansing of a woman's hair took place for the benefit of society and if the result could not be achieved by use of a medicinal lotion. Since Islam prohibited that a man cut the hair of a woman strange to him, only women in the sanitary services were allowed to do so.[102] Thus the CDPH ordered that the shaving of Muslim women be performed by female doctors, nurses, female social workers, midwives and, in cases where there was a lack of female medical personnel, by female teachers or other female attendants.[103]

It is quite possible that the idea of the total eradication of lice among Muslims was influenced by the example of the Third Reich. Again, it is apparent that the sanitization measures were not employed primarily to ensure the health of the Muslims but in order to protect the Bulgarian majority from diseases allegedly originating from these minorities. This neatly fits with the claimed demographic threat, namely that the higher birth-rates among the Turkish population would, in the long run, make Bulgaria vulnerable to foreign aspirations.[104] Similar to the Nuremberg Laws of 1935, the 1940 Bulgarian "Law for the Protection of the Nation" prohibited marriages between ethnic Bulgarians and Jews, but also between Bulgarians and Roma or Turks.[105] Consequently, with growing ethnic ghettoization, typhus would become a medical codeword for the designation of relations between an imagined clean Bulgarian *Volkskörper*, and the verminous Muslim minorities living within it.

However, the big delousing action had to wait. Unexpectedly, in autumn 1940 the CDPH was forced to establish a plan for the mass medical examination of 65,000 Bulgarian migrants from northern Dobrudja to

102 "№ 14874 ot 26. VII. 1940 g.,"422–423.

103 "№ 14874 ot 26. VII. 1940 g.," 420.

104 Stefan Konsulov, "Nay-golyamata opasnost za balgarskiya narod – negovoto stopyavane," *Prosveta* 3, 7 (1938): 794–808.

105 Neuburger, *The Orient Within*, 51–52; Elena Marushiakova and Vesselin Popov, "Die bulgarischen Roma während des Zweiten Weltkriegs," in Donald Kendrick, ed., *Sinti und Roma unter dem Nazi-Regime*, vol. 2: *Die Verfolgung im besetzten Europa* (Berlin: Centre de Recherches Tsiganes – Edition Parabolis, 2000), 94.

Bulgaria when southern Dobrudja was returned to Bulgaria in the Treaty of Craiova. However, such mass examinations proved quite difficult due to the congestion at the border checkpoints, a problem not anticipated by the authorities.[106] Bulgaria's occupation of Yugoslav Macedonia and Western Thrace in 1941 additionally strained the limits of the CDPH, which was also responsible for public health in the occupied territories. Two anti-epidemic detachments were sent to Macedonia, which had long been a hotbed of typhus, with louse infestation of the population standing at about 80%.[107]

Over 1941–1942 the cases of typhus tripled from 267 to 815, while fatalities increased fivefold from 13 to 67.[108] The reasons for such an increase have to be sought in difficulties of the supply of materials. In the mountainous regions, a lack of soap, fuel and clothing meant that the population had no means to attend to their hygiene. Of the 980 cases registered between 1 January 1941 and 1 July 1942, 44.7% (438 cases) were Turks, 36.9% (362 cases) were Roma, 7.4% (73 cases) were Pomaks and 18% (176 cases) were Bulgarians.[109]

In the hinterland every doctor therefore had to be on watch. Those district and communal physicians and surgeons who incorrectly performed delousing had their wages cut.[110] Although the master plan for the eradication of the lice could not yet be realized, the results of an initial compulsory delousing campaign—from the beginning of 1941 until halfway through 1942—were impressive. During this period 169,727 civilians were deloused, 80,000 alone in the Turkish areas of northeastern Bulgaria. For this fight against lice, 101 disinfections machines and 504 medical personnel (among them 130 doctors) were deployed.[111] The most striking problem was the lack of motor vehicles for transporting disinfection machines to remote and inaccessible regions—and even a lack of military boots was noted.[112]

---

106 TsDA (f-372k, op-1, ae-2195, l-146).

107 Sia Nikoforova, "Zdravna politika vav Vardarska Makedoniya," *Makedonski pregled* 28, 4 (2005): 5.

108 "Ostri zarazni bolesti v stranata prez 1944 godina," *Izvestiya na Glavnata direktsiya na narodnoto zdrave* 30 (286) (1945): 392.

109 TsDA (f-372k, op-1, ae-2178, l-9-10).

110 "Zapovyad № 4409 ot 20. XI. 1941," *Izvestiya na Glavnata direktsiya na narodnoto zdrave* 26, 231 (1941): 684; "Nakazaniya," *Izvestiya na Glavnata direktsiya na narodnoto zdrave* 27, 254 (1942): 438–439.

111 TsDA (f-372k, op-1, ae-2178, l-10-12).

112 TsDA (f-372k, op-1, ae-2178, l-11); TsDA (f-372k, op-1, ae-2199, l-80); (TsDA (f-372k, op-1 ae-2200, l-388).

The ordinary procedure, identical to the one used in World War I, was that individuals were given a deadline for self-delousing after which, if lice were found, compulsory delousing took place. However, in the case of the Roma, compulsory delousing was conducted principally.[113] The reputation of past delousing campaigns showed effects, as the head of a detachment reported about its arrival in a village: "Turkish and Roma women, thinking that they would be shaved, abandoned their work in the field, put large kettles on the fires and as one party started to wash, a second party started to boil and a third party with combs in the hands and kerosene annihilated the lairs of the nits and lice."[114]

Around the same time the CDPH raised another issue, that of whether the threat to the Bulgarian population represented by the Muslims and Roma could be explained in terms of immunology. Lethality among Bulgarians was 9% and 9.5% among Turks, but only 4.4% among Roma. This was explained in terms of "a relative racial immunity" of the Roma population. [115] The argument proceeded from the assumption that since the first massive onset of typhus in the Balkan Wars three decades previously, typhus had become endemic among Pomaks and Turks. During this time it appeared that these two groups had developed partial immunity, often showing atypical forms of typhus that the doctors were unable to detect easily.[116] Atypical forms were also found among the various types of migrant populations—Roma, beggars, peddlers, tinkers, polishers, basket makers, minstrels, owners of performing bears and their families—so that "there is no way to discern it from customary febrile diseases."[117] "From these two groups—Mohammedans and Roma—the infection spreads to the other population groups. But among the latter, who were not immune, the clinical forms are apparent and typical, and the course of disease is more serious."[118]

It appears that the concept of partial immunity among Muslims and Roma was influenced by a recent scientific article by Lyubomir Tsvetkov

113 TsDA (f-372k, op-1, ae-2178, l-13).
114 TsDA (f-372k, op-1, ae-2199, l-80, l-84).
115 TsDA (f-372k, op-1, ae-2178, l-9-10).
116 TsDA (f-372k, op-1, ae-2178, l-5-6).
117 "№ 3713 ot 14. II. 1941 g.," *Izvestiya na Glavnata direktsiya na narodnoto zdrave* 26, 221 (1941): 51–52.
118 TsDA (f-372k, op-1, ae-2178, l-6).

(1899–1984), lecturer at the Hygienic Institute, University of Sofia on *inapparent* forms of typhus (i.e. those with no symptoms). Tsvetkov had encountered such forms during a small epidemic in Sofia in spring 1938. There the disease from people with Muslim names, who lived in extremely poor quarters, and were probably Roma, had spread to those with Bulgarian names. In the course of serological tests of the diseased and their contacts, Tsvetkov discovered that in some cases the intensity of the serological Weil-Felix reaction was higher among apparently healthy persons than among some of the diseased. Tsvetkov furthermore proved the existence of these, with respect to their clinical profile inapparent forms of typhus in experiments with guinea pigs. It is important to mention that contrary to the view of the CDPH, which wanted only to see atypical forms of the disease among Muslims and migrants, Tsvetkov ably demonstrated that inapparent forms did not discriminate. Indeed, Tsvetkov was not concerned with the ethnic or religious origin of the diseased but instead was troubled by their living conditions, which were described as "worse than those of animals."[119]

But the CDPH had its own way of handling the matter. After the campaign of 1941–2 it prepared a second "extraordinary" assault. Although this campaign was ostensibly directed at the whole population, the Muslims, once again, were the primary target. Dimitar Topalov, head inspector for the CDPH, was in charge of the action planned for 1943.[120] Once again resources and manpower had to be found (see fig. 4), the health services not having sufficient staff and financial resources of their own for the exhaustive and systematic delousing of the whole population. The CDPH therefore asked the Bulgarian Red Cross as well as *Brannik* [Defender], the official youth organization of the regime, for their support.[121] The district physicians were ordered to make "extraordinary" tours through all settlements in their territory with instructions to assess the lice infestation of the population. Subsequently the CDPH would establish a general plan for delousing with provisions for the compulsory delousing of every Roma every second month.[122]

119 Tsvetkov, "Petnist tif", 464–471.

120 TsDA (f-372k, op-1, ae-2200, l-277).

121 TsDA (f-372k, op-1, ae-2200, l-512, l-526-527).

122 "№ 507 ot 9. II. 1943 g.," *Izvestiya na Glavnata direktsiya na narodnoto zdrave* 28, 255 (1943): 81–83.

Fig. 4. Graduates from a disinfection course who were supposed to serve in anti-epidemic detachments (city of Kazanlak, late 1943). Source: Bulgarian Central Archives [TsDA (f-372k, op-1, ae-2199, l-28)]

The district physicians had to lay the foundation for establishing local committees based in the communes. They were expected to support delousing, control the sanitary situation as well as instruct the population in hygienic measures whilst ensuring that Roma were not billeted in the communities. These committees were to be headed by the mayor and consisted of teachers, religious men of all faiths, medical staff and representatives of all cultural and economic organizations.[123] Although the committees were reminiscent of the hygiene councils during World War I, they did not have the same freedom of action and were under direct supervision of the CDPH. Dimitar Topalov asked the committees to isolate louse infested families by erecting signs in front of their houses with the inscription "Watch out! Infested family!" Furthermore, their children were to

[123] TsDA (f-372k, op-1, ae-2190, l-7-9, l-15, ibid., ae-2200, l-466); "№ 537-II ot 12. II. 1943 g.," *Izvestiya na Glavnata direktsiya na narodnoto zdrave* 28, 255 (1943): 80–81.

be excluded from school and young unmarried people expelled from folk dances and from inns. "In general, they should be treated as citizens of lower quality than the clean ones."[124]

After a great deal of persuasion by the CDPH, the military medical service finally deployed twenty anti-epidemic detachments to southeast Bulgaria in December 1942. This region was close to the border with Turkey and had a large concentration of Turkish residents.[125] Furthermore, it was the location for the Bulgarian Army's *Covering Front* [*Prikrivashtiya Front*], opened in 1939 to offset the risk of military conflict with Turkey. Therefore, it was imperative to cleanse the operational zone of typhus in order to protect Bulgarian soldiers.[126] After touring the *Covering Front* in early 1943, Topalov concluded that it was possible to rid the local Turkish population of lice, but only if they were under the continuous control of the authorities.[127]

With respect to the Roma population, Topalov had different plans. For this particular group he suggested "special measures, so that this population stops being a burden for society and is tied into labor, useful for society and economy, compelling them to stop migrating and to lead a sedentary life."[128] His plan was segregation, the total prohibition of Roma migration, the establishment of ghettos, "settlements," along with the exploitation of their labor.[129] As early as 1941, the head of an anti-epidemic detachment had considered it legitimate to suggest the segregation of the Roma population in separate districts and settlements. He genuinely believed that segregation would "provide them with autonomy and the essential conditions for life as well as for the development towards a civilized way of life."[130] Compulsory labor for Roma was decreed in May 1942. However, it appears that the provision was not implemented to its full extent. In August 1943 the Ministry of the Interior ordered the restriction of Roma movements within Bulgaria under the pretext that they were spreading typhus

124 TsDA (f-372k, op-1, ae-2200, l-311-312).

125 TsDA (f-372k, op-1, ae-2200, l-423, ibid., l-471; ibid,. ae-2203, l-25-26).

126 TsDA (f-372k, op-1, ae-2200, l-277).

127 TsDA (f-372k, op-1, ae-2200, l-312).

128 TsDA (f-372k, op-1, ae-2200, l-312).

129 TsDA (f-372k, op-1, ae-2190, l-2).

130 D. Bratovanov, "Epidemiologichno prouchvane na petnistiya tif v s. Lyubimets, Svilengradska okolya," *Izvestiya na Glavnata direktsiya na narodnoto zdrave* 26, 230 (1941): 584.

and other contagious diseases. It would appear that the authorities sought to deport a considerable proportion of Bulgarian Roma together with Bulgarian Jews. The Bulgarian public's successful opposition and resistance against the deportation of the Jews from "Old Bulgaria" (i.e. without the occupied territories of Macedonia and the Aegean coast) also saved the Roma from deportation.[131]

In spite of these draconian measures, the mobility of the Roma population was seen to to be increasing rather than declining. For the CDPH the reason was patently clear: "as unproductive elements" they were the most time engaged in foraging.[132] The CDPH forgot to note, however, that by that time food rations for Roma had been drastically reduced in comparison to ethnic Bulgarians.[133] The challenge of increased mobility was not just limited to the Roma population; recruitment for military service, conscription for fatigue duty, the search for food and evacuation measures (after the first Allied air raids in November 1943) ensured that large quantities of Bulgarians were forced into motion. Under these conditions epidemiological control was difficult to manage.[134] Already in 1943, typhus had doubled to 1,794 cases (154 deaths). Statistics suggest that this remained steady in 1944 with 1,846 cases (162 fatalities). Among the infected persons the ethnic division showed the same bias as in 1941–1942: 43.6% of the infected were Roma, 32.5% Turks, 4% Pomaks and 13.6% ethnic Bulgarians.[135]

In the meantime the second delousing campaign, which had started in late 1942, continued unabated. It would last until the Communist-dominated "Fatherland's Front" [*Otechestven Front*] took over control on 9 September 1944. There are no complete figures available but a careful estimation of individual reports would allow for at least 200,000 cases of delousing, whereby the focus lay on areas with Turkish populations. Aside from the civilian population, delousing measures were also undertaken among sol-

131 Marushiakova Popov, *Gypsies (Roma) in Bulgaria*, 32–33; and Marushiakova Popov, "Die bulgarischen Roma," 95–97. On the relationship between Roma and Jews with respect to their treatment and classification by the Nazis, see Sevasti Trubeta, "'Gypsiness,' Racial Discourse and Persecution: Balkan Roma during the Second World War," *Nationalities Papers* 31, 4 (2003): 498–501.

132 TsDA (f-372k, op-1, ae-2192, l-4).

133 Marushiakova, Popov, "Die bulgarischen Roma," 95.

134 TsDA (f-372k, op-1, ae-2192, l-4).

135 "Ostri zarazni bolesti v stranata prez 1944 godina," *Izvestiya na Glavnata direktsiya na narodnoto zdrave* 30, 286 (1945): 392–393.

diers, persons enrolled in fatigue duty and interned Serbs.[136] In early 1943 the Jewish inmates of concentration camps in the occupied territories of the Aegean coastal strip and of Macedonia (and initially southeastern Bulgaria) were also subjected to compulsory delousing, which they had partly to organize themselves with physicians from their own ranks.[137] Unlike the Jews of "Old Bulgaria" they had no lobby and were simply handed over to the SS for transport to Auschwitz-Birkenau.

## Conclusion

Statistical data from the 1940s indicates that typhus was disproportionately prevalent among Turks, Roma and Pomaks. But what do these figures really mean? We can assume that the over-representation of Muslims was partly a consequence of the official scrutiny they found themselves under; in the archives we rarely find reports about ethnic Bulgarians, although lice and endemic typhus were certainly common among them as well. The over-representation of Muslims also had much to do with poverty, which was more common among them. Finally, the relatively high figures may have indeed been the result of their traditional lifestyle, which was retained not least, because the Bulgarian nation-state was reluctant to offer effective models of integration for its Muslim population. It is notable that positive Bulgarian voices about Muslim hygiene and the Turkish custom of bathing and washing, which were still heard in the 1920s, appeared to cease shortly thereafter. Instead, the "Central Directorate of Public Health" launched compulsory delousing measures in order to control and discipline the Muslim population. Medical practice was thus among the instruments of the nation-state to impinge furthest into the intimate sphere of the Bulgarian Muslims, into their very bodies.

Typhus was understood as a war epidemic and a racist language of military action was applied in a crusade against the enemy, the louse and those infested by it. This language ultimately became the accepted way of speak-

136 The value is based on individual reports on delousing from various districts from Bulgaria, collected in TsDA (f-372k, op-1, ae-2199, ae-2200, ae-2205, ae-2206 and ae-2208).

137 TsDA (f-372k, op-1, ae-2200, l-379; (TsDA (f-372k, op-1, ae-2202, l-1-4).

ing about Muslim minorities. This is in line with the arguments provided by Shula Marks and Neil Andersons about the health policies applied to the black population of South Africa and Paul Weindling's observations about the development of German epidemiological policies vis-à-vis the Jews.[138] There is one essential difference, however: Bulgarian and South African medicine in their sanitization attempts stopped at the stage of segregation of the human carriers of epidemics, while the German doctors went further in their eradication. The Bulgarian case shows that the line between cultural and biological racism was not as decisive with respect to the applied methods as was suggested in the introduction. In the case of the Pomaks, because of their Slavic language the cultural distance between them and a Bulgarian national culture appeared to be surmountable within an assimilationist model of integration. In the case of Turks and Roma the public health administration with its models of control, containment and segregation—developed gradually after the passing of the "Law for Public Health" in 1929—behaved as if irreconcilable differences existed. However, explicit racist arguments were only used in the 1940s when the concept of partial immunity against typhus among Muslims and Roma was discussed.

From the viewpoint of individual rights, compulsory delousing was a severe infringement, one which was justified by an alleged state of imminent danger. In the case of Turks who had no symptoms of typhus, but were infested with lice, compulsory delousing was applied when the order for self-delousing was not adhered to. In the case of the Roma, who occupied the lowest wrung in Bulgarian society and who were classified as obstinate, only compulsory delousing came into question. A delicate issue that demanded particular attention was the delousing of Turkish women. In this case religious principles were challenged. To take the hair of a Muslim woman was in Muslim eyes the ultimate encroachment, committed by the secular nation-state. In this respect the representatives of Bulgarian public health at least, had to come to terms with Muslim cultural values. To be sure, the issue was not a matter which was discussed between women and men. On the one hand we have the male Bulgarian doctors who saw Muslim

138 Shula Marks and Neil Anderson, "Typhus and Social Control: South Africa, 1917–50," in Roy MacLeod and Milton Lewis, eds., *Disease, Medicine, and Empire. Perspectives on Western Medicine and the Experience of European Expansion* (London: Routledge, 1988), 257–283; and Weindling, *Epidemics and Genocide.*

women as caring for the sick but as being ignorant of hygienic norms; they wore traditional dress, the *shalvari*, the traditional pants being hotbeds for lice, and what is more, they hid their head lice under their veils. "Here and elsewhere," Mary Neuburger observed, "it is evident that Muslim women with their hidden faces, bodies, and sentiments heightened the sense of Muslim menace."[139] On the other hand we have the patriarchal advocates of mute Muslim women, their fathers and husbands, and ultimately, the Islamic hierarchy guarding the principles of the *sharia*. The whole issue was a negotiation among males of different cultural backgrounds. If women's interests were addressed, then involuntarily. The situation afforded the employment of a third party, namely female doctors whom their male colleagues had to grant a broad field of activity among Muslim women.

With respect to the complex of Orientalism and medicine it is essential to point out that Neuburger's unterstanding of the "Orient within" as she calls the presence of rural Muslim communities in Bulgaria, only describes a localist variant of a general Western discursive formation about full-fledged Muslim civilization(s) of the lands of the "Orient," as they have been described by Edward Said.[140] The relative simplicity of the discursive structures of Orientalisms in the Balkans facilitates their use for descending hierarchies of moral ascriptions—"nesting Orientalisms," as they have been called by Milica Bakic-Hayden.[141] Neuburger claims that because "the Balkans themselves were Orientalized or cast as the Eastern other by Western Europe and later by the United States, [...] Balkan Christians intensified such frameworks as they focused them inward and on Muslim minorities in their midst."[142] Neuburger even speaks of an "inverted postcoloniality" thus: "Specifically, Bulgarian national thinkers had to deal with a colonial past in which they, as Europeans, were subject to the rule of a theoretically inferior non-European power, the Ottoman Empire."[143] That is why in this case it was imperative that the description of filthiness among Turks should be thwarted by the reference to the European character of the Bul-

139 Neuburger, *The Orient Within*, 130.

140 Edward Said, *Orientalism* (London: Routledge and Kegan Paul, 1978).

141 Milica Bakic-Hayden, "Nesting Orientalisms: The Case of Former Yugoslavia," *Slavic Review* 54, 4 (1995): 917–931.

142 Neuburger, *The Orient Within*, 5.

143 Neuburger, *The Orient Within*, 7–8.

garian peasant, as was claimed by the aforementioned 1939 dissertation from Leipzig. Consequently, the Bulgarian position was Janus-faced. Bulgarian intellectuals truly looked askance at their Muslim contemporaries. Nevertheless, some Bulgarian physicians, and among them eugenicists in particular, also warned of dangers to health coming from the West—diseases of civilization, venereal diseases and degeneration in general.[144] As we have seen, in a rare example a Bulgarian doctor even praised the Bulgarian Turks for their healthy living conditions on the basis of their refusal to accept a modern lifestyle. But such an idyllic account was only a rose-colored, sub-variant of the general Orientalist tune in Bulgarian medicine.

Finally, it should be noted that compulsory delousing did not stop immediately after the Communist takeover in September 1944. Up until the early 1950s the Communist government explicitly considered the Turkish population as a reactionary obstacle to the visions of socialism and therefore instigated the emigration of about 150,000 of them to Turkey.

[144] See Christian Promitzer, "Taking Care of the Nation Body: Eugenic Visions in Interwar Bulgaria, 1905–1940," in Marius Turda and Paul J. Weindling, eds., *Blood and Homeland. Eugenics and Racial Nationalism in Central and Southeast Europe, 1900–1940* (Budapest: Central European University Press, 2007), 223–252.

# HEALTH POLICY AND PRIVATE CARE

## MALARIA SANITIZATION IN EARLY TWENTIETH CENTURY GREECE

*Katerina Gardikas*

"Alongside tuberculosis, alcoholism and syphilis, many believe malaria to be one of the four wheels of the vehicle precipitating man towards degeneration."[1] Thus spoke Spyridon Livieratos, professor of medicine at the University of Athens, at a 1914 conference in Athens held to celebrate the fiftieth anniversary of the accession of the Ionian Islands to Greece. In this political context, Livieratos's statement also suggests a serious concern for the biological future of the Greek nation. Furthermore, in its language of decay, this metaphor vividly illustrates the fear of a looming, yet preventable, breakdown in public health.

I shall here focus on malaria, one of Livieratos' sources of fear, arguing that, in the absence of a public health system to deal with such a threat in early twentieth century Greece, the disease itself partially shaped the practice of medical care. Indeed, as Charles Rosenberg has noted, within a given social setting, biological realities shape "the variety of choices available to societies in framing conceptual and institutional responses to disease."[2] While it is arguable that malaria, the most prevalent disease throughout Greece, fuelled a nationwide demand for medical services, it is nonetheless

---

1 Σπυρίδων Γ. Λιβιεράτος, "Η ελονοσία και η φυματίωσις εν Επτανήσω," *Πρακτικά του εν Κερκύρα Πρώτου Πανιονίου Συνεδρίου (20–22 Μαΐου 1914) και αι εν αυτώ ανακοινώσεις* (Athens: n.p., 1915), 174–175; all translations from the original are my own unless otherwise cited.

2 Charles E. Rosenberg, "Framing Disease: Illness, Society and History," in Charles E. Rosenberg and Janet Golden, eds., *Framing Disease: Studies in Cultural History: Health and Medicine in American Society* (New Brunswick: Rutgers University Press, 1992), xv.

true that socio-economic, political and geographical limitations affected access to such services. "Perhaps [...] the relationship between medicine and the laity entailed wider interactions between self, society and knowledge," reflected Roger Cooter, "all according to competing priorities and the different material constraints of everyday life."[3]

Once alerted to this health crisis, by specifically targeting malaria, state policy adopted a vertical approach to malaria control, a strategy suitable for countries like Greece during this period—and for decades to come—that lacked a comprehensive system of public health.[4] This policy sanctioned the extensive use of quinine, the old febrifuge that now promised not only to cure the individual patient but also to benefit the wider social environment; to sanitize the nation. With the development of an international space of public health after World War I, rural health in particular became the focus of public health policies of the League of Nations in the 1930s as a social response to socialist medicine.[5] Before the Great War, however, the Greek medical profession discussed health matters in terms of material concerns impacting an apparently socially undifferentiated nation.

Malaria control had been a national concern since the early days of Greek state-building; for instance it was taken into account in the first town planning and land reclamation projects. Moreover, even in the early days of modern Greek statehood, malaria was studied intensely by individual physicians as well as by the Greek Medical Board and the Athens Medical Society.[6] In 1898, when the transmission mechanism of the disease (through the anopheles mosquito) was discovered, unanswered questions over how humans contracted malaria were resolved—questions that

3 Roger Cooter, "'Framing' the End of the Social History of Medicine," in Frank Huisman and John Harley Warner, eds., *Locating Medical History: The Stories and Their Meanings* (Baltimore: Johns Hopkins University Press, 2004), 314.

4 John Farley, *To Cast Out Disease. A History of the International Health Division of the Rockefeller Foundation (1913–1951)* (Oxford: Oxford University Press, 2004), 285.

5 Paul Weindling, "From Moral Exhortation to the New Public Health, 1918–45," in Esteban Rodriguez-Ocaña, ed., *The Politics of the Healthy Life. An International Perspective* (Sheffield: European Association for the History of Medicine and Health Publications, 2002), 127; Lion Murard, "Health Policy Between the International and Local: Jacques Parisot in Nancy and Geneva," in Iris Borowy and Wolf D. Gruner, eds., *Facing Illness in Troubled Times. Health in Europe in the Interwar Years, 1918–1939* (Frankfurt: Peter Lang, 2005), 238–239; Lion Murard and Patrick Zylberman, "French Social Medicine on the International Public Health Map in the 1930s," in Ocaña, ed., *The Politics of the Healthy Life*, 198–199.

6 *Ασκληπιός* 1 (1836): iv; on the purpose of the official publication of the Society, B. Roeser and N. Kostis to ministry of interior, Athens 15 January 1836, State Archives, Ministry of the Interior, f. 204, doc. 004.

Greek physicians had been speculating about for almost two decades, since Alphonse Laveran had first identified the pathogen causing malaria.[7]

At the time of the scientific discoveries of laboratory-based Western medicine, Greece was effectively without a system of public health. The country's early public health system had been designed by the Bavarian bureaucrats who had accompanied King Otho to Greece in 1833. It had been modeled on late eighteenth century theory, which introduced the institution of the medical police. This system was hierarchical in structure and quite suitable for an absolutist regime like the one established in Greece in the 1830s. At the head of this hierarchy was the Medical Department of the Ministry of the Interior, with an associated Medical Council acting in an advisory capacity. At the lower level of the system, the medical police took orders from and answered to the county and departmental medical authorities.[8]

Officially, this skeleton of a public health system was retained after the overthrow of Otho's regime in 1862, although it essentially disintegrated thereafter for lack of state funding. State funds were instead diverted towards a costly system of state-funded patronage, one controlled by liberal political elites. Even so, the executive branch of the decaying system of medical policing remained in the hands of municipal authorities, who would occasionally appoint a local doctor as a municipal physician to benefit the urban poor.[9] In effect, however, private medicine was the only health service available to civilians. This private care extended to services offered by private railway and mining companies to their personnel.

---

7 Felix von Niemeyer, *Ειδική νοσολογία και θεραπευτική: συντεταγμένη κατ' αναφοράν ιδίως προς την φυσιολογίαν και παθολογικήν ανατομίαν*, trans. Γεώργιος Καραμήτσας (Athens, 1879 [-1885]), 2, 728–804. The translator Georgios Karamitsas, a professor of medicine, wrote the supplementary chapter on malaria included in the second volume published in 1885. Karamitsas was particularly puzzled by the presence of malaria in dry areas (732), by the apparent reproduction of the infecting parasite outside the human body (733) and its mechanism of entry into the human body (737).

8 Peter Baldwin, *Contagion and the State in Europe, 1830–1930* (Cambridge: Cambridge University Press, 1999), 559–63; Marion Maria Ruisinger, *Das griechische Gesundheitswesen unter Koenig Otto (1833–1862)* (Frankfurt: Peter Lang, 1997); James C. Riley, *The Eighteenth Century Campaign to Avoid Disease* (London: Palgrave Macmillan, 1987), 41; Αριστοτέλης Κούζης, "Αι μετά την ίδρυσιν του βασιλείου της Ελλάδος πρώται παρ' ημίν αρχαί υγιειονομικής πολιτικής και οργανώσεως της δημοσίας υγείας επί τη βάσει των ανεκδότων χειρογράφων πρακτικών του Ιατροσυνεδρίου," *Πρακτικά της Ακαδημίας Αθηνών* 21 (1946): 87–89.

9 Λυδία Σαπουνάκη-Δρακάκη, *Οικονομική ανάπτυξη και κοινωνική προστασία. Η γέννηση των υπηρεσιών υγείας στον Πειραιά και το Τζάνειο Νοσοκομείο* (Athens: Διόνικος, 2006); and Φωκίων Κοπανάρης, *Η δημοσία υγεία εν Ελλάδι* (Athens: n.p., 1933).

In comparison to civilian society, the Greek army was in a more advantageous position on account of its own medical corps, a medical journal, military hospitals and infirmaries. In the early years of the twentieth century, medical and, moreover, national statistics drew on material from 43 such establishments. Still, conscripted men were certainly more exposed to infectious diseases. For instance, the men mobilized to fight in the 1897 war against Turkey suffered from the hardships of malaria for months at the banks of the Spercheios River. It appears that their mortality nevertheless remained lower than that of the civilian population owing to their better access to care—and especially to quinine as medication.[10] Poor health in the army, however, developed into a serious military and indeed social problem, particularly after the 1880s, when the army grew in numbers thanks to a program of military reforms.[11]

Despite its medical services, overall sanitary conditions in the army were in fact a source not just of disease but of disgrace as well. More than two in every three recruits fell ill while serving their country.[12] Figures for the peaceful first decade of the twentieth century suggest that 35% of all army personnel contracted malaria. At precisely the same time, malaria also adversely affected the army from a demographic perspective, as emigration to the United States removed able-bodied men from the country; as a result, officers on tour to enlist recruits frequently encountered malaria stricken young men too weak either to emigrate or to serve in the army.[13] Within army ranks, both malaria and tuberculosis spread to healthy conscripts, who then went on to contaminate their native towns and villages.[14] Thus, besides being a hotbed of disease affecting its overall performance, the army also became an important source of infection for civilian society.

10 Ιωάννης Π. Καρδαμάτης, *Πραγματεία περί ελειογενών νόσων* (Athens: n.p., 1909), 275, 383; Γρηγόριος Σκαμπαρδώνης, et al., *Ιατρική Εφημερίς του Στρατού (1890–1897). Κατάλογος περιεχομένων και επιλογή άρθρων* (Athens: Παρισιάνου, 2001).

11 On malaria morbidity in the Greek army see Ιωάννης Π. Καρδαμάτης, *Πραγματεία περί ελειογενών νόσων*, 383; and Κωνσταντίνος Γ. Σάββας and Ιωάννης Π. Καρδαμάτης, eds., *Σύλλογος προς Περιστολήν των Ελωδών Νόσων* (Athens: n.p., 1907), 383–411.

12 As late as 1928, that is, on the eve of the public health reforms in Greece, the official general mortality rate in the army was high at 8.12%, but the mean general morbidity rate in 1927 and 1928, at 68.2%, was alarming.

13 *Ακρόπολις* 22, 8954 (6 April 1907): 1.

14 Ιωάννης Π. Καρδαμάτης, *Πραγματεία περί ελειογενών νόσων*, 64; Πρωθυπουργός (Γραφείον Τύπου), *Το έργον της κυβερνήσεως Βενιζέλου κατά την τετραετίαν 1928–1932* (Athens: n.p., 1932), 245. On the spread of malaria see Σάββας and Καρδαμάτης, eds., *Σύλλογος προς Περιστολήν των Ελωδών Νόσων*, 255. On the spread of tuberculosis in the army see Ιωάννης Π. Καρδαμάτης, *Πραγματεία περί ελειογενών νόσων*, 480.

In May 1901, the first national medical conference was convened by the Athens Medical Association. The main aim was to highlight the public health aspects of infectious diseases. Additionally, the conference underlined the importance of the private specialist among medical practitioners and surgeons. With this broad agenda in mind, the organizing committee, consisting of university professors and leading Athens specialists,[15] selected specific themes of priority to the medical profession and to public health in Greece; these themes related to tuberculosis, malaria, leprosy, echinococcus, syphilis, ophthalmic infections, asepsis and antisepsis, appendicitis, miscarriages, alcoholism and school hygiene.[16]

The conference—with around 460 participants, of whom approximately 25% had come from Greek communities abroad—approved unanimously eight resolutions on these pre-defined issues. The first of these resolutions created a National League against Tuberculosis. With regard to malaria, a second unanimous resolution established a committee that would study the disease and its causes, compile a nosological map and make policy recommendations for its eradication.[17] Thus, by placing the major public health problems, namely tuberculosis and malaria, in the hands of specialists, the conference effectively launched a vertical approach to disease control.

The malaria committee sent out a questionnaire to physicians around the country. Beyond their scientific significance, both the conference itself and the questionnaire became an important publicity event, establishing the leading figures of Greek malariology in a position of authority. Foremost among them was Ioannis Kardamatis, in his early 40s at the time. Furthermore, the conference and the distribution of the questionnaire resulted in civilian physicians assuming the lead from their army col-

15 The organizing committee was chaired by Rigas Nikolaidis, professor of physiology. The other committee members were Nikolaos Pezopoulos, professor of pathology, Georgios Vafas, professor of forensic medicine, Michail Mangakis, ORL specialist, Dimitrios Kokkoris, surgeon, Michail Kairis, obstetrician and Dionysios Gouzaris. Λ. Βλαδίμηρος and Χ. Φραγκίδης, "Το 'Πανελλήνιον Ιατρικόν Συνέδριον' του 1901. Ιατρικά δρώμενα, υγειονομικά προβλήματα και κορυφαίοι ιατροί πριν από έναν αιώνα," *Αρχεία Ελληνικής Ιατρικής* 19, 6 (2002): 700–709.

16 Βλαδίμηρος and Φραγκίδης, "Το 'Πανελλήνιον Ιατρικόν Συνέδριον' του 1901."

17 Βλαδίμηρος and Φραγκίδης, "Το 'Πανελλήνιον Ιατρικόν Συνέδριον' του 1901"; Λήδα Παπαστεφανάκη, "Δημόσια υγεία, φυματίωση και επαγγελματική παθολογία στις ελληνικές πόλεις στις αρχές του 20ού αιώνα: η αντιφατική διαδικασία του αστικού εκσυγχρονισμού," *Ελευθέριος Βενιζέλος και ελληνική πόλη: πολεοδομικές πολιτικές και κοινωνικοπολιτικές ανακατατάξεις* (Athens: Εθνικό Ίδρυμα Ερευνών και Μελετών 'Ελευθέριος Κ. Βενιζέλος', Τεχνικό Επιμελητήριο Ελλάδας, and Σχολή Αρχιτεκτόνων Μηχανικών ΕΜΠ, 2005), 160.

leagues. The conference thus brought about a restructuring of the elite in the medical profession, together with the development of a nationwide network of empowered medical practitioners. It was therefore a turning point in Greece's medical history.

Nonetheless, this first nationwide malaria survey elicited responses from only 76 physicians. These were then published in the first volume of the *Proceedings of the Anti-malarial League*, an association founded with royal patronage in 1905 as a further result of the social awareness generated by the 1901 conference. The questionnaire covered issues relating to the nature of malaria, in addition to its distribution and the geographical, economic and climatic conditions for spreading infections. The League carried out two further nationwide surveys with questionnaires sent out to physicians in 1905 and 1906. A fourth investigation in 1907 surveyed the country's municipal authorities in addition to the physicians.[18] Kardamatis, secretary-general of the League, subsequently coordinated a fifth nationwide malaria survey during World War I, drawing data from state engineers, medical and municipal sources.[19]

With public health low on the list of priorities, successive Greek governments consistently failed to act on the evidence provided on Greek public health. However, the impact of the survey initiative of the Anti-malarial League was substantial. By transmitting up-to-date scientific knowledge, the surveyors doubtless shaped the collective perception of the medical practitioners in a meaningful way. Beyond the value of this material to health planners, the exchanges involved in the vetting of questionnaires emerged into an extensive network of private physicians—most of whom were seriously engaged in treating malaria patients throughout Greece—exchanging experiences, and generally interacting with a broader circle of colleagues in their respective districts, freshly empowered by a meaningful grasp of the disease they were combating. Some of these physicians also

18 "Απαντήσεις των κ. ιατρών επί των υποβληθέντων αυτοίς ερωτημάτων υπό της επιτροπής," in Κωνσταντίνος Γ. Σάββας and Ιωάννης Π. Καρδαμάτης, eds., *Η ελονοσία εν Ελλάδι και τα πεπραγμένα του Συλλόγου. Έτος Πρώτον και Δεύτερον*, (Athens: Σύλλογος πρός Περιστολήν των Ελωδών Νόσων, 1907); Κωνσταντίνος Σάββας and Ιωάννης Καρδαμάτης, "Απαντήσεις δημάρχων του Κράτους ως προς τα εν τη περιφερεία του δήμου των έλη", in Κωνσταντίνος Γ. Σάββας and Ιωάννης Π. Καρδαμάτης, eds., *Η ελονοσία εν Ελλάδι και τα πεπραγμένα του Συλλόγου*, vol. 4–5 (Athens: Σύλλογος προς Περιστολήν των Ελωδών Νόσων, 1908–9).

19 Ιωάννης Π. Καρδαμάτης, *Στατιστικοί πίνακες των ελών και της συχνότητος της ελονοσίας εν Ελλάδι* (Athens: Υπουργείον Συγκοινωνίας. Τμήμα Υγειονομικόν, 1924).

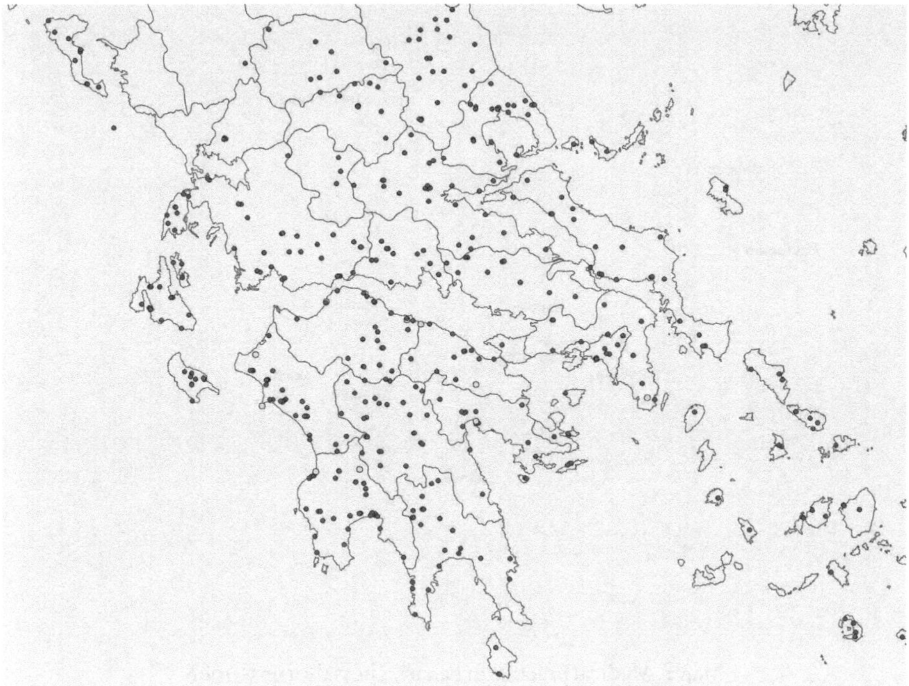

Map 1. Locations of physicians who responded to the malaria surveys, 1901–1907.

served the local needs of large business enterprises, mostly in the construction and operation of Greek railroads.

Eventually as many as 777 physicians, or approximately one third of all Greek doctors, replied to the four surveys held between 1901 and 1907. These doctors represented most of the 443 total Greek municipalities[20] [see map 1]. Note in the map above that the limited presence of doctors in the border regions should not be attributed to a freedom from malaria but rather to their extreme poverty and remoteness. The local doctor from the district of Malakasi, in the area of Kalabaka in Thessaly, for instance,

20 Βασίλειον της Ελλάδος. Υπουργείον των Εσωτερικών. Γραφείον Απογραφής, *Απογραφή του Πληθυσμού κατά την 27 Οκτωβρίου 1907. Μέρος Δεύτερον Πίνακες αριθμητικοί. Α'- Πληθυσμός πραγματικός κατά νομούς, επαρχίας, δήμους, πόλεις και χωρία* (Athens; s.n., 1908); "Λογοδοσία των πεπραγμένων του Συλλόγου κατά το 1905, πρώτον έτος της ιδρύσεως αυτού," in Κωνσταντίνος Γ. Σάββας and Ιωάννης Π. Καρδαμάτης, eds., *Η ελονοσία εν Ελλάδι και τα πεπραγμένα του Συλλόγου. Έτος Πρώτον και Δεύτερον* (Athens: Σύλλογος προς Περιστολήν των Ελωδών Νόσων, 1907), 57. On the per capita distribution of doctors see Βλαδίμηρος and Φραγκίδης, "Το 'Πανελλήνιον Ιατρικόν Συνέδριον' του 1901," 701 and Σαπουνάκη-Δρακάκη, *Οικονομική ανάπτυξη και κοινωνική προστασία*, 125–26.

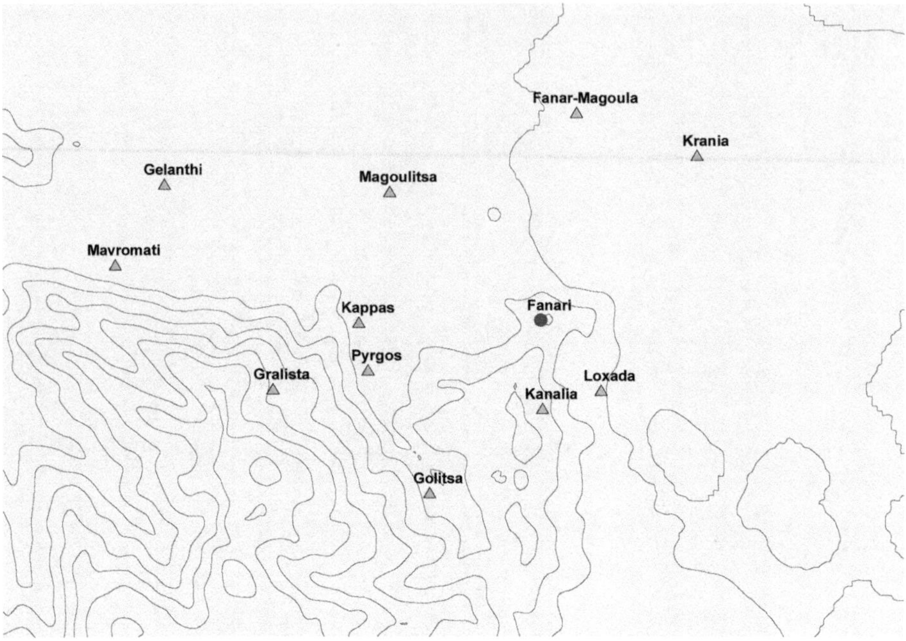


Map 2. Medical practice in Fanari, Thessaly, 1905–1906.

reported a malaria mortality rate of 2‰ only to protest: "Beyond Malakasi, most inhabitants in the other villages unfortunately die without a doctor!"[21] Conversely, and predictably, there existed a high density of physicians in urban centers.[22]

The following two cases offer some further insight into the social context and the medical practice of the rural doctor. In one location at the western edge of the rolling plain in Karditsa (also a town in Thessaly), the doctor, Nikolaos Zoumbos, resided in Fanari, a town of 2,000 inhabitants, the largest and healthiest community in the region [see map 2]. He also practiced medicine in eleven surrounding villages in a radius of ten kilometers, thus serving a total population of 15,000 potential patients.

21 K. [Σάββας, I. Καρδαμάτης, and Σπ.] Δάσιος, "1906. [Αι περί της κατά τόπους ελονοσίας πληροφορίαι των ιατρών]," in Κωνσταντίνος Γ. Σάββας and Ιωάννης Π. Καρδαμάτης, eds., *Η ελονοσία εν Ελλάδι και τα πεπραγμένα του Συλλόγου. Έτος Πρώτον και Δεύτερον*, (Athens: Σύλλογος προς Περιστολήν των Ελωδών Νόσων, 1907), 612.

22 The distribution should rather be read as the product of morbidity and relative prosperity. Judging by their surnames, occasionally relatives in the same profession appear to succeed one another or share patients in neighboring districts.

Fanari and Kanalia, both located at some slight elevation from the Thessalian plain, were healthier than the other villages of the region. In 1905, the malaria pandemic appears to have overstretched Zoumbos's ability to assist the sick in the plain, who were at any rate unable to pay for his services. As he admitted, the doctor lacked detailed information about conditions in all but the two more fortunate communities, themselves plagued by 20% malaria morbidity and 19 malaria deaths. As for the villages in the plain, Zoumbos vaguely estimated that "40, not to say 50%" of the local population came down with malaria, offering an equally vague observation that mortality was higher than in previous years, and highest among children between the ages of three and fifteen. A drought in the following year helped reduce overall malaria incidents considerably. Gralista, another village in the region served by Zoumbos, was even spared completely, thanks to its elevated position.[23] With the crisis over, his villages returned to lower morbidity rates by 1907.[24]

A second case relevant here concerns a particularly malaria-infested district in southern Peloponnese, on the Western shore of the Messinia Bay. The local doctor, Dionysios Markopoulos, resided in Petalidi, a small town of 2,000 inhabitants [see map 3]. Petalidi had a total of three physicians, serving a regional population of approximately 4,000. One of the few doctors responding to all four surveys, Markopoulos was first mobilized by the 1901 initiative, that is, merely three years after the implication of the mosquito in the malaria cycle and only two years after setting up his medical practice in Petalidi. Malaria morbidity varied significantly across the area. He reported that on the whole, 80% came down with malaria, mostly in September and October, except for in the town itself, which typically only suffered from a malaria morbidity of 30%. Nearly all of the inhabitants in the Miska and Karakasilion villages were chronic malaria cases (despite the absence of marshes), probably, in his view, owing to poor water absorption in the clay soil. Of 150 patients that Markopoulos treated between August

23 "1905. Αι περί της κατά τόπους ελονοσίας πληροφορίαι των ιατρών," in Κωνσταντίνος Γ. Σάββας and Ιωάννης Π. Καρδαμάτης, eds., *Η ελονοσία εν Ελλάδι και τα πεπραγμένα του Συλλόγου. Έτος Πρώτον και Δεύτερον* (Athens: Σύλλογος προς Περιστολήν των Ελωδών Νόσων, 1907), 460; [Σάββας, Καρδαμάτης, and Δάσιος, "1906. Πληροφορίαι ιατρών", 616 Altitudes are as follows: Fanari, 450 m.; Kanalia, 350 m.; Loxada, 100 m.; Pyrgos, 180 m.; Gralista, 640 m.; Gelanthi, 115 m.

24 The same district recorded an average malaria morbidity rate of 15% between 1915 and 1919. Ιωάννης Π. Καρδαμάτης, *Στατιστικοί πίνακες*, 183.

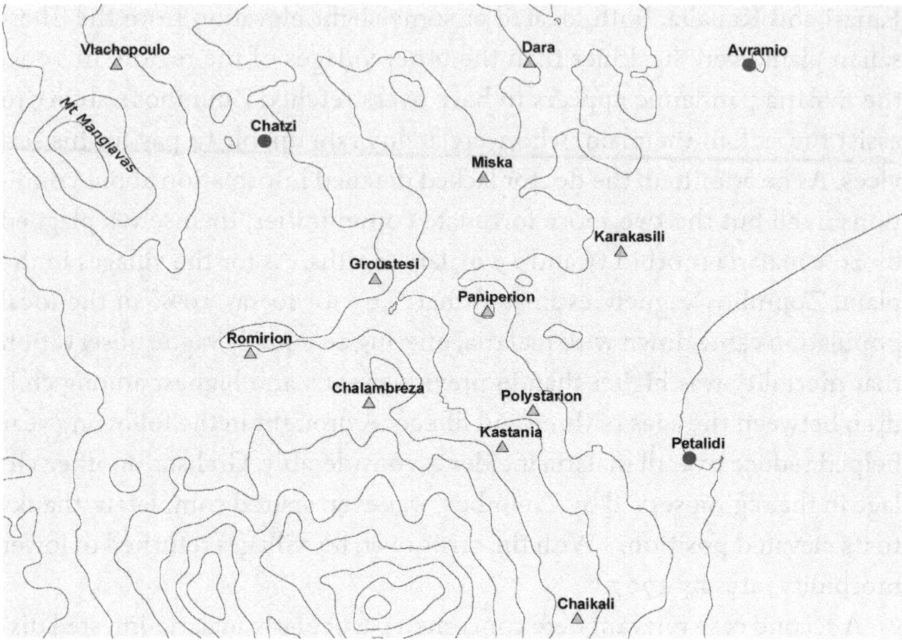


Map 3. Medical practice in Petalidi, Peloponnese, 1901–1907.

and October 1905, 125 suffered from malaria while out of 280 patients in the following year, 160 were malaria cases.

Encouraged by the Anti-malarial League's guidelines, Markopoulos performed spleen examinations on 80 primary school boys in October 1907; he found 37% enlarged spleens, with 85% of the boys having reported contracting malaria in that year. Like his colleague Zoumbos in Fanari, he noted that the disease seriously affected children under the age of 15, particularly those under the age of five. By 1907, Markopoulos had begun to collect information from his patients on the local history of malaria. According to his informants, only 15 to 20 years previously, Mount Manglava, which rose to the west of Petalidi, had been entirely free of malaria. However, when the villagers cleared the forest to cultivate the slope, the disease advanced to higher altitudes from the plain below. In Markopoulos's words, "the infection is carried and spread by man, who infects the area's healthy mosquitoes." Meanwhile, by the time of the 1907 survey, malaria had also infected inhabitants in the village of Vlachopoulo, at an altitude of 450 meters, one of the largest in the region. Further downhill, at an altitude of about 260 meters, the villages of Kastania, Bali, Polystarion and Panyperi were all seriously affected by malaria in the same year. Yet many malaria cases were never treated by a doctor and, presumably, after multiple attacks over the years, became asymptomatic carriers of the disease. Whenever untreated cases sought medical help, these "simple-hearted people" refused to believe that their complaints were symptoms of malaria complications.[25]

Evidently, eight years of rural medical practice alongside the influence of the League transformed the young provincial doctor into a dedicated scientific and social observer. In turn, this empowerment was forged and elaborated by the vertical strategy of the elite medical profession in Athens. Neither the success of its national surveys nor the authority earned by the Anti-malarial League is surprising given that Greece suffered its most severe malaria pandemic in 1905, when 40% of all Greeks fell ill.[26] Until the eve of World War II, extant medical statistics pointed consistently to the fact that, even in normal years, malaria was the single most frequent disease in both rural and urban milieus.[27] Clearly, malaria chiefly drove the demand for medical services and, even though self-healing was common practice in the countryside, malaria ensured that private medical care thrived across urban and rural Greece alike. The Anti-malarial League's ability to mobilize private medical practitioners in large numbers thus points to the success of its leadership.

With respect to malaria in prewar Greece, a physician's role was to diagnose the disease and administer quinine. To keep himself up-to-date, the doctor relied on the Anti-malarial League in Athens: its publications, its educational events and its guidelines. Specifically regarding the use of quinine for treatment and prevention, the League defined its policy recommendations by adopting the views of Ioannis Kardamatis, its secretary-general, from the outset. Quinine was, however, soon to become the basis for

---

25 "Απαντήσεις των ιατρών," 233–34; "1905. Πληροφορίαι των ιατρών," 444; [Σάββας, Καρδαμάτης, and Δάσιος, "1906. Πληροφορίαι ιατρών," 582–83; "1907 [Απαντήσεις ιατρών και δημάρχων]," in Κωνσταντίνος Γ. Σάββας and Ιωάννης Π. Καρδαμάτης, eds., *Η ελονοσία εν Ελλάδι και τα πεπραγμένα του Συλλόγου. Έτος Τρίτον* (Athens: Σύλλογος προς Περιστολήν των Ελωδών Νόσων, 1908), 372, 375, and 530–531.

26 Mortality was 6‰ of those infected. I Καρδαμάτης, "Η ελονοσία εν Ελλάδι κατά το 1905," *Η ελονοσία εν Ελλάδι και τα πεπραγμένα του Συλλόγου. Έτος Πρώτον και Δεύτερον*, 191.

27 As late as 1937, that is, even after the first measures of reform in public heath, malaria still ranked first among all infectious diseases treated at Evangelismos Hospital, a general hospital in Athens. Θεραπευτήριον "Ο Ευαγγελισμός," *Έκθεσις των πεπραγμένων του έτους 1937 του Θεραπευτηρίου "Ο Ευαγγελισμός", του Νοσοκομείου "Γ. Σταύρου & Γ. Φούγκ" και του Ταμείου Συντάξεων του Προσωπικού* (Athens: n.p., 1938).

a national malaria sanitization program, not just for therapeutic treatment. Rather than treating the entire population through the preventive administration of quinine—as advocated by the rival theory—the League opted for a scheme proposed by Robert Koch that appeared less utopian and more practical. Koch, as well as Giovanni Battista Grassi in Rome, expected that, if all infected humans were treated with quinine throughout the infection-free winter months, and if this practice was carried out over an adequate number of years, the number of infecting individuals would drop each spring until eventually all human sources of infection would be eliminated; thus, it was hoped, putting an end to the annual cycle of infection.[28]

This approach to malaria control suffered from serious drawbacks. Even in the highly unlikely event that some ideal medical service could identify and reach all sick persons in a given country, Koch's and Grassi's scheme still could never work as it was founded on exaggerated scientific assumptions about the curing powers of quinine.[29] Nonetheless, owing to the influence of its Anti-malarial League, Greece—unlike Italy, which supplemented malaria control with social reform programs—adopted the quinine distribution scheme on a national scale as its principal, state-assisted strategy. A large part of the country's malaria control program was thereafter based upon the administration of quinine, partly on the basis of a free distribution scheme and partly by private physicians. Thanks to the social and scientific leverage afforded to the Anti-malarial League, Greek authorities committed state resources to the League's doctrine of quinine sanitization, Greece's equivalent to Grassi's *bonifica umana*.

Greece's early anti-malarial campaign was largely modeled on the Italian experience. Indeed, before Italy's entry into the Great War in 1915, the

28 "Λογοδοσία των πεπραγμένων του Συλλόγου κατά το 1905, πρώτον έτος της ιδρύσεως αυτού," 42–43; Ιωάννης Π. Καρδαμάτης, *Πραγματεία περί ελειογενών νόσων*, 521, 582; Κ. Σάββας and Ι. Καρδαμάτης, "Η εν Μαραθώνι εξυγιαστική εργασία του προς περιστολήν των ελωδών νόσων Συλλόγου κατά το θέρος του 1907," in Κωνσταντίνος Γ. Σάββας and Ιωάννης Π. Καρδαμάτης, eds., *Η ελονοσία εν Ελλάδι και τα πεπραγμένα του Συλλόγου*, Vol. 3 (Athens: Σύλλογος προς Περιστολήν των Ελωδών Νόσων, 1908), 60–63; Ιωάννης Π. Καρδαμάτης, *Στατιστικοί Πίνακες*, 250–54; and G. MacDonald, *The Epidemiology and Control of Malaria* (Oxford: Oxford UP, 1957), 150.

29 This is not the place to discuss the effectiveness of quinine in public health. I will merely cite two recent works on the subject: David A. Warrell, William M. Watkins, and Peter A. Winstanley, "Treatment and Prevention of Malaria," in David A. Warrell and Herbert M. Gilles, eds., *Essential Malariology* (London: Arnold, 2002), 269–293; and Martin Eichner, et al., "Genesis, Sequestration and Survival of *Plasmodium falciparum* Gametocytes: Parameter Estimates from Fitting a Model to Malariatherapy Data," *Transactions of the Royal Society of Tropical Medicine and Hygiene* 95 (2001): 497–501. For an historical perspective, see also MacDonald, *The Epidemiology and Control of Malaria*, 150.

country was leading in the field of malaria research and control. As Frank M. Snowden has shown in his authoritative *The Conquest of Malaria: Italy, 1900–1962*, malaria had become a national issue that drew state attention already in 1878, while since the end of the nineteenth century the Roman School of malariology had developed into one of the leading scientific centers of both basic research and clinical application, Italy being the only Western European country for which malaria was a domestic rather than a colonial public health problem. Snowden also explains the difference between Giovanni Battista Grassi, who advocated *bonifica umana,* or sterilizing the blood of infected individuals with the administration of quinine, and another Italian pioneer in malariology and republican deputy, Angelo Celli, who supported the prophylactic use of quinine by all Italians living in malarious areas, a practice to be enhanced by policies of educational reform. Thanks to Celli's political influence, it was his policy that prevailed before World War I; it was abandoned, however, in favor of Grassi's *bonifica umana* after 1918.[30]

As a consequence of its heavy reliance on quinine, by the interwar years Greece was purchasing one fifth to one quarter of the world's annual quinine production, distributing some five tons of free quinine to its citizens each year—that is, beside the amount sold on the free market. Yet even these quantities were insufficient. While it is true that the policy brought relief to a large number of suffering individuals, its effect on malaria control clearly proved limited. Quinine sanitization was a popular but costly failure often articulated in the egalitarian and patriotic rhetoric of the period.[31]

As the state failed in its distributive role, malaria control lay at the mercy of international and local market forces.[32] Thus, oscillations in the interna-

30 Frank M. Snowden, *The Conquest of Malaria. Italy, 1900–1962* (New Haven: Yale University Press, 2006) 9, 38-39, 47, 50-52, 168.

31 Leften Stavros Stavrianos, *Greece: American Dilemma and Opportunity* (Chicago: H. Regnery, 1952), 192; World Health Organisation, *The Chronicle of the World Health Organisation* (Geneva: World Health Organisation, 1959), 138; comments by G. Alivisatos in Ιατρική Εταιρία Θεσσαλονίκης, *Πρακτικά του Πρώτου συνεδρίου των ιατρών της Βορείου Ελλάδος με μόνον θέμα "Η ελονοσία εν Μακεδονία και Θράκη,"* Συνέδριον Ιατρών Βορείου Ελλάδος κατά της ελονοσίας, 16–17–18 Σεπτεμβρίου 1933 (Thessaloniki, 1933), 236–37. For a criticism of quinine distribution, see also Randall M. Packard, *The Making of a Tropical Disease: A Short History of Malaria* (Baltimore, Md.: Johns Hopkins University Press, 2007), 122-25.

32 The literature on European and international malaria control programs in the interwar period is considerable. It includes contemporary studies such as L.W. Hackett, *Malaria in Europe: An Ecological Study* (Oxford: Oxford University Press, 1937) and M.C. Balfour, "Malaria Studies in Greece. Measurements of Malaria, 1930–33," *American Journal of Tropical Medicine* 15 (1935): 301-30.; comprehensive accounts like Leonard

tional price of quinine imports affected the local distribution of the drug, not to mention its quality. Preferential access to medical care revolved around access to the distribution of cheap, even free, unadulterated quinine. Familiar with the signs of malaria, patients would often purchase whatever quinine the village grocery was selling without even consulting a doctor.[33] Falling far short of national demand, the drug was constantly being adulterated, even after 1908, when the state began importing "state quinine" under strict government controls from Italy's state quinine factory in Turin. This was the outcome of an intensive campaign by the Anti-malarial League, one significantly influenced by the Italian experience.[34]

Like other scarce resources in Greece, access to quinine was affected by political influence. According to one malariologist, free quinine went largely to the politically connected.[35] Steep price rises, especially in the period immediately after the First World War, were also a contributing factor to quinine shortages impacting its distribution, and therefore the rise in morbidity in some, less privileged communities.[36] Geographic isolation was also a factor. Evidence from the mid-1940s suggests that, owing to their

---

Jan Bruce-Chwatt and Julian De Zulueta, *The Rise and Fall of Malaria in Europe: A Historico-Epidemiological Study* (Oxford: Oxford University Press, 1980); critical reviews such as H. Evans, "European Malaria Policy in the 1920s and 1930s: The Epidemiology of Minutiae," *Isis* 80 (1989): 40–59; G. Corbellini, "Acquired Immunity Against Malaria as a Tool for the Control of the Disease: The Strategy Proposed by the Malaria Commission of the League of Nations in 1933," *Parassitologia* 40:1–2 (1998): 109–15; and national and regional case studies, including Margaret Humphreys, *Malaria: Poverty, Race, and Public Health in the United States* (Baltimore: Johns Hopkins University Press, 2001); Esteban Rodríguez-Ocaña, "The Birth of the Anti-Malaria Campaign in Spain During the First 30 Years of the 20th Century: Scentific and Social Aspects," *Parassitologia* 47, 3–4 (2005): 371–77 and "International Health Goals and Social Reform: The Fight Against Malaria in Interwar Spain," in Borowy and Gruner, eds., *Facing Illness in Troubled Times*, 247–76; D. H. Stapleton, "Internationalism and Nationalism: The Rockefeller Foundation, Public Health and Malaria in Italy, 1923–1951," *Parassitologia* 42, 1–2 (2000): 127–34; Sandra M. Sufian, *Healing the Land and the Nation: Malaria and the Zionist Project in Palestine, 1920–1947* (Chicago: University of Chicago Press, 2007); Patrick Zylberman, "Mosquitoes and the Komitadjis: Malaria and Borders in Macedonia (1919–1938)," in Borowy and Gruner, eds., *Facing Illness in Troubled Times*, 305–43.

33 "Απαντήσεις των ιατρών," 218.

34 Snowden, *The Conquest of Malaria. Italy, 1900–1962*, 52–54; also James L.A. Webb, Jr., *Humanity's Burden: A Global History of Malaria*, Studies in Environment and History (Cambridge: Cambridge University Press, 2009), 135–36, 139–40. In 1924, Kardamatis estimated the country's annual needs at 70 tons. Ιωάννης Π. Καρδαμάτης, *Στατιστικοί πίνακες*, 252. On a similarly costly program of quinine distribution in Spain promoted by Gustavo Pittaluga, see Esteban Rodríguez-Ocaña, "International Health Goals and Social Reform," 265–276.

35 Ιατρική Εταιρία Θεσσαλονίκης, *Πρακτικά του Πρώτου συνεδρίου των ιατρών της Βορείου Ελλάδος με μόνον θέμα "Η ελονοσία εν Μακεδονία και Θράκη*," 326–27.

36 Ιωάννης Π. Καρδαμάτης, *Τα πεπραγμένα προς περιστολήν της ελονοσίας εν τω κράτει κατά το 1921* (Athens: Υπουργείον Συγκοινωνίας. Τμήμα Υγειονομικόν, 1922); Ιωάννης Π. Καρδαμάτης, *Τα πεπραγμένα προς περιστολήν της ελονοσίας εν τω κράτει κατά το 1922* (Athens: Υπουργείον Συγκοινωνίας. Τμήμα Υγειονομικόν, 1923).

remoteness, some villages in northern Greece (which had become part of Greece after 1912) had never seen any form of malaria control whatsoever before being contacted by postwar foreign aid workers.[37]

Finally, before the emergence of a modern public health system in twentieth century Greece, malaria, an endemic disease of high infection rates but low adult mortality, sustained a system of private health care, one driven by an acute nationwide demand for medical services given the limitations of state networks. The Anti-malarial League, a medical body earning a position of authority in prewar Greece, drew much of its inspiration from contemporary Italy, albeit selectively, and promoted a costly policy of malaria control of limited effectiveness, namely, quinine sanitization. With annual malaria morbidity rates as high as those revealed in the surveys, it is not surprising that the vertical strategy of disease control remained unchallenged and that no horizontal alternative—one encouraging social reform—was offered prior to Greece's prolonged humanitarian crisis of the 1920s. In effect, this policy lasted until the early 1930s when the new liberal Greek government inaugurated public health reforms and the Rockefeller Foundation's International Health Division representatives introduced to Greece a modern, research-based approach to malaria control.[38]

For its part, the Anti-malarial League empowered rural medical practitioners through an old cure, namely quinine, but with a new and meaningful conceptual frame for a disease with which they already had extensive experience.[39] As the state failed to extend its provision, it fell upon the League to mediate between the state on the one hand and private medical practitioners on the other, for both designing and implementing sanitary policy.

---

37 PAG-4/4.2:36, folder Greece #38, Region EG, Health Division Final Report United Nations Relief and Rehabilitation Administration, Health Division, Region EG, History, Chapter I: General Report on area before onset of UNRRA operation, p. 67, United Nations Archives.

38 Malaria control in interwar Greece and the involvement of the League of Nations and the Rockefeller Foundation's International Health Division are covered extensively in Maria Vassiliou, "Politics, Public Health, and Development: Malaria in 20th Century Greece," D.Phil. Oxford, Oxford University, 2005. For an account of the work of the Greek anti-malarial service in the 1930s, see Gregory A. Livadas and John C. Sphangos, *Malaria in Greece (1930–1940)* (Athens: Pyrsos Press, 1941).

39 In the sense used by Rosenberg, "Framing Disease: Illness, Society and History," xvi.

## Note on maps

The maps were compiled with the help of ArcGIS, version 8.3, using information from the four surveys cited in the text and data from *Κεντρική Ένωση Δήμων και Κοινοτήτων Ελλάδας. Λεξικό διοικητικών μεταβολών των δήμων και κοινοτήτων* (Athens, 2002), a database available on CD-ROM. I wish to thank Mr. Michalis Spyridakis, cartographer of the Greek Archaeological Service, for his valuable assistance with the GIS software. Any errors are my own responsibility.

# COMBATING INFANT MORTALITY IN BULGARIA

## WELFARE ACTIVITIES, NATIONAL PROPAGANDA, AND THE ESTABLISHMENT OF PEDIATRICS, 1900–1940

*Kristina Popova*

At the beginning of the twentieth century, infant mortality made its first appearance as a medical and social problem in Bulgarian medical periodicals. These tentative publications were not based upon systematic research but instead were isolated attempts to place this problem in a Bulgarian context, in order to compare the Bulgarian situation to that of other countries where infant mortality had been a research topic for much longer. The authors of these early publications pointed to the unavailability of data concerning infant mortality.[1] Although statistical data had been collected since 1881, the validity of such information was highly questionable. At this time, as laid down by statute, statistics were collected and systematized by local priests. About ten years later, however, medical officers took over the role. Unfortunately, the data remained inaccurate, primarily due to general difficulties in the Bulgarian infrastructure as well as poor communication between geographically isolated parts of the country.

Interestingly, infant mortality was not discussed in the central medical media as perhaps we would expect. In 1903 the subject was introduced by a medical journal published in the town of Vidin and two years later it was

1 Krum Kambosev, "Dvizhenie na naselenieto i smartnostta na detsata v Balgaria," *Zdrave. Mesechno populyarno medicinsko spisanie* 2, 5 (1903): 300–306; and Alexander Nedyalkov, "Dve dumi za detskata smartnost v Balgaria i borbata sreshtu neya (Parvata meditsinska izlozhba v Balgaria)," *Letopisi na lekarskiya sayuz* 3, 10–11 (1905): 119–121.

discussed in a periodical as well as in the first Bulgarian medical exhibition, both in the town of Tarnovo. A few years later the tuberculosis specialist Dr. Aleksandar Nedyalkov (1879–1967) explained that he had expected this introduction to provoke discussion in the medical literature, which however, did not occur. Therefore he continued to collect more data, which formed the basis for his book *Infant Mortality in Bulgaria and the Means to Fight Against It* [*Detskata smartnost v Balgaria i sredstvata za borbata s neya*] published in 1911.[2]

In the limited contemporaneous publications available, infant mortality was considered a purely medical problem. Nonetheless, there was a growing collective interest in child protection and child care that took into consideration the different social groups of children who needed society's protection. At the end of the nineteenth century the problems of school hygiene and child malnutrition had come to the forefront of concerns. In 1897 the first "Society for the Establishment of Free Soup-Kitchens for Schoolchildren in the Capital" [*Druzhestvo za ustroyvane na bezplatni uchenichenski trapezarii v stolnitsata*] was founded in Sofia, and a year later the Women's Society "Evdokia" (named after the recently born daughter of the Bulgarian ruler, Prince Ferdinand) was established to care for abandoned babies. At this juncture it is also pertinent to mention the 1905 parliamentary discussions concerning the "Law on Child and Women's Labor," which would prohibit child labor in industry for under twelve and in some exceptional cases for those under eleven years of age. The working hours of children aged between 12-16 years, of pregnant women and of nursing mothers were also regulated. In some instances the law prohibited child and women's labor in particular industries regarded as a threat to health.[3]

Another initiative was the introduction of school physicians, the so-called teacher-doctors [*uchiteli-lekari*], in 1904. They systematically observed both the health and development of the pupils, introducing standardized health cards to regularly record the physical condition of each individual child. By the end of the nineteenth century the first reports on

2 Alexander Nedyalkov, *Detskata smartnost v Balgaria i sredstvata za borbata s neya* (Veliko Tarnovo: Pechatnitsa Pryaporets, 1911), 136.

3 See Iliya Yanulov, *Razvitie na sotsialnoto zakonodatelstvo v Balgariya* (Sofia: Pridvorna Pechatnitsa, 1939), 96–103. In 1917 the law was replaced by the "Law for Hygiene and for Work Safety," which limited the age of children working in factories and granted certain rights to working mothers. See Yanulov, *Razvitie*, 123–132.

the health condition of boys were published. These were necessary in order to prepare boys for military service.[4] Street children and childhood delinquency became increasingly important topics in the media. The first societies for the care of young criminals—dealing mainly with the development of educational institutions for them—were called into being by judges in 1906. The professionals employed in these institutions were expected to have knowledge and understanding of both European and transnational discourses with regard to the etiology of juvenile delinquency.[5]

With his 1911 book, Alexander Nedyalkov tried to provoke a discussion about infant mortality among his colleagues. In order to give this problem of "great social importance" more relevance, he cited foreign authors and attempted to place the issue of infant mortality in a wider international, scientific context. Following medical discussions and interpretations of the causes of infant mortality, Nedyalkov emphasized the importance of natural child nutrition. Women's labor in industry hindered them from nursing and the ensuing use of artificial milk was considered a precondition for the high infant mortality rate. Other prerequisites included the paucity of hygiene in the poor households as well as alcoholism and sexually transmitted diseases. Nedyalkov agreed with the findings of his cited foreign authors, insisting that infant mortality rates depended on "cultural factors."[6] At the end of his book he published a detailed program for combating infant mortality, including recommendations for the opening of gynecological consultation stations, the provision of assistance by midwives, health promotion, the overcoming of traditional practices of childrearing, the introduction of individual health cards for newborn children and the development of institutes for child health research.[7]

Until the Balkan Wars of 1912 and 1913, the importance of infant mortality rates were simply considered a social medicine problem and only a small circle of Bulgarian physicians and child welfare reformers regularly published their research on child health in which they also referred to inter-

4 Stefan Georgiev Tonchev, "Opit za statisticheski svedeniya varhu zdraveopaznoto sastoyanie na nabornata (22 voenen priziv) mladezh ot Dobrichka okoliya prez 1898 godina," *Meditsinski napredak* 1 (1900): 1–3.

5 See Todor Mirkov, "Druzhestvo za pokrovitelstvuvane na nepalnoletni prestapnitsi v Plovdiv," *Uchilishten pregled* 5 (1906): 471–490.

6 He cited the German author Hugo Bernheim's, *Die Sterblichkeit der Kinder im ersten Lebensjahre* published in Würzburg in 1891. See Nedyalkov, *Detskata smartnost*, 134.

7 Nedyalkov, *Detskata smartnost*, 131.

national debates. These physicians adopted prevention methods and practices initially developed in Europe and North America including research institutions, milk stations, children's health exhibitions and education for mothers. In this way they managed to highlight infant mortality as a social problem.[8]

During the years of war between 1912 and 1918, the need for an improved Bulgarian health system together with the better organization of child care became an urgent imperative. The war created large groups of children needing social help such as children of refugee families, war orphans as well as those children judged as neglected. The specter of large numbers of suffering children and mothers provoked numerous international initiatives already during the war. In 1919 the "Save the Children Fund" initiated activities in the field. In the same year the "International Labor Organization" passed a "Maternity Protection Convention" while in 1920 the "International Child Protection Union" was founded. Finally, in 1924, a "Declaration of Children's Rights" recommending common standards for the child protection movement was proclaimed in Geneva. This declaration gave new impetus to the struggle against infant mortality.

After World War I the number of publications relating to infant mortality increased. Bulgarian publications defended the necessity of a struggle against infant mortality by stressing the growing difference between infant mortality rates in Bulgaria and Central and Western European countries. The prevention of high infant mortality was considered a major component in the modernization of Bulgarian society.[9]

The modernization pressure caused by this comparison with the so called "cultured countries" played an important role in the struggle against infant mortality in Bulgaria. In all publications, the high Bulgarian rates were contrasted unfavorably with the lower rates found in Western Europe

---

8 According to the Canadian author Terry Cop "The most important single issue raised by child welfare reformers was the very high rate of infant mortality which the Mid-Victorian fascination with statistics brought to light." See Terry Cop, "The Child Welfare Movement in Montreal to 1920," in D. C. M. Platt, ed., *Social Welfare 1850–1950, Australia, Argentina and Canada Compared* (Basingstoke: Macmillan, 1989), 45–60. With reference to the organization of measures against infant mortality in early twentieth century Germany, see also Larry Frohman, "Prevention, Welfare, and Citizenship: The War on Tuberculosis and Infant Mortality in Germany, 1900–1930," *Central European History* 39, 3 (2006): 431–481.

9 See Kristina Popova, *Natsionalnoto dete. Blagotvoritelnata i prosvetna deynost na Sayuza za zakrila na detsata v Balgariya 1925–1944* (Sofia: LIK 1999), 44–64.

**Decrease in infant mortality in Europe (per 1000 live births)[10]**

| | 1891–1900 | 1926–1928 |
|---|---|---|
| Germany | 217 | 96 |
| Switzerland | 149 | 56 |
| England | 153 | 68 |

**Infant mortality rates in Bulgaria[11]**

| Year | births | deaths in the first 12 months | mortality per 1000 |
|---|---|---|---|
| 1921 | 196,442 | 31,076 | 157.8 |
| 1922 | 202,602 | 31,401 | 155.0 |
| 1923 | 192,381 | 31,702 | 164.8 |
| 1924 | 207,111 | 31,074 | 150.0 |
| 1925 | 196,312 | 29,865 | 152.1 |
| 1926 | 202,730 | 25,825 | 127.4 |
| 1927 | 183,334 | 30,796 | 168.0 |
| 1928 | 185,189 | 27,512 | 148.6 |

**Infant mortality rates per 1000 live births (1936)[12]**

| | |
|---|---|
| Austria | 93 |
| Belgium | 79 |
| Bulgaria | 144 |
| UK | 59 |
| Germany | 66 |
| Greece | 114 |
| Italy | 100 |
| Portugal | 149 |
| Romania | 180 |
| Sweden | 43 |
| USA | 56 |

10 Dimitar Mishaykov, *Osnoven kurs po demografiya* (Sofia: Universitetska Pechatnitsa, 1941), 201.

11 See Zahari Zahariev, *Zashto umirat detsata (Prinos kam izuchavane na detskata smartnost v Balgariya)* (Sofia: Druzhestvo na narodnoto zdrave, 1939), 15.

12 Mishaykov, *Osnoven kurs*, 200. See also Iliya Yanulov, *Sotsialna zashtita na detsata* (Sofia: author's edition, 1932), 12–13.

and the United States.[13] The fall in the birth rate after 1924 prompted discussions about the problem of child care and infant mortality. This trend was interpreted as evidence of the degenerate population's balance between birth and mortality rates.

In 1924 the "Bulgarian Red Cross"[14] was invited to participate in the "Fourth International Congress for Child Protection" organized in Vienna and Budapest. Two Bulgarian representatives, the pediatricians Prof. Dr. Stefan Vatev (1866–1946) and Dr. Dimitar Kirov (1872–1957), took part in the congress. Key figures from the new child protection movements in the Balkans appeared at the conference—Dr. Safie Ali and Prof. Besim Omer Pasha from Turkey, Ms. Andreades and Ms. Kennedy, Chief Administrator of the "Save the Children Fund" representing Greece and Dr. Miloš Popović (1876–1954) from the Kingdom of Serbs, Croats and Slovenes.[15] With support from the "International Union for the Protection of Children" [*UIPE—Union Internationale de Protection de l'Enfants*] these individuals were able to present and assess the situation of children in the Balkans. They discussed the problems caused by the vast increase in numbers of refugee children together with the extraordinarily high infant mortality rate in Turkey, which stood at about 70–80% in some regions according to Turkish reports. The congress ended with an appeal for both international support as well as participation by the national state authorities.

The Vienna Congress initiated an exchange between the national "Child Protection Unions" of the Balkan states. From the end of the 1920s they began to meet at international conferences and in turn organize regional meetings: the "International Congress for Social Work" (Paris, 1929), the "Pediatric Panel of the Slavic Medicine Congress" (Sofia, 1936), the "First Balkan Congress for Child Protection" (Athens, 1936) and the "Second

[13] Elena Malcheva-Georgieva, "Detskata smartnost u nas," *Trimesechno spisanie na Glavnata direktsiya na statistikata* 1, 2–3 (1929): 204–211; Yanulov, *Sotsialna zashtita na detsata*; Ivan Nazlumov, *Plovdiv ot zdravno gledishte* (Plovdiv: Pechatnitsa Hr. G. Danov, 1934)., Stefan Vatev, *Kakvo tryabva da znae vsyaka mayka po othranvaneto, otglezhdaneto i vazpitanieto na detsata* (Sofia: Balgarsko druzhestvo "Cherven crast," 1927); Pencho Penchev, "Razhdaemostta v Balgaria," *Trimesechno spisanie na Glavnata direktsiya na statistikata* 1, 1 (1929): 13–32 and Slavcho Zagorov, *Namalenieto na razhdaemostta v Evropa i negovite prichini* (Sofia: Universiteska Pechatnitsa, 1931).

[14] The Bulgarian Red Cross Society was organized in 1885. Its network developed during the war period (1912–1918) and involved more than 10,000 volunteers.

[15] Stefan Vatev, "IV mezhdunaroden kongres za blagoto na detsata, stanal vav Viena, na 6-8 i v Budapeshta na 9-11 oktomvri 1924 g.," *Higienichesko spisanie i izvestiya na Balgarskoto druzhestvo "Cherven Krast"* (9 February 1925): 243–253.

Balkan Congress for Child Protection" (Belgrade, 1938). The participation of the Bulgarian Red Cross in the 1924 Congress together with the recognition of the Geneva Declaration of Children's Rights encouraged organizations to launch new initiatives. On 28 May 1924, the first "Child Health Consultation Station" opened in Sofia. A further eight stations for refugee children opened in 1926/27 with the support of both international organizations and the Bulgarian Red Cross. These stations were founded in towns with high refugee populations. During the 1930s similar health stations opened in virtually every town, supported by the municipalities. The activities of these Consulting Stations were recorded by Stefan Vatev, who published a quinquennial report replete with statistics and photographs.[16]

A regular course on public health for visiting nurses was established in 1926. It was held by Boyana Hristova, the first trained visiting nurse in Bulgaria, who had finished the Bedford School in London in 1923. Visiting nurses, who were supposed to have regular contacts with mothers and their children, to hold consultations and to prepare young mothers for modern childrearing, started to play an increasingly important role. Soon they were regarded as the "most powerful institution in the struggle against infant mortality."[17] Consequently, pediatricians, visiting nurses and female teacher–advisors together with their clients—families and young mothers—established a network which promoted the norms of home hygiene and modern childrearing.

In 1924 and 1925 the leading Bulgarian hygienist Toshko Petrov (1872–1942) published his *Letters to the Mothers and Fathers* in the journal of the Bulgarian Red Cross.[18] He recommended children be considered partners by their parents and that both parents care for children in order to overcome some of the superstitions found in childrearing. The journal of the Red Cross and other hygiene journals reported the child mortality rates in other countries, arguing that the correct way to combat child mortality was by opening consulting centers and introducing visiting nurses.[19]

16 See Stefan Vatev, *Deynostta na zdravnite saveshtatelni stantsii za mayki i detsa v Sofia prez 1924–1929* (Sofia: Pechatnitsa T.T. Dragiev, 1930).

17 F. Kisov, *Znachenieto na sestrata – posetitelka v borbata s detskata smartnost, doklad pred sestrinskiya sayuz* (1927), Bulgarian Central State Archive – *Tsentralen Durzhaven Archiv* (f-360k, op-2, ae-46, l, 24-26).

18 Toshko Petrov, "Kam maykite i bashtite, pismo II," *Higienichesko spisanie i izvestiya na Balgarskoto druzhestvo "Cherven krast"* (9 January 1925): 230–231.

19 Petrov, "Kam maykite."

In 1927 Professor Stefan Vatev and his assistants at Sofia's Pediatric Clinic published a book *What Every Mother Should Know about Child Nutrition, Childrearing and Child Education* [*Kakvo tryabva da znae vsyaka mayka po othranvaneto, otglezhdaneto i vazpitanieto na detsata*]. In the preface Vatev wrote, "Annually in Bulgaria 230 of 1,000 live births die, different from Sweden where only 76 die... The reason for the low rate of infant mortality in Sweden is that mothers are educated and instructed on how to rear their children."[20] Nurses explained childrearing techniques during home visits, and also instructed mothers on how best to follow their physician's advice. The authors identified the most dangerous diseases—diphtheria, tuberculosis, malaria—mentioning both their scientific and popular names. The authors not only advised mothers, but also described different aspects of children's social situation. After each topic they used the opportunity to address the wider community by issuing short appeals urging people to take part in the struggle against infant mortality: "Organize Milk Kitchens!", "Become a Samaritan Woman!", "Organize Soup Kitchens for All Pupils!", "Organize Summer Camps for All Children!", "Organize Hospitals for All Children Suffering from Tuberculosis."[21] Free copies of the book were given to each visitor to the traveling "Child Health Exhibition" [*Detska zdravna izlozhba*], organized by the new "Child Protection Union." In 1926 Vatev himself accompanied the exhibition, visiting six Bulgarian towns and giving lectures about childrearing.[22] A year later, delegations of both female and male physicians from the University Hospitals visited small towns and villages in order to give lectures about proper childrearing.[23] Furthermore, these physicians organized a committee to offer lectures to local municipalities. Unfortunately, local authorities were not interested in such initiatives.[24]

In the following years the reduction of infant mortality became an important target of official state health policy. According to the 1929 "Public Health Law," it was defined as a key element in the "care for next gen-

20 Vatev, *Kakvo tryabva da znae vsyaka mayka*, 3.

21 Vatev, *Kakvo tryabva da znae vsyaka mayka*, 3.

22 *Otchet za deynostta na nastoyatelstvoto na Sayuza za zakrila na detsata v Balgariya ot 1 yanuari 1926 do 25 fevruari 1927)* (Sofia: Pechatnitsa S. M. Staykov, 1927).

23 *Otchet za deynostta na nastoyatelstvoto 1927*, 11.

24 *Otchet za deynostta na nastoyatelstvoto na Sayuza za zakrila na detsata v Balgariya ot 8 yuli 1928 do 31 dekemvri 1929)* (Sofia: Pechatnitsa S. M. Staykov, 1930), 15.

eration," together with the organization of maternity and child welfare services. This meant that the struggle against infant mortality formed a separate issue both with respect to other demographic tendencies and other issues of child health care.[25] Child protection as a whole was proclaimed a state responsibility, with local municipalities taking responsibility for the organization of consulting stations, midwives services and supporting mothers in poverty.

Experts were sent to Plovdiv by the "Central Directorate of Public Health," the central authority for public health in Bulgaria, to study the reasons behind the particularly high infant mortality rates recorded in that region.[26] The Central Directorate recommended a special research project to find out the reasons for the conspicuous difference between infant mortality rates.[27] According to the demographers, physicians and economists, the main reasons were poverty and ignorance. These experts insisted on a detailed examination of social, ethnic and religious groups to uncover the reasons for infant mortality and to decide how best to fight it. Statistical data was presented separately for towns and villages, but the results were ambiguous and had to be interpreted. During the first years of investigation, no significant difference between infant mortality in the towns and villages was discovered,[28] but by the end of 1929, one could detect that—with exceptions—infant mortality on average was slightly lower in the towns than in the villages. The experts explained this difference as consequence of the 1917 "Law of Hygiene and Security of Labor" which had ensured some rights to pregnant women and mothers working in industry, rights that had not been extended to village women.[29] Therefore it is no accident that the rural population became the main target of the activities of the Bulgarian "Child Protection Union."

25 Ivan Balkanski, ed., *Arhiv na Glavnata direktsiya na narodnoto zdrave. T. 1. Godishnik za 1930 g. (v. svrazka s desetletieto 1921–1930)* (Sofia: Glavna direktsiya na narodnoto zdrave, 1932), 118–124.

26 Balkanski, ed., *Arhiv na Glavnata direktsiya*, 62.

27 Balkanski, ed., *Arhiv na Glavnata direktsiya*, 63.

28 See Malcheva-Georgieva, "Detskata smartnost u nas," 204 – 211.

29 Ivan Balkanski, "Zakonodatelni grizhi za deteto po Zakona za narodnoto zdrave," *Den na deteto*, May 4, 1930, 11–12.

Infant mortality (rural and urban) per 1000 live births 1926–1930[30]

| | Rural | Urban |
|---|---|---|
| 1926 | 125.9 | 135.9 |
| 1927 | 170.2 | 155.9 |
| 1928 | 147.9 | 152.3 |
| 1929 | 156.2 | 152.3 |
| 1930 | 138.2 | 128.7 |

## *'Our child dies more than the others...'*[31]—The Child Protection Union's Health Promotion and Networks

In May 1925 the Bulgarian "Child Protection Union" was founded with the support and influence of the "International Child Protection Union." The activities of the Bulgarian branch were furthermore influenced by the Geneva "Declaration of Children's Rights" and by the acknowledgment of the international standards laid down with respect to treating children. Returning from the 1924 "International Congress for Child Protection" (Vienna and Budapest), the Bulgarian delegates, Stefan Vatev and Dimitar Kirov, called together representatives from all institutions active in child welfare to discuss "the conditions of official and private care for children in Bulgaria. They recommended a permanent "Union of Child Protection Rights" as a branch of the "International Union" to be founded on the basis of the Geneva "Declaration of Children's Rights."[32] Stefan Vatev was elected as the first Chairman of the newly founded Union. To support the first initiatives of the Union, Frederica Freund, a representative of the "International Union," came to Bulgaria (from Vienna) and remained there for more than four years. According to the "Statute of the Child Protection Union," the struggle against infant mortality was considered a priority. However, other activities were also deemed important and thus recommended, such as the extension of legislation for the protection of mother

30 Balkanski, *Arhiv na Glavnata direktsiya za narodno zdrave,* 118.

31 Stefan Vatev, "Spasete Detsata," *Nasheto dete* 1, 1 (1928): 1.

32 *Otchet za deynostta na nastoyatelstvoto na Sayuza za zakrila na detsata v* Balgariya ot 8 yuli 1928 do 31 dekemvri 1929.

and child together with additional safeguards for orphans, illegitimate and abandoned children.

The Union planed to promote health knowledge for pregnant women and young mothers, among other things hygiene for pregnant women, child nutrition, household hygiene and childhood diseases. To disseminate its ideas among a wider public the Union since 1926 organized an annual "The Day of the Child" [*Den na deteto*] on the day of the Holy Mayrrhbearing Women, i.e. the second Sunday after Easter. A magazine with the same name started in 1927 with an article written by Stefan Vatev on "infant mortality."[33] The author delineated the social reasons for the very high infant mortality rates and furthermore outlined the most common diseases responsible: diarrhea, scarlatina, diphtheria and bronchitis. He stressed the need to open health consulting stations as well as offering health advice to young mothers, particularly in rural areas.

Encouraged by the success of these first propaganda actions, the Union started the monthly publication *Our child* [*Nasheto dete*] in 1928. Infant mortality was also a prominent topic in this magazine and it was stressed that in order to bridge the gap between the Bulgarian situation and the "normal" situation in the "cultured countries," child protection and the "normalization" of child care were necessary. "Our child is getting sick, it is degenerating and dying more often than normal," Vatev wrote in the first issue of *Our Child*. He continued:

> There are social and individual reasons for this: the lack of health promotion in the family, of knowledge about proper childrearing, [and] of social child care. In such countries the cultural progress is not sufficient to make people aware of the need for rational childrearing and social care. Our child is disadvantaged by the family life situation, the school and labor system, [and] by street life.[34]

*Our Child* regularly published articles about different childhood diseases, offering advice for the appropriate treatment as well as guidance with regard to child nutrition and childrearing. Many issues contained photographs which presented poor village children within their families and schools and showed the lack of food and clothes. *Our Child* greatly contrib-

33 Stefan Vatev, "Smartnostta u detsata," *Den na deteto* (8 May 1927): 4.

34 Stefan Vatev, "Spasete Detsata," 1.

uted to raising awareness of poverty, of the dire conditions of child nutrition and more generally to issues of child care. This campaign of health promotion and the information exchanged between local and central branches of the Union soon destroyed the romantic myth of the village as a "healthy place." Hence, Bulgarian peasant children and their mothers became a priority group for the activities of the Union. During the next years the articles, letters and photographs created a symbolic image of "our child" who "dies and suffers" more than other children. Local reports about the difficulties of child care, details and photographs of successful soup kitchens for children and the presentation of the annual celebrations of the "Day of the Child" ensured that the conditions of children in Bulgaria were represented in detail.

The Union established a wide network of local organizations, but to be truly effective required specially trained personnel, particularly volunteers who were in regular contact with young mothers and children. Research on literacy in rural areas revealed that about a half of the young mothers in the villages were illiterate or had difficulties reading. It was therefore, impossible to offer written texts as education material. Instead, the Union was forced to mobilize the existing modernization agents within the villages: teachers, (especially women) who had permanent access to the young mothers. A new type of Union activist was introduced, the so-called female teacher-advisor [*uchitelka-savetnichka*]. In order to do so, the Union began annual courses for female teachers, to prepare them for voluntary work, principally in the villages. Usually such teachers were young and came from the local village schools. In the course they were instructed in various disciplines—school hygiene, symptoms of tuberculosis and ways of combating it, symptoms of malaria, infant mortality, eugenics, children's rights, anatomy and physiology of the female body, child dental care, sport and tourism.[35] These courses were intended to provide teachers with some elementary medical knowledge. The trainees also visited children's hospitals and other children's institutions to enable their understanding of existing hygiene practices and standards. More than a thousand female teach-

[35] Penka Georgieva-Petkova and Todor Zubov, eds., *Rakovodstvo za uchitelki-savetnichki* (Sofia: Sayuz za zakrila na detsata v Balgariya, 1933).

ers were trained in this way in the next ten years.[36] In the villages the female teacher-advisors were expected to collect data about the living conditions, hygiene and nutrition of families and children. They asked schoolchildren about their breakfast, lectured young mothers and girls about hygiene and health and shaved the heads of children to prevent the spread of typhus. They were supposed to organize hygienic soup kitchens for schoolchildren and to decrease the influence of the old untrained midwives who defended traditional practices of childrearing.

Annual reports for all these activities were required by the central body of the Union, some of which were published in *Our Child*. According to the archived reports, many teachers were enthused with the idea and practice of child protection activities. The reports offer information about their support and control functions. Rayna Boydeva was an elementary school teacher in the village of Vresovo in the region of Burgas. She completed her teacher-advisor course in 1930 and in the same year started work in the village. In her annual report for 1930/31 she wrote that after school she had visited 37 families in their homes. Mothers were more ready to accept her instructions than fathers. She lectured village women on several topics including "Child Nutrition," "Childrearing," "Hygiene for Pregnant Women and Mothers," "The Care for the Future of Our Child" and "More Care for Our Child." To improve the individual hygiene of the schoolchildren she supplied the school with "cheap soap" and made them wash their hands. She also directed sick children to the appropriate physician.[37] In her annual report for 1932/33 Rayna Boydeva wrote that she had given three lectures on "The Village Women as a Housewife, Working Women and Mother," "Care for Sick Children" and "The Significance of the Fight against Alcoholism." She organized a free school kitchen for 14 children and visited 33 homes: "Initially I was not invited into the families; now I am invited more often. My suggestions support the physicians' work, they contribute to the faster recovery of sick children, and the physicians are accepted with more confidence than the old women."[38]

36 In the 1930s there were 5,712 elementary schools in the Bulgarian villages. See Konstantsa Lyapcheva, "Zakrila na selskoto dete v Balgaria," *Nasheto dete* 15, 3 (1942): 9–10.

37 Central State Archive, Sofia (TsDA) (f-365k, op-1, ae-30, l-88).

38 TsDA (f-365k, op-1, ae-30, l-89).

Rada Bakardzhieva, another teacher-advisor working in the village of Chadurlii in the region of Sevlievo confirmed that "the mother accepts my instructions more easily than the father, and according to her ability and time she tries to improve the situation in the house."[39] In her previous report for 1929/30 she had expressed the conviction that every village woman needed some knowledge of hygiene. She gave 10 lectures to women—about the anatomy and physiology of the human body, the woman's situation and her central place within the family, sexual hygiene, childrearing and child nutrition. When visiting families she observed both the hygiene of the house and nutrition. To improve the hygienic situation she also supplied the school with "cheap soap" direct from the producer and also provided a small mechanic shaver for cutting the children's hair.[40]

For the period of 1930/31 Radka Chervenkova from the village of Golyamo Yalare (now known as Rusalya) in the region of Veliko Tarnovo reported that she had visited ten young mothers and instructed them on how to maintain modern hygiene, "I was successful; our village women accepted suggestions at once."[41] Female teacher-advisors introduced health cards for every child to compensate for the lack of school doctors in most schools. In the 1930s the teacher-advisers succeeded in creating a wide network in which to observe child health conditions, to spread modern standards of childrearing and to exert social control over the rural population. In an attempt to support teacher-advisors and to teach village mothers, Vatev wrote a small book *The Mother—The Mistress of the Family*[42] [*Maykata—gospodarka na semeystvoto*] in simple language and explained the basic rules of childrearing together with the advantages of contacting medical staff and health stations instead of simply tolerating the dangerous traditional practices of untrained midwives.

At the end of the 1930s Zahari Zahariev (1905–1959) published another book titled *Why Do Children Die?* [*Zashto umirat detsata*] as part of the struggle against infant mortality. Abundantly equipped with illustrations, the book was based on the records of village physicians and presented the existing dangerous traditional practices and beliefs as mere superstitions. The

39 TsDA (f-365k, op-1, ae-30, l-71).
40 TsDA (f-365k, op-1, ae-30, l-72).
41 TsDA (f-365k, op-1, ae-30, l-108).
42 Stefan Vatev, *Maykata – gospodarka na semeystvoto* (Sofia: Sayuz za zakrila na detsata v Balgariya, 1936).

Fig 1. "Putting salt on the newborn body." Source: Zahari Zahariev, *Zashto umirat detsata (Prinos kam izuchavane na detskata smartnost v Bulgariya)* (Sofia: Druzhestvo na narodnotozdrave, 1939).

texts and the illustrations—the latter done by Stoyan Venev (1904–1989), one of the famous young artists of the 1930s—depict the rural mothers and babies as innocent victims of old illiterate village women. He stigmatized traditional childrearing as a "dark labyrinth" where babies die.[43] Like *Our Child,* this book presented villages as homogenous social spaces united by common traditional practices that should be blamed for bad sanitary conditions and childhood diseases. Zahariev revealed just some of the practices and beliefs surrounding child birth—the use of the mother's hair or the thread of the father's clothes to bind the umbilical cord, a practice which could cause tetanus, the cutting the umbilical cord with a sickle, which could cause infection, or cutting it immediately after birth, which could cause anemia, powdering the new born baby with salt and various other ingredients (see fig. 1). He described the characteristics of the new born organism and explained how each dangerous practice could cause disease.

A mobile "Child Health Exhibition" was created by the "Union of Child Protection" to visit villages and small towns, promoting norms of modern hygiene and childrearing. During the ensuing years the exhibition became

43 Zahariev, *Zashto umirat detsata.*

a centerpiece for health promotion, especially for the rural public. The exhibition explored models for childrearing supported by instructions: "Your child needs to have a bath at least once weekly if not twice," "Your child needs clean clothes," "You have to cut every child's hair daily," "There is no health without hygiene."[44] Pictures of the different stages of pregnancy, embryonic and child development together with illustrations of the causes of infant disease were also presented. The exhibition presented examples of how a modern mother should take care of her baby and explained what mothers needed to do in order to ensure that their baby remained healthy. Many real-life examples were also included to illustrate where the risks were, for example dirty, untidy rooms or babies growing up together with domestic animals. In this way, promoted scenarios of childrearing were contrasted favorably with negative models. The arrival of the exhibition was celebrated by the village as a great event that brought local authorities, physicians, teachers, young mothers and children together. Young mothers and schoolchildren wore their best clothes and often let themselves be photographed. By 1935 the exhibition visited 85 towns and 122 villages.[45]

## *"The Care for the Bulgarian Child is a Guarantee for the Health and Welfare of the Bulgarian State Tomorrow"*[46] —Infant Mortality as National Problem

The agenda of the "Child Protection Union" together with annual celebrations for the "Day of the Child" and other activities embraced a national rhetoric of child care and the recognition of the struggle against infant mortality that promoted ethnocentric views of childhood in Bulgaria. The principles of the Geneva Declaration were translated into nationalistic terms. In his 1935 book about infant mortality, Lazar Binder (1888–?) wrote: "In order to fulfill the requirements of the Geneva Declaration the starving Bulgarian child should be fed, the neglected Bulgarian child should be protected and sheltered and the sick Bulgarian child should be cured. By the Bulgarian

44 Popova, *Natsionalnoto dete*, 150–165.

45 Popova, *Natsionalnoto dete*, 150–165.

46 "Negovo Velichestvo Tsarat za deteto," *Den na deteto* (8 May 1927) (s. l., s. d. [1927]), 3.

child for the Bulgarian state and nation!"[47] He pointed out that the struggle was, "one of the most important questions for the future of the nation."[48]

After the establishment of the military government in 1934, the state intensified its support for the health promotion network of the "Child Protection Union" and for the modernization efforts it conducted particularly in rural areas. The "International Child Protection Union" celebrated and promoted its achievements, which were even exported as best practice for modernization in child care to other Balkan countries. Village children and young mothers were a special focus of interest for the "International Child Protection Union." which, in turn, supported the exchange of ideas and practices in Balkan child protection. This resulted—as previously mentioned—in the first "Balkan Congress for Child Protection" (Athens, 1936) with the second one held in Belgrade in 1938. These congresses encouraged the development of training courses for village teachers and priests to fully prepare them for child protection activism. They also recommended the collection of information on children's living conditions with the respective criteria for processing the data in order to establish common criteria according to League of Nations rules. Furthermore, the congress in Athens stressed the need to improve "puericulture," understood as a system of skills and measures of care for pregnant and puerperal women, together with children of all age groups. It strongly recommended the implementation of the health care system, especially in rural areas.[49] The Balkan congresses stimulated official attention toward child care. The Bulgarian journal *Our Child* presented child care institutions with the respective statistical data of the neighboring Balkan countries and additionally introduced publications by authors from these countries, like the Romanian hygienist Gheorghe Banu (1889–1957)[50] and the Serbian social reformer Slobodan Vidaković, President of the "Yugoslav Union of Child Protection."[51]

---

47 Lazar Binder, *Detskata smartnost u nas, prichini i borba s neya* (Sofia: Roditelsko-uchitelski komitet na 12 Sofiysko narodno uchilishte "Tsaritsa Yoanna," 1935), 29.

48 Binder, *Detskata smartnost u nas*, 29.

49 Popova, *Natsionalnoto dete*, 125–127

50 Gheorghe Banu, *Mortalitatea infantilă în România: studiu de igienă socială şi de statistică* (Bucharest: Tipografia Curţii Regale F. Göbl, 1930); the book was donated to Sofia's National Library by the famous Romanian historian and politician Nicolae Iorga. Gheorge Banu took part in the "First Balkan Congress for Child Protection" in Athens in 1936.

51 Slobodan Vidaković, *Komunalna zaštita dece* (Belgrade: Knjižara S. B. Cvijanovića, 1937).

## The Establishment of the Pediatric Society

The network of child health stations together with the system of school physicians, the "Child Protection Union" and the 1929 "Public Health Law," contributed to the advancement of pediatrics as an important part of the infrastructure of health sciences in the 1930s. Aldready in 1895 a pediatric section was opened in the ward for Internal Diseases in Sofia's Alexander Hospital, the main hospital in the country. The section became a ward of its own in 1905 and later it became a pediatric clinic. The "Bulgarian Pediatric Society" [*Balgarsko pediatrichno druzhestvo*] founded in 1929 published the *Pediatric Review* [*Pediatrichen pregled*] in 1931. The first issue proclaimed that the journal was concerned with "childhood in a medical, social and pedagogical sense." The periodical focused not only on child health but also on wider issues concerning children. Members of the Editorial Board included such luminaries as the philosopher Dimitar Mihalchev (1880–1967) and the renowned biologist Metodi Popov (1881–1954).[52] The editor of the journal was Persiado Rahamimov (1905–?), author of several popular books on childrearing.[53] In the introductory article of the journal, Stefan Vatev outlined the reasons for producing such a periodical: an increased social interest in childhood, the academic research on childhood diseases with the subject's inclusion in the university curriculum, the need for medical internship and the introduction of state exams in pediatrics.[54] This periodical contributed greatly to the establishment of academic pediatrics in Bulgaria; it called for the establishment of preparation courses for mothers as a general measure, it informed its readers of new foreign findings and publications dealing with diseases affecting infant mortality or the requirements of proper child nutrition[55] and it published the minutes of the society's discussions on different scientific issues. The editors emphasized the necessity of the *Pediatric Review* for rural physicians, who were particularly "hungry for knowledge" and

52 *Pediatrichen pregled* 1 (1) (1931).

53 Persiado Rahamimov, *Upatvane za hranene na karmacheto,* (Sofia: Pechatnitsa S.M. Staykov, 1936), 41; Persiado Rahamimov, Lyuben Rachev, *Razvitie, higiena i hranene na malkoto dete* (Sofia: Pechatnitsa Rila, 1938), 229.

54 Stefan Vatev, "Nashite zadachi," *Pediatrichen pregled* 1, 1 (1931):1.

55 See Persiado Rahamimov, "Zahranvane na karmacheto," *Pediatrichen pregled* 5, 3 (1935): 180–190; and Mihail Adzharov, "Detskata smartnost i neynite prichini," *Pediatrichen pregled* 5, 2 (1935): 81–82.

"devoured" it.[56] In 1932 the journal published an article about infant mortality in Sofia. The author related the mortality rates to particular diseases and months and underlined the social reasons for such high mortality rates. He argued that every case needed separate research and home visits according to the recommendations of the "Health Section of the League of Nations."[57]

In 1935 the Pediatric Society officially appealed to the head of the state-run "Central Directorate of Public Health" to take measures against the high infant mortality rate in Bulgaria.[58] The Pediatric Society also requested improvements in the practical preparation of physicians by including at least a year practical work in a pediatric clinic together with the establishment of pediatric wards in every major hospital.[59] In these years many female physicians joined the profession and took positions of responsibility in the organization: Dr. T. Zlatkova and Dr. Sophia Epelbaum became secretaries of the Pediatric Society and Dr. Bogdana Vinarova the editor of the Society's *Pediatric Review*.

In providing a separate panel for pediatricians, the "Fifth Slavic Medical Congress," held in Sofia in 1936 acknowledged the achievements of Bulgarian pediatricians. Pediatricians from Yugoslavia, Bulgaria and Czechoslovakia took their places on the panel. The different experiences of the pediatric hospitals in Belgrade, Zagreb and Prague were presented.[60] The panel's discussions focused on three main diseases: tuberculosis, scarlet fever and diphtheria. Dr. Marija Gajić (1899–?) and Dr. Miloš Petrović spoke about tuberculosis and infant mortality in Belgrade, Dr. Dragoljub Ilić (1889–1974) spoke about diphtheria and Dr. Smilja A. Kostić-Joksić (1895–1981) talked about scarlet fever and the immunization of children in Belgrade.[61] The pediatric panel of the Fifth Congress passed two important resolutions concerning "Diphtheria in the Slavic Countries" and "Tuberculosis in the Slavic Countries," insisting on the necessity of a regular immunization program for children.[62]

56 "Protokoli ot zasedanie na pediatrichnoto druzhestvo," *Pediatrichen pregled* 6, 1 (1936): 60.

57 Simeon Gunchev, "Detskata smartnost v Sofia," *Pediatrichen pregled* 2, 2 (1932): 90–92.

58 "Do direktora na Narodnoto zdrave," *Pediatrichen pregled* 6, 2 (1936): 111. In the late 1930s the Pediatric Society had about 150 members, the majority of whom lived in Sofia.

59 "Do direktora na Narodnoto zdrave," 111.

60 The fifth issue of the fifth volume of the *Pediatric Review* was dedicated to the pediatric panel of the "Fifth Slavic Medicine Congress" in Sofia. See *Pedriatrichen pregled* 5, 5 (1936).

61 See the issue of *Pedriatrichen pregled* 5, 5 (1936).

62 "Iz zasedaniyata na pediatrich. D-vo," *Pediatrichen pregled* 7, 1 (1937): 57–58. According to the decision of the Supreme Medical Council, immunization against diphtheria became compulsory in 1940.

We should nevertheless not overemphasize the influence of the Pedriatric Society. After all, of the circa three thousand physicians in the whole of Bulgaria in 1940, only 150 were members of the Pediatric Society, of which 104 from Sofia and 46 from the rest of the country.

A few words should be said about the connections between Bulgarian pediatric and eugenic thinking. Up until the end of the 1930s there does not appear to be any essential discussion of eugenics in the *Pediatric Review*; infant mortality tended to be discussed in a universalistic way as caused by insufficient hygiene, lack of health stations and immunization. In 1940 Dr. Asparuh Panov (1902–?) visited the pediatric clinic in Berlin. Upon his return he gave a lecture that was also published in the *Pediatric Review*.[63] Panov was impressed by the order, cleanliness and high level of the scientific research but he disapproved of German eugenic practices, especially the replacement of individual care with the care of "the nation" as the main objective for treatment. He was highly critical of the division of children and elderly into groups of normal and mentally weak. He was afraid that innocent people diagnosed as mentally weak could become victims of such practices, and warned physicians, explicitly of the danger of such an approach.[64]

## Conclusions

The struggle against infant mortality placed special emphasis on child care and raised social respect for young mothers and children. It contributed to the establishment of a nationwide network that included medical professionals, civic activists, local child protection societies, health consulting centers and professional and popular media. This highly developed network was supported by the International Red Cross, the "International Union of Child Protection" and other international health organizations. Its activities raised the awareness of children's situations, especially in the countryside, where the focus was on abandoned, illegitimate or orphaned children. This new field of public activity gathered together many people who were

63 Asparuh Panov, "Vpechatleniya ot dneshna Germaniya," *Pediatrichen pregled* 10, 1 (1941): 22–29.
64 Panov, "Vpechatleniya."

inspired by modern ideas of child protection. In the fields of hygiene and medicine the struggle against infant mortality was realized as a program of instruction, supervision and, essentially, controlling mothers in their child-rearing practice. Modern norms of child care were initiated as processes of modernization following the example of other European countries.

Discussions about infant mortality in Bulgaria had various goals. Popular media engaged in child protection and health exhibitions while books stressed the national importance of the struggle against infant mortality. They spoke about protecting "our child" or the "Bulgarian child" in an attempt to mobilize a wide public audience. Their ultimate goal was the survival and health of children in the name of a better national future. "Puericulture" was considered the main tool for improving the culture of childrearing in the families.

In the Bulgarian case the struggle against infant mortality was interpreted as care for the "progeny" as well as for the "national future." The movement exhausted itself in the improvement of health promotion, hygiene, child nutrition, childrearing and regular consultative treatment of pregnant women and young mothers. The work of pediatricians was assisted by women who held new professions or practiced voluntary vocations: visiting nurses and teacher-advisers. Eugenics was involved in the preparatory courses first as an element of health promotion but later as a guiding principle when prevention of the consequences of malnutrition, of diseases like tuberculoses or of alcoholism and venereal diseases were discussed.

The Pediatric Society had its own aims in the struggle. Pediatricians used the opportunity to emphasize their professional and social importance in an effort to become a more visible part of the national health policy. They strengthened the authority of pediatrics as an academic discipline and by maintaining their international contacts shared medical knowledge about childrearing as well as treatment of the main childhood diseases. Both popular and scientific discourses were inspired by a series of international conferences and congresses on child protection and pediatrics in which Bulgarian representatives not only took part but also found recognition.

# POLITICS, MODERNIZATION AND PUBLIC HEALTH IN GREECE

## THE CASE OF OCCUPATIONAL HEALTH 1900–1940

*Leda Papastefanaki*

The process of modernization undertaken by Greece in the early twentieth century, mainly by liberal governments, was prompted by fundamental transformations in economic and social life. At the time, these were perceived chiefly as "Europeanization" and concerned urban planning and education, the re-organization of public health and labor policy. In the field of social policy, attempts to pass legislation on public health and the creation of public health institutions were one pole of the modernization of the Greek state. The specific undertaking to re-organize public health on the basis of a rational treatment of people was linked with the gradual formation of the welfare state. In Western Europe the development of public health from the nineteenth century onwards was looked upon as a kind of "social reformation," a response to the acute social and health problems created by urbanization and industrialization.[1] The other pole of modernization in the sphere of Greek social policy was the introduction of labor legislation and the creation of institutions such as the Labor Inspectorate, ensuring the mediation of the state in labor relations so as to safeguard the elementary conditions for the reproduction of the labor force.[2]

---

1 Elizabeth Fee and Dorothy Porter, "Public Health, Preventive Medicine and Professionalization: England and America in the 19th century," in Andrew Wear, ed., *Medicine in Society. Historical Essays* (Cambridge: Cambridge University Press, 1992), 249–275; and Ann F. La Berge, *Mission and Method. The Early Nineteenth Century French Public Health Movement* (Cambridge: Cambridge University Press, 1992).

2 For a general overview of Greek modernization, see Γιώργος Μαυρογορδάτος, Χρήστος Χατζηιωσήφ eds., *Βενιζελισμός και Αστικός εκσυγχρονισμός* (Heraklion: Πανεπιστημιακές Εκδόσεις Κρήτης, 1988).

In the context of the economic crisis, social unrest and political ferment on the international stage during the interwar years, the International Labor Organization and its secretariat, the International Labor Office (ILO), played an important role—and in many cases lanced decisive interventions—in the regulation of labor relations, especially through a *sui generis,* tripartite collaboration of state, employers and workers. The process of economic modernization specific to Greece unfolded within the particular operations of the political system. Parliamentary democracy alternated, after 1909, with military coups, and even when the parliament was functioning, functionaries from the leading bourgeois parties were almost always willing to take measures to undermine constitutional guarantees of freedom, adopting authoritarian measures against their opponents—above all labor unions and communists.[3]

Public health movements in Greece involved not only an improvement in infrastructure and a reduction in the population's morbidity and mortality rates but raised more general issues for the rejuvenation of Greek society—both literally and metaphorically. In public debates on these issues the incorporation of the population, forms of social exclusion, supervision of social behavior and the confinement of the "unhealthy" were hotly discussed. The views expressed were not neutral, neither from the point of view of social class nor gender; on the contrary, they promoted new social and gender values in addition to renegotiating dominant social relations of power. As such, medical knowledge —and, more generally, scientific knowledge—is not approached here as an objective body of 'facts'; instead, my view is that scientific knowledge engages in a dialogue with the social environment in which it is produced and by which it is affected.[4]

3 Gunnar Hering, *Die politischen Parteien in Griechenland 1821–1936* (Munich: R. Oldenbourg, 1992).

4 For this discussion, see Paul Weindling, ed., *The Social History of Occupational Health* (London: Croom Helm, 1985); Richard Gillespie, "Accounting for Lead Poisoning: The Medical Politics of Occupational Health," *Social History* 15, 3 (1990): 303–331; Robert Gray, "Medical Men, Industrial Labour and the State in Britain, 1830–50," *Social History* 16, 1 (1991): 19–43; and Catherine Omnès, Anne Sophie Bruno, eds., *Les mains inutiles. Inaptitude au travail et emploi en Europe* (Paris: Belin, 2004).

## Modernization of Public Health System

When Constantinos Savvas (1861–1929), university professor and court physician who had studied in Athens, Vienna and Berlin, delivered his inaugural address in 1900 on the subject of hygiene and microbiology at the University of Athens, he was animated by great optimism. This feeling derived, no doubt, from the significant progress made in medicine by the end of the nineteenth century, from the gradual decrease in mortality observed in Greece during the last quarter of that century and to the medical students' "well-known conscientiousness and thirst for knowledge."[5] Savvas claimed that in order to improve public health, measures that needed to be taken included the organization of an efficient statistical service to study population movements, the legal enforcement of vaccination and re-vaccination to annihilate smallpox, the mandatory reporting and isolating of infectious diseases, the popularization of hygiene by every means possible and the introduction of an extensive state health care system.[6] Driven by the conviction that "the state of health of a people constitutes the criterion of its civilization," Savvas further hoped:

> Not only shall we greatly reduce illness and mortality but we shall also reinforce the people's resistance, thus rendering the population denser, healthier and therefore more hard-working. And hence an increase in the economic prosperity of the nation, constantly becoming more robust through ceaseless and advancing assumption of new life and well-being, will swiftly be able to rise to the appropriate degree of true civilization and of that perfection which is feasible for humanity.[7]

Hygiene, this new branch of medicine, the teaching of which Savvas inaugurated at the university (in conjunction with microbiology), seemed to be the science best able to bring about "social improvements" at the dawn of the twentieth century. With the sciences of hygiene and microbiology it was now possible to prevent infectious diseases and to realize a significant reduction in the Greek mortality rate. Savvas continued:

---

5 Κωνσταντίνος Γ. Σάββας, *Λόγος εναρκτήριος εις το μάθημα της Υγιεινής και Μικροβιολογίας εκφωνηθείς τη 15 Νοεμβρίου 1900* (Athens: Εστία, 1900), 26; all translations from the original are my own unless otherwise cited.

6 Σάββας, *Λόγος εναρκτήριος εις το μάθημα της Υγιεινής και Μικροβιολογίας*, 21–24.

> This gain in terms of life and vital power saved has not only a boundless moral and philanthropic value but also an incalculable economic one [...] because a very large human population is kept alive and engages in work. But apart from the profit arising from saving and lengthening the life of so many people, another most important economic factor is the lowering of the mortality rate necessarily occurring as a result of the reduction of the number of diseases.[8]

Interestingly, the latter, according to the calculations of the German chemist and hygienist Max von Pettenkofer, brought about significant economic benefits.[9] Modern hygiene, therefore, served as an essential political economy of health since it secured the necessary labor force for industrial capitalism while at the same time reducing sick days and the expenses of medical treatment. For these reasons the strict implementation of modern hygiene had to be the principal concern of Greek governments, like in the "civilized countries." Savvas concluded—following the British Prime-minister Benjamin Disraeli (1804–1881)—that "an improvement in the state of public health of the people is the primary social aim on which every politician must first and foremost fix his attention."[10]

In Savvas' 1907 university textbook on "Hygiene"—which had gone through four editions by 1928—a definition of the science of hygiene is given: its object is the recommendation and implementation of any measure that contributes to the maintenance and promotion of human health. As a medical science, hygiene thus examines the significance of all factors influencing human health and formulates rules by which humankind should organize life—not only so that it can develop in a healthy manner physically and mentally but so that it can maintain its health and increase its resistance and ability to work. The object of hygiene is therefore a broad one: the study of the environment, both natural and artificial.[11] Hygiene was represented as a unifying science that supplied the greatest benefit to society since "it does not belong only to medical science but is also an

7 Σάββας, *Λόγος εναρκτήριος εις το μάθημα της Υγιεινής και Μικροβιολογίας*, 25.

8 Σάββας, *Λόγος εναρκτήριος εις το μάθημα της Υγιεινής και Μικροβιολογίας*, 13.

9 Quoted by Constantinos Savvas in Σάββας, *Λόγος εναρκτήριος εις το μάθημα της Υγιεινής και Μικροβιολογίας*, 14–16.

10 Quoted by Constantinos Savvas in Σάββας, *Λόγος εναρκτήριος εις το μάθημα της Υγιεινής και Μικροβιολογίας*, 24.

11 Κωνσταντίνος Γ. Σάββας, *Εγχειρίδιον Υγιεινής* (Athens: Εστία, 1907), 8.

important branch of public economy whose main aim is to combat social malaise."[12]

By 1910 Constantinos Savvas was no longer an optimist, quite the contrary. He had realized the dispiriting state of public health in Greece. Since 1845, no legislative activity whatsoever had addressed public health issues and no modern health care system had been established. At the same time, Savvas was shocked by the relatively high mortality rate of the Greek population as well as by its high fatality rate owing to infectious diseases, this at the exact same time that Western European countries were taking extensive measures to significantly reduce mortality, fatality and morbidity rates due to infection.[13]

Undoubtedly, conditions for public health and its infrastructure in Greece were in an embryonic stage before the outbreak of World War I. The "Council of Physicians" [*Ιατροσυνέδριον*], the committee responsible for the medical organization of the state since the reign of King Otho (r. 1833–1862), had left the task uncompleted. The reasons for this failure were largely associated with the state's indifference to public health, the lack of funds, the shortage of medical and nursing personnel and the absence of an organized program for public health intervention.[14] Greece's health system was based mainly on the institutional health services of the army, municipal authorities and private charities. Since the nineteenth century, epidemics of every kind had afflicted the population—and chiefly among it the weaker and poorer strata in the growing cities. The Balkan Wars (1912–1913), World War I and Greece's campaign in Asia Minor (1919–1922) with the concomitant large-scale movements of armies and refugees only intensified the problem of morbidity.[15] In the rural areas especially, malaria manifested a high rate of frequency and fatality. The disease's frequency remained high until 1928 and no central anti-malaria service was ever established, the state refusing responsibility for combating the disease in an effective manner.[16]

---

12 Σάββας, *Εγχειρίδιον Υγιεινής*: 18.

13 Κωνσταντίνος Γ. Σάββας, *Σχέδιον νόμου περί επιβλέψεως της δημοσίας υγιείας* (Athens: Εθνικόν Τυπογραφείον, 1910).

14 Αριστοτέλης Κούζης, *Αι μετά την ίδρυσιν του βασιλείου της Ελλάδος πρώται παρ' ημιν αρχαί υγιειονομικής πολιτικής και οργανώσεως της Δημοσία υγεία*, Λόγος εκφωνηθείς κατά την πανηγυρικήν συνεδρίαν 28ης Δεκεμβρίου 1946 (Athens: Ακαδημία Αθηνών, 1947).

15 Φωκίων Κοπανάρης, *Η δημόσια υγεία εν Ελλάδι* (Athens: τυπογραφείον Χρήστος Β. Χρονόπουλος, 1933).

16 Constantinos Savvas and Jean P. Cardamatis, *Ligue antimalarienne hellénique (1905–1928)* (Athens: Imprimérie Progrès, 1929).

Tuberculosis had spread quickly since the late nineteenth century and had in the cities taken on unprecedented dimensions with a high fatality rate. The official initiatives in the organizing of the "anti-tuberculosis struggle" had limited financial resources at their disposal while it was generally recognized that poor living conditions, diet and working conditions among the laboring strata contributed to the constant spread of this disease.

The problems of public health in Greece were linked directly to the quality and conditions of life, including the high population density in urban centers and the effectively non-existent infrastructure and medical treatment network. In the two largest cities, Athens and Piraeus, working class houses and communities were described as sunless, cramped, dirty, unhealthy "caves," gloomy basements, "pits" lacking ventilation and rudimentary infrastructure.[17] Furthermore, the water supply, sewage and drainage in urban centers were a major problem for public health. Before 1930, very few towns had "healthy" water with even fewer having elementary drainage. Nearly all Greek cities' aqueducts were old and derelict, while catch-basins were usually uncovered and primitive; drainage systems were clearly inadequate while the facilities of most slaughterhouses had major defects. Given these conditions, pollution in the water was frequent and extensive while urban typhus epidemics made their appearance each year.[18]

With the influx of refugees from Asia Minor in 1922, these public health problems were intensified. The continuing expansion of cities, especially Athens and Piraeus, took place without infrastructural development, meaning that "the simplest rules of hygiene are ravaged."[19] The inadequacy of infrastructure, the lack of a health system of any kind and the low standard of living among the working strata already observable before 1922 contributed to the constant spread of tuberculosis and other infectious diseases. The lack of any statistics on fatality rates and disease impeded the effective

17 Υπουργείον Εθνικής Οικονομίας / Επιθεώρησις Εργασίας, *Έρευνα επί των συνθηκών της εργατικής κατοικίας των πόλεων Αθηνών – Πειραιώς κατά το 1921* (Athens: Εθνικόν Τυπογραφείον, 1922).

18 Αναστάσιος Δαμβέργης and Κωνσταντίνος Γ. Σάββας, *Έκθεσις περί της υγιεινής εξετάσεως του ύδατος των Αθηνών υποβληθείσα εις το επί των Εσωτερικών υπουργείον* (Athens: τυπογραφείον Α. Καλαράκη – Ν. Τριανταφύλου, 1895), 25–27; Κοπανάρης, *Η δημόσια υγεία εν Ελλάδι*, 343–396.

19 *Θάρρος* (Piraeus) (3 February 1931); Κοπανάρης, *Η δημοσία υγεία εν Ελλάδι*, 357–359. On urbanization, see Lila Leontidou, *The Mediterranean City in Transition. Social Change and Urban Development* (Cambridge: Cambridge University Press, 1990), 47–88.

treatment of problems related to public health in Greece. This occurred in a period in which in other European countries and in the US, medicine—bacteriology in particular—were making remarkable progress, so much so that even the classification system of patients changed, providing evidence for the causes of diseases and a new medical semiology.[20] Given this population explosion and corresponding lack of health services and infrastructure in urban centers over the early decades of the twentieth century, the modernization of the cities, the redesigning of urban space, the development of the public transportation infrastructure, the installation of infrastructural networks (water supply, drainage, gas, public lighting) and the re-organization of the public health services were all undertaken by successive liberal governments (in 1910–1914, 1917–1920, 1928–1932).

Infectious diseases continued to be treated according to the guidelines of the antiquated Royal Decree "On Impeding Transmission of Infectious Diseases" dated 1836, resulting in little or no application of measures to combat infection or epidemics.[21] According to Constantinos Savvas, these deficiencies were also due to the fact that the Sanitary Services of the Ministry of the Interior was staffed by administrative employees rather than physicians, reflecting the overall disarray of Greek health services. In the early 1920s the experience of general mobilizations and mass population movements showed the increasing need for the reorganization of public health services in Greece under a single central ministry, one able to confront the public health care problem with a unified approach facilitated by state intervention. At any rate, the substantial public health problems in Greece were as much connected to an inadequate and obsolete public health infrastructure—which had been commented upon by Savvas as early as 1910—as to the quality and standards of life throughout the country.[22]

---

20 Κωνσταντίνος Γ. Σάββας, *Υπόμνημα περί ιδρύσεως Υπουργείου Υγείας και Κοινωνικής Προνοίας υποβληθέν εις τον κύριον Πρόεδρον της Κυβερνήσεως κατά μήνα Δεκέμβριον 1920* (Athens: Αρχεία Ιατρικής, 1922); and Paul Weindling, "From Infectious to Chronic Diseases: Changing Patterns of Sickness in the 19th and 20th Centuries," in Andrew Wear, ed., *Medicine in Society,* 303–316.

21 Σάββας *Σχέδιον νόμου περί επιβλέψεως της δημοσίας υγιείας*, 3; Σάββας, *Υπόμνημα περί ιδρύσεως Υπουργείου Υγείας*, 3–5.

22 Λήδα Παπαστεφανάκη, "Δημόσια υγεία, φυματίωση και επαγγελματική παθολογία στις ελληνικές πόλεις στις αρχές του 20ου αιώνα: η αντιφατική διαδικασία του αστικού εκσυγχρονισμού," Proceedings of the Conference *Ελευθέριος Βενιζέλος και ελληνική πόλη. Πολεοδομικές πολιτικές και κοινωνικοπολιτικές ανακατατάξεις* (Athens: Εθνικό Ίδρυμα Ερευνών και Μελετών 'Ελευθέριος Κ. Βενιζέλος' – ΤΕΕ – ΕΜΠ/Σχολή Αρχιτεκτόνων Μηχανικών, 2005), 155–170.

The Ministry of Hygiene and Social Welfare was finally established in 1922 while throughout the 1920s, experts from the League of Nations Health Organization assisted the Greek government in confronting public health problems generated by the resettlement refugees from Asia Minor. Financial and technical aid from the League of Nations to Eleftherios Venizelos' government grew in 1929, when a committee of public health experts from the Health Organization was summoned to discuss reforming the country's public health services. The committee conducted extensive research in the cities and countryside and was astonished by the absence of health services and social welfare. This was unparalleled not only by European standards and often rivaled contemporaneous developments in several non-industrialized nations. The committee's plan called for the creation of a National Health Service to coordinate the scattered state services and private activities concerning the population's health. According to the International Labour Office, which worked closely with the Venizelos government in the period 1928–1932, reform of public health needed to be combined with establishment of social security (in order to appease labor unions and the working class).[23] Until 1932 the Sub-Ministry for Hygiene concerned itself systematically with (a) passing the necessary legislative measures for the re-organization of public health, (b) combating infectious diseases and safeguarding public health generally by extending the hospital system and creating a network of public surgeries and (c) a national program of vaccination and provision of pharmaceuticals to deal with malaria, trachoma, smallpox and so on. Yet implementing this new public health program met with financial and administrative difficulties.[24]

Despite problems encountered by state intervention in the organization of public health, debates about 'hygiene' and ways to regenerate Greek society gained ground in the early twentieth century. Scientific hygiene was elevated to a science and given scope for general intervention in the name of public health, while the number of hygiene associations increased and popular medical publications multiplied. One of the first such publications was entitled *Diseases and Microbes* [*Νοσήματα και μικρόβια*], written by Vassileios Patrikios (1847–1926),[25] a distinguished figure in the medical world,

[23] Αντώνης Λιάκος, *Εργασία και πολιτική στην Ελλάδα του μεσοπολέμου. Το Διεθνές Γραφείο Εργασίας και η ανάδυση των κοινωνικών θεσμών* (Athens: Ίδρυμα Έρευνας και Παιδείας της Εμπορικής Τράπεζας, 1993), 326–333.

[24] Venizelos Archive, Ministry of Economics (1932), Files 74, 134.

[25] Βασίλειος Πατρίκιος, *Νοσήματα και μικρόβια* (Athens: Σύλλογος προς Διάδοσιν Ωφελίμων Βιβλίων, 1901).

a public health expert who had studied in Athens and Paris, General Secretary of the Medical Congress and a close associate of Constantinos Savvas. Patrikios' book was published in 1901 by the *Association for the Dissemination of Useful Books* [*Σύλλογος προς Διάδοσιν Ωφελίμων Βιβλίων*] with the aim of enlightening the public about infectious diseases, their carriers, the ways in which the organism becomes infected and the means of preventing illness. In a chapter called "Housing–Diet–Work," emphasis was placed on the importance of well-aired and sunlit houses, a healthy diet and work suited to the ability of the individual. Here special attention was paid to alcoholism as "one of the greatest evils of modern societies" since alcohol abuse gave rise, on the one hand, to various diseases while on the other, "heavy drinkers produce degenerate, neurasthenic and imbecilic children or are themselves prone to disease, incapable of begetting children and, finally, easily susceptible to phthisis. It is not only individuals who are harmed by the abuse of alcohol but the nation [...]. Those who do not drink spirits have more robust health. [...] Alcohol does not bestow powers. On the contrary, it destroys them."[26] The destructive power of alcohol, weakening the organism and rendering it prone to sicknesses, was a repeated motif in publications on the subject of public health in early twentieth century Greece. Alcoholism was also to fuel the intensifying (scientific and social) control over the use of leisure time—particularly the leisure time of the working class.

In the chapter on "Work–Fatigue," the physician stressed that weak constitutions gave rise to illnesses and that fatigue that was harmful chiefly to working women, children and adolescents. He further argued that special care should be taken to regulate the working hours of women and children in factories, as was already the case in the rest of Europe.[27] The medical prognoses advanced here treated working women and children as "special categories" that needed extra protection and was thus in full accord with wider discourses by social reformers (both socialists and liberals) of the same period who also sought special protection for certain categories of workers. Savvas appears to have been of the same view in maintaining that work should be "proportionate to the physical and intellectual ability of the workers and should not be shared out equally between workers of unequal

26 Πατρίκιος, *Νοσήματα και μικρόβια*, 46–48.
27 Πατρίκιος, *Νοσήματα και μικρόβια*, 52–54.

strength."[28] At another point in his own study Savvas makes a distinction between soldiers, who constitute "the healthiest persons in society by reason of selection and by reason of age," and the various social classes, including "in them old men, women, and children."[29]

In the sphere of hygiene, the general and private interests were interlinked. Hygienic measures could contribute to an upgrading of public health only with the input of individuals: "through the strict observance of the precepts of hygiene, each person will contribute not only to the maintenance of his own individual health and that of his family, but, further, to the prevention of the spread of infectious diseases."[30] In consequence, the infectious nature of certain illnesses was used by public health experts as a tool to intervene in private life, in work and in leisure time. The object of hygiene was, in the end, to deal with the 'social question' by means of social reform. Tools for scientific intervention into the social question extended to the popularization and diffusion of hygienist ideas among the public at large, particularly among the working class. In this mission of enlightenment, members of the working class were perceived as uneducated, prodigal and drunken "others" in whom health reformers needed to inculcate principles of a healthy, peaceful, regular life—family life. For most physicians, members of the middle-class, the family was a place of moral regulation contributing critically to the larger pacification of the social question. In this framework of new values, men and women, children and adults were the object of enlightenment and moralizing over expected family and work roles. These roles were invariably founded on pre-existing gender power relations.

## Tuberculosis: A Meeting-Point for Public Health and Labor Policies

Among the varied public health problems facing physicians and hygienists in the early twentieth century, tuberculosis occupied a special place. Attempts to quantify the disease statistically in Greece remained incom-

28 Σάββας, *Εγχειρίδιον Υγιεινής*, 570.
29 Σάββας, *Εγχειρίδιον Υγιεινής*, 17.
30 Πατρίκιος, *Νοσήματα και μικρόβια*, 123.

plete until the interwar period. When this was finally completed, more than 10,000 deaths annually due to tuberculosis were recorded in the period between the two World Wars,[31] although in other estimates deaths reached 20,000 per annum.[32] On the basis of this evidence, fatality from tuberculosis in Greece exceeded that of most other European states.[33] Medical studies stressed the high fatality rates and infectiousness of the disease, particularly in industrialized cities, and therefore urged countering of it through organized and systematic preventive measures, the reinforcement of the population's health and the establishment of "Rural Clinics" and sanatoria where tuberculosis patients could be treated. The First Panhellenic Medical Congress convened in Athens in 1901 devoted a large part of its work to tuberculosis. It commissioned a special committee to set up the "Panhellenic Anti-Tuberculosis Association" [*Πανελλήνιος Σύνδεσμος κατά της Φυματιώσεως εν Ελλάδι*] in the summer of 1901 and tasked it with combating the disease. The Association was active in locating funds needed for the anti-tuberculosis campaign, in publicizing the problem and canvassing public opinion, in creating a tuberculosis surgery in Athens in 1907 and in holding the First Panhellenic Medical Congress to Counter Tuberculosis in 1909.[34]

The first tuberculosis clinic in Athens, "Sotiria" [*Σωτηρία*, meaning "Salvation"], opened upon the initiative of philanthropists in 1905. By the end of the 1930s, the "Sotiria" tuberculosis clinic was not really serving general hospitalization purposes but was used chiefly as an isolation asylum for lower class patients. Of the 3,586 tuberculosis patients admitted to the "Sotiria" clinic in the period 1924–28, most belonged to the working class,

31 Μάνος Αλ. Τζατζάνης, *Αι κοινωνικαί ασφαλίσεις αλλαχού και εν Ελλάδι*, vol. 2 (Athens: Τύποις Ν. Απατσίδη, 1937), 411.

32 According to the physician Metallinos' calculations as cited by Τζατζάνης, *Αι κοινωνικαί ασφαλίσεις*, 411.

33 Βασίλειος Πατρίκιος, *Η φθίσις εν Ελλάδι* (Athens: τυπογραφείον 'Νομικής' Λ Χ Βεργιανίτου, 1903), 64, 68, 73; and Τζατζάνης, *Αι κοινωνικαί ασφαλίσεις*, 412.

34 Αθανάσιος Ε. Τσακαλώτος, *Περί της δημοσίας, υγείας εν Σύρω και ιδία της φυματιώσεως* (Athens: Εθνικόν Τυπογραφείον, 1914); Ευάγγελος Αρφάνης, *Η φυματίωσις εν Σύρω και ο κατ' αυτής αγών* (Ermoupoli: Σύλλογος Υγιεινής Σύρου, 1918); Χρ. Ν. Αποστολίδης and Ιφικράτης Δ. Χατζηεμμανουήλ, *Συμβολή εις την μελέτην της οργανώσεως του αντιφυματικού αγώνος* (Athens, 1936); Γεώργιος Δ. Παπακώστας, "Οργάνωσις αντιφυματικού αγώνος," *Αρχεία Υγιεινής* 9 (1937): 333–383; Κοπανάρης, *Η δημοσία υγεία εν Ελλάδι*, 231–252; Πανελλήνιος Σύνδεσμος κατά της φυματιώσεως (φθίσεως), *Τα υπό του Συνδέσμου Πεπραγμένα 1922* (Athens: n.p., 1923); Πανελλήνιος Σύνδεσμος κατά της φυματιώσεως (φθίσεως), *Τα υπό του Συνδέσμου Πεπραγμένα (1.1.1924-31.12.1926)* (Athens: n.p., 1927).

"living from their daily labor."[35] The infectiousness of the disease caused something akin to a social panic and it was the perceived inability to control the living conditions of the working strata that set in motion these confinement and institutionalization procedures.

In order to prevent and fight tuberculosis, physicians like Vassileios Patrikios stressed, among other measures, the need to study the workers' living and working conditions. According to Patrikios, a leading representative of the struggle against tuberculosis in early twentieth century Greece, it was imperative that the reasons behind the spread of tuberculosis among the working class become known. The question of whether the spread of the disease was due to internal migration and continuing urbanization also needed examination, given "the ceaseless shift from the fields into the cities, to the replacement of agricultural tasks carried out in the open air by others performed in a dusty and confined atmosphere at dizzying speed, where the spread of infection is easy since conditions of hygiene are not observed."[36] It was also imperative to investigate the extent to which "the various occupations or the kind of industry concerned contribute to spreading the disease."[37] At the same time, morbidity, mortality and fatality statistics for varying professions became an essential medical tool. In popular publications on tuberculosis precautions, physicians deemed certain conditions necessary, such as having nutritious, good-quality food and sunny, well-ventilated homes in addition to taking rest, meticulously avoiding any kind of physical abuse (like alcohol), exercising outdoors and so on.[38] In the accompanying medical discourse, factors relating to the population's living conditions (like nutrition and accommodation) and to the intensive work pace dictated by wage labor were interwoven with "cultural" recommendations bearing upon leisure time and morality, especially among the working classes. The physician Anastasios Aravantinos (1879–1948) for example, in a popular study published in 1910, drew attention to

35 Τζατζάνης, *Αι κοινωνικαί ασφαλίσεις*, 412; Πανελλήνιος Σύνδεσμος κατά της φυματιώσεως (φθίσεως), *Τα υπό του Συνδέσμου Πεπραγμένα 1922* (Athens: n.p., 1923); Πανελλήνιος Σύνδεσμος κατά της φυματιώσεως (φθίσεως), *Τα υπό του Συνδέσμου Πεπραγμένα (1.1.1924-31.12.1926)* (Athens: n.p., 1927), 5–6.

36 Βασίλειος Πατρίκιος, *Η πορεία της φθίσεως εν Ελλάδι: από του 1890–1905* (Athens: Εστία, 1906), 22.

37 Πατρίκιος, *Η πορεία της φθίσεως εν Ελλάδι*, 23.

38 Βασίλειος Πατρίκιος, *Οδηγίαι περί προφυλάξεως από της φθίσεως* (Athens: Εστία, 1902), 14–15; Παναγιώτης Παμπούκης, *Ο αγών κατά της φθίσεως. Προφύλαξις - θεραπεία* (Athens: Εθνικόν Τυπογραφείον, 1927), 8–10 and 25–26.

the poor living conditions of the working class, where "poverty, little and inadequate food, deprivations in general and the abuse of wine and alcohol predispose the body to phthisis."[39]

Similarly, the last chapter of Savvas's 1907 university textbook on hygiene was devoted to "occupational hygiene." Here, he argued that it was difficult to compile accurate statistics showing the developmental effects of the various diseases and the rate of morbidity and fatality for each profession not least because other factors also contributed to morbidity and fatality in Greece at the time. It was a fact established by medical research that workers were more exposed than other people "to the effects of their occupation," that is, to the unfavorable consequences of certain jobs.[40] Nevertheless, in a striking passage, Savvas points out:

> It should not be thought that the great morbidity and fatality observable in the case of workers should be attributed only to the harmful effect of their particular occupation because *poverty* and *living in adverse hygienic conditions* contribute to a great extent to them. The sum of money that each worker earns daily is not always sufficient for the upkeep of his family [....] Thus, from close cohabitation [in unsuitable housing] and the prevailing *dirty condition* of clothing, the body and the house, infectious diseases very frequently develop that very easily attack the organism of the worker, weakened by inadequate food, excessive work and perhaps *alcoholism* and other reasons, and that of the members of his family. For this reason we very often observe tuberculosis, diphtheria, acute rashes and other infectious diseases developing in the case of such dwellings.[41]

Indeed, in both Savvas's textbook and in Aravantinos's study, ranking causes for fatality, morbidity and mortality in certain professions is reminiscent of the French physician Louis René Villermé (1782–1863), despite the fact that the latter is never explicitly mentioned. For Villermé, a defender of the view that specific occupations have only an indirect effect on workers' health, the working classes' poverty and the associated moral and physical

39 Αναστάσιος Ι. Αραβαντινός, *Τι είναι η φθίσις; Πώς ειμπορούν οι υγιείς να προφυλαχθούν από την φθίσιν και πώς πρέπει οι φθισικοί να προσπαθούν να γίνουν καλά και να μην είναι επικίνδυνοι εις τους υγιείς. Βιβλίον δια τον λαόν* (Athens: Τύποις Αθαν. Γ Δεληγιάννη, 1910), 16.

40 Σάββας, *Εγχειρίδιον υγιεινής*, 563.

41 Σάββας, *Εγχειρίδιον υγιεινής*, 564–565.

"degradation" that followed were due to uncontrolled diseases; these were social causes that could be dealt with by means of appropriate regulatory interventions. To this end an improvement in standards of living and the moral instruction of laborers were the means by which to eradicate those basic factors giving rise to illness.[42]

European socialists and communists stressed the direct relation between work and worker's health through recourse to Bernardino Ramazzini's (1633–1714) methodology. Ramazzini, professor at the School of Medicine of the University of Padua, had published his 1713 study *Diseases of Workers* [*De morbis artificum diatriba*], he laid the foundations for scientific approach to occupational pathology, that is, the effects of work on the health of workers. In Ramazzini's work, occupational diseases were causally connected to their unhealthy occupations for the first time together with a scientific method for their analysis, namely, the observation of labor conditions and the detection of harmful health factors in the workplace.[43] On the other hand, the liberal bourgeois in Europe tended to dissociate work and health. They claimed that the actual reasons behind workers' diseases were exogenous, meaning they were not directly connected to the work itself but rather to the conditions under which labor was carried out: excessive fatigue, mandatory overtime, insufficient nutrition, alcoholism, poor living conditions, lack of personal hygiene and low wages. Based on their social research into the state of Greek public health, physicians like Constantinos Savvas, Vassileios Patrikios and Athanasios Aravantinos were close to nineteenth century liberal medical traditions in Europe in their claims that poverty and living conditions contributed to the working population's higher rates of mortality.

Nevertheless, alongside the moral counsel to avoid indulgence and excess, Greek physicians also drew attention to conditions prevailing in the

42 Louis René Villermé, *Tableau de l'état physique et moral des ouvriers employés dans les manufactures de coton, de laine et de soie* (Paris: Renouard, 1840).

43 A strong indication of Ramazzini's impact can be seen in his works' continual translation and publication (since the eighteenth century). See Bernardino Ramazzini, *De morbis artificum diatriba* (3 Works) (Ultrajecti van de Water, 1703); Greek translation as *Οι ασθένειες των εργατών* (Athens: ΕΛΙΝΥΑΕ, 2001). On Ramazzini's influence see, Francesco Carnevale, ed., *Bernardino Ramazzini (1633–1714). Le malattie dei lavoratori. I testi delle edizioni del 1700 e del 1713* (Florence: Libreria Chiari, 2000). For discussion among French physicians see, Bernard-Pierre Lécuyer, "Les maladies professionnelles dans les *Annales d' hygiène publique et de la médecine légale* ou une première approche de l' usure au travail," *Le Mouvement Social* 124 (1983): 45–69; and Gérard Jorland, "L'hygiène professionnelle en France aux XIXe siècle," *Le Mouvement Social* 213 (2005): 71–90.

workplace, stressing the need to improve them. "Eliminate from industrial establishments every source of danger (dust, smoke, etc.)," urged Secretary of the Medical Congress and the General Secretary of the Anti-Tuberculosis Association, Vassileios Patrikios. He continued: "dirt, absence of air and light, a sedentary life and, particularly, spending time on dirty premises where the atmosphere is dusty (unhealthy occupations, etc.) and where there are many people, predispose the organism to phthisis."[44] With increasing frequency, physicians connected tuberculosis fatalities with specific professions since "among the harm caused by occupations, the most frequent is that caused by dust [ ... ]. Dust, when inhaled, irritates and damages the respiratory organs or is a means of entry for germs and the consequent inflammations. Phthisis is frequent among stonemasons and stonecarvers. Dust from smoke, cotton, wood, cereals, etc. also have a deleterious effect."[45] In addition, "those occupations are dangerous where individuals work bent over in a small and enclosed space, such as printers, shoemakers and tailors."[46] Conditions in the workplace thus became objects of medical study:

> Because large sites, large rooms, high ceilings, many windows call for expenditures, factory-owners, merchants and bankers, etc., particularly when they are unscrupulous and if the state does not exert strict, organized, legislated supervision, cram many workers into a small space. A person who works on such premises has a greater chance of being infected with phthisis. A small space has less light and less air, which swiftly becomes unfit to breathe, and the workers soon are almost inhaling each other's breath. This situation is even worse if the premises are damp, if the ventilation is bad, as when there are few windows or they are small or look out on to a narrow, sunless, damp yard, if the sun cannot enter at all, if the nature of the work produces smoke, dust, fumes.[47]

Faced with the danger of high fatality rates and general infection from tuberculosis, state intervention in matters of occupational health was becoming a matter of urgency:

44 Πατρίκιος, *Οδηγίαι περί προφυλάξεως από της φθίσεως*, 14, 16.

45 Πατρίκιος, *Νοσήματα και μικρόβια*: 54; Αραβαντινός, *Τι είναι η φθίσις*, 16.

46 Αραβαντινός, *Τι είναι η φθίσις*, 16.

47 Αραβαντινός, *Τι είναι η φθίσις*, 21–22.

> Those who have one of those occupations, which we have seen to predispose them to phthisis, must be protected by the state. Factory-owners and merchants must be obliged by law to build or to rent healthy factories and to provide the workers with masks specially made to prevent dust passing into the lungs but without hindering breathing [...]. The times and duration of the breaks in work in the light of our climate and the nature of the job should be set by law. It should be prohibited for children under the age of 12 to work and under the age of 16 to do any hard labor. Ventilators should be required so that ventilation is good for as long as work takes place.[48]

Patrikios concluded that "the state should have specially trained people to supervise the strict observance of the law and the hygiene of the workers."[49] Such demands for state mediation as an antidote to the increase in tuberculosis had already been made by Patrikios and Savvas a year earlier in 1909, at the first Greek conference on combating tuberculosis.[50] As previously mentioned, this conference's resolutions, along with those made by Aravantinos, are reflected in the labor legislation instituted by the Venizelos government between 1910 and 1914.

With the creation of the first insurance funds for various occupations, the occupational identity of tuberculosis-sufferers began to emerge: mill workers, cigarette-makers, tobacco workers, printers, hairdressers, employees in hotels and restaurants, banks, post offices and telegraph offices all showed high rates of tuberculosis.[51] The collection of morbidity data (from tuberculosis or other causes) in specific professions was rendered possible by these new occupational insurance funds, although the total absence of such funds in some occupations (in textile mills, for example) made the collection of such data there more difficult. The general extension of social security—notably through the Agency for Social Insurance [*Ίδρυμα Κοινωνικών Ασφαλίσεων*; IKA] in 1937—was the point of departure for more systematic recording of morbidity in other workplaces where, to that point, no professional insurance funds had been established.

48 Αραβαντινός, *Τι είναι η φθίσις*, 30.

49 Αραβαντινός, *Τι είναι η φθίσις*, 23.

50 Βάσω Θεοδώρου, "Οι γιατροί απέναντι στο κοινωνικό ζήτημα. Ο αντιφυματικός αγώνας στις αρχές του 20ου αιώνα (1901–1926)," *Μνήμων* 24 (2002): 145–178.

51 Κοπανάρης, *Δημοσία υγεία εν Ελλάδι*, 232, 248; Αποστολίδης and Χατζηεμμανουήλ, *Συμβολή εις την μελέτην της οργανώσεως*, 26–27; Παπακώστας, *Οργάνωσις αντιφυματικού αγώνος*, 335.

It was slowly being realized that a variety of harmful factors in occupational health had an unfavorable effect on public health, rendering workers vulnerable to tuberculosis and other infectious diseases. Most Greek physicians, apart from suggesting other measures (such as the improvement of housing, avoidance of alcohol, open-air exercise by workers and so on), proposed the strict implementation of a national labor law, the supervision of occupational hygiene in the workplace (particularly in certain jobs having a deleterious effect on the respiratory system with dust, smoke, etc.), the combating of "industrial fatigue," the general application of the eight-hour day and greater control over domestic labor.[52]

Early twentieth century Greek physicians considered that in order to deal with tuberculosis, legislative measures needed to be taken, addressing the operation of factories and the maximum working hours. They also pointed to the need for existing labor laws to be enforced and for occupational hygiene to be monitored in order to reduce morbidity from tuberculosis in the workplace. As with the general issue of the improvement of public health, so in the more particular but crucial case of tuberculosis did the medical discourse have targets: on the one hand it was addressed at the state in the form of demands for legislating intervention in the improvement of working conditions. On the other hand, it was addressed to the working people themselves, urging them to take adequate nourishment, to live healthy lives, not to indulge in abuses and, in short, to adopt a quiet family life in accordance with the new social values of the nascent middle class.

## Labor Relations and Modernization

As previously noted, the foundations laid for labor legislation in Greece—meaning the legal provisions meant to intervene in wage labor and shape the terms of exchange in the Greek labor market—were established by the liberal government between 1910 and 1914. This legislation adopted the Sunday holiday, a decrease in working hours for certain occupations, prohibition on the employment of children under 12 years of age, the pro-

[52] Παπαστεφανάκη, *Δημοσία υγεία*, 155–170.

tection of children and women's labor and work accident compensation.[53] Until the early twentieth century, sporadic strikes and the rising demands of the labor movement were met by both the state and the employers with a paternalistic approach. Yet widespread strikes in the years 1906–1909 taking place across the country seemed to have threatened social harmony and revealed that the opposition between capital and labor may be inherent to the system. Subsequent proposals put forward for state intervention and social reforms for the improvement of labor conditions were connected with social balance and the avoidance of class confrontations.

Critically, the introduction of labor legislation in 1910–1914 under Venizelos led to the legislative normalization of labor relations. This legislation, a product of the dynamics of Greek class relations, was a basic axis of the economic and social policy of the liberal governments although, at the time of their introduction, the laws were regarded by the "Association of Greek Industrialists and Craftsmen" [*Σύνδεσμος Ελλήνων Βιομηχάνων και Βιοτεχνών* – ΣΕΒΒ] and other representatives of the bourgeoisie as "premature" for Greek reality, even "socialist" (although in 1914 Venizelos accepted that Greek legislation lagged behind that of other European countries). Nevertheless, it reinforced the political hegemony of the liberals, while at the same time expanding the field of preventive intervention by the state.[54]

Furthermore, the Labor Inspectorate was instituted in 1912 as part of the Ministry of the National Economy, intended as a state instrument for controlling the operation and enforcement of labor legislation. The data registered by Labor Inspectors in the Inspectorate's *Reports* published between 1913 and 1934 provide systematic information on illnesses among the working population as well as more general occupational pathology. The first *Report* from 1913 observed that every factory inspected, without exception, failed to comply with conditions of occupational hygiene and could thus "undoubtedly be described as hazardous and unsanitary."[55]

53 See transcripts of the laws in Μιχαήλ Κρητικός and Ιωάννης Ζάρρας, *Εργατική και κοινωνική νομοθεσία* (Athens: Ερμής, 1929).

54 George Leon, "The Greek Labor Movement and the Bourgeois State, 1910–1920," *Journal of the Hellenic Diaspora* 4, 4 (1978): 5–28. See also Efi Avdela, "'To the Most Weak and Needy': Women's Protective Labor Legislation in Greece," in Ulla Wikander, Alice Kessler Harris, and Jane Lewis, eds., *Protecting Women. Labor Legislation in Europe, the United States, and Australia, 1880–1920* (Urbana-Champaign: University of Illinois Press, 1995), 290–317.

55 "Γενικαί Εκθέσεις των Επιθεωρητών Εργασίας του έτους 1913," *Δελτίον του Υπουργείου Εθνικής Οικονομίας*, vol. 2 (Athens: n.p., 1914), 163.

During the ensuing two decades, although some steps were taken towards improving the working conditions in factories, the situation remained for the most part unsatisfactory. Among the diverse cases of occupational disease detected by Inspectors between 1913 and 1934, tuberculosis and the diseases associated with it were most frequent.[56]

The Inspectorate often lamented its inability to direct the application of relevant Greek legislation. The League of Nations committee that studied Greece's public health organization between 1928 and 1929 came to similar conclusions about factory hygiene and working conditions. In his report, Adrien Tixier (1893–1946), Director of the Social Security Department of the League's International Labor Office, established that Greek labor legislation was not implemented with respect to working hours, occupational health and safety or the protection of women and minors, while child labor had reached scandalous proportions. Yet the Labor Inspectorate had extremely limited jurisdiction and remained powerless to enforce the stoppage of work in places judged to be unfit and unsanitary. Inspectors did not even have the right to directly impose fines and their jurisdiction was all the more diminished by the courts to which factory-owners were summoned.[57] Thus, the need to reform the country's health system indirectly encountered still another aspect of modernization, the institution and enforcement of employment legislation. In the eyes of the International Labor Office and Venizelos' liberal government, reform of public health needed to be combined with the establishment of social security.

Labor Inspectors, educated civil servants who largely belonged to the circle of social reformers, compiled their reports in the light of the scientific criteria common at the time (using systematic description and statistical documentation), imbued with a positivist faith in the objectivity of their data. As has been noticed in the German case, Labor Inspectorate *Reports* contributed to a better comprehension of the "ideological assumptions and hopes of part of the ruling classes and the extent to which these became institutionalized in the emerging industrial society" by the implementation of labor legislation and state intervention in the formation of new social

56 Λήδα Παπαστεφανάκη, *Εργασία, τεχνολογία και φύλο στην ελληνική βιομηχανία. Η κλωστοϋφαντουργία του Πειραιά (1870–1940)* (Heraklion: Πανεπιστημιακές Εκδόσεις Κρήτης, 2009), 334–340.

57 Venizelos Archive, Ministry of Economics, File 72, Adrien Tixier, "Les conditions de travail des salariés de l' industrie et du commerce et leurs conséquences au point de vue sanitaire" [1928–29].

relations more generally.[58] In the Greek Labor Inspectorate *Reports*, workers are represented as incapable of understanding and pursuing their interests in implementing labor legislation. More specifically, Inspectors were unable to explain the refusal of workers to use protective safety measures, their indifference towards the employers' provisions for better hygienic conditions in the workplace and, *a fortiori*, their consenting to violation of the working hours.[59]

Efi Avdela has proposed that the Labor Inspectorate *Reports* be read as references to "the latent and inarticulate meanings that 'others', that is, men and women workers, might have given to their acts."[60] The reevaluation of these official sources thus document "informal responses of men and women workers to the provisions of protective legislation and interpret these responses according to their own values and norms."[61] From this perspective, the workers' refusal to use the protective legislative provisions expressed certain forms of resistance to collective exploitation. In 1921, for example, workers in a large metallurgical firm refused to use the wash-basins with running water, an attitude seized upon by Labor Inspectors as proof of the lack of the "instinct and knowledge of sanitation and cleanliness" not only for the workers but for Greek society as a whole.[62] Yet, in actual fact the workers' refusal was connected with their employers' insistence that they should clean the wash-basins in their leisure time. Moreover, workers' rejection of protective devices, such as gloves and masks, may be related to the corresponding reduction in productivity in work paid piecework, to maintaining a distinct working class identity (in terms of skill and pride) or the desire to exercise some control over the labor process. In general, then, the refusal of workers to adopt the protective legislation may be interpreted as not merely as a matter of "backwardness" but as unspoken resistance to the "civilizing" process imposed from above by social reformers.

58 Jean H. Quataert, "A Source Analysis in German Women's History: Factory Inspectors' Reports and the Shaping of Working-Class Lives, 1878–1914," *Central European History* 16 (1983): 100.

59 Efi Avdela, "Contested Meanings: Protection and Resistance in Labour Inspectors' Reports in 20th c. Greece", *Gender and History* 9, 2 (1997): 310–332.

60 Avdela, *Contested Meanings*, 312

61 Avdela, *Contested Meanings*, 312.

62 Υπουργείον Εθνικής Οικονομίας/Επιθεώρησις Εργασίας, *Εκθέσεις του προσωπικού επιθεωρήσεως εργασίας επί της εφαρμογής των εργατικών νόμων, έτος 1921* (Athens: Εθνικόν Τυπογραφείον, 1923): 18. See also Avdela, *Contested Meanings*, 324.

## Compensation for Accidents at Work and Occupational Hazards

Work-related accidents were another basic occupational health concern. Until 1914, Roman law requiring labyrinthine judicial procedures in which the employer's responsibility had to be established had long governed accident in the workplace issues. In effect, employees were not insured against any occupational hazards and, in the event of an accident, were completely dependent upon the goodwill of their employers. Law 551 of 1914, "Concerning the Responsibility to Compensate Laborers and Office Workers Injured by Accident While at Work,"[63] based on a similar French law from 1898 and an Italian law from 1904 was, in effect, the first general law in Greece to regulate workplace accidents according to the principle of occupational hazards. According to the French definition of 1898, occupational hazards were the risks "inherent in a certain occupation irrespective of any fault on the part of employer or employee."[64] Since it was accepted that industrial production exposed the worker to certain risks, it followed that in the event that these risks were realized, compensation to the injured party ought to be made by those profiting from the enterprise (namely the employer), irrespective of any culpability on the part of the latter. To put it differently, all accidents "except for those induced by malicious intent" by the worker were thereafter considered "inherent to the workplace and the responsibility of the employer" as owner and main beneficiary of the enterprise. Through this law, the employer was first rendered occupationally responsible in Greece and therefore liable for compensating the injured worker.[65]

---

63 See transcript of the law 551/1914 (Government Gazette , 8 January 1915) as codified in 1920 by the Royal Decree "Concerning the Codification of Laws on Compensation Liability for Accidental Death..." in Κρητικός and Ζάρρας, *Εργατική και κοινωνική νομοθεσία*, 285–295. Law 551 was amended several times, mostly with respect to compensation limits, namely through the 1920 Royal Decree 'Concerning the Codification....', and through Law No. 4705 "Concerning the Amendment of the Royal Decree..." in 1930. See Γεώργιος Λ. Ροϊλός, *Περί επαγγελματικού κινδύνου και της σχετικής ελληνικής νομοθεσίας (Εργατικά ατυχήματα)* (Athens: Δημοσθ. Θεοφιλόπουλος, 1928), 7–18 and 23–24; and Άγγελος Τρουπάκης, *Η θεωρία του επαγγελματικού κινδύνου* (Athens: Έκδοσις Βιβλιοθήκης Σιδηροδρόμων Κράτους, 1929), 15.

64 Ροϊλός, *Περί επαγγελματικού κινδύνου*, 29.

65 Ροϊλός, *Περί επαγγελματικού κινδύνου*, 7–18, 23–24; Τρουπάκης, *Η θεωρία του επαγγελματικού κινδύνου*: 15–18; Αγνή Ο. Ρουσοπούλου, *Πρώτα στοιχεία κοινωνικής πολιτικής και εργατικής νομοθεσίας* (Athens: Εκδόσεις 'Φλάμμα', 1937): 65–68; Σωκράτης Κλαδάς, *Η εργατική μας πολιτική και νομοθεσία* (Athens: 'Το Νομικόν' Σάκκουλας, 1945), 40–42.

The 1914 law on the principle of occupational hazards did not, however, include occupational diseases, meaning occupational illness manifesting themselves gradually over time due to "progressive, slow and often latent effects caused by the workplace, the bad climate, the overstraining of limbs or, mostly, by the use of caustic or poisonous substances."[66] According to both French and Greek law, occupational illnesses were excluded from compensation. Although this exclusion was acknowledged as unjust at the time, the difficulty in verifying occupational illnesses and the employers' responses as well as the inadequate insurance system caused some delay in adding occupational illnesses to the list of insurable occupational health risks. Yet only a few years later, the country's first insurance law, Law 2868 of 1922 "Concerning the Mandatory Insurance of Workers and Private Employees" included—albeit in a rather general formulation—occupational diseases among the risks against which workers were to be insured: "it is obligatory that [the worker] be insured against illness while at work, death, old age and inability to work in general."[67] In all cases, including those of occupational illnesses, the employer was held ultimately responsible for compensation under common criminal law, when proper working conditions were met (e.g., lead poisoning of the miners due to bad construction of galleries, indicating fault on the part of the employer).

In Greece, the concept of occupational illness was introduced in the 1920s by way of the gradual institution of social security and the principle of occupational hazards. These started to become accepted, if only in certain social circles, with "the essential defining characteristic of a risk to be covered [being] the underlying causal connection between the injury and the work, as opposed to the nature of the injury itself."[68] In 1928 a Labor Inspector named Nikolaos Saliveros admitted for the first time that "the world of labor is plagued by professional illnesses which [...] without doubt lead to the inability to work and to tuberculosis; this disease does not only afflict tobacco workers but also workers in textile, tannery and so many other industrial enterprises."[69]

66 Ροΐλός, *Περί επαγγελματικού κινδύνου*, 45–46.

67 Law 2868/1922, article 1. See the law's transcription in Κρητικός and Ζάρρας, *Εργατική και κοινωνική νομοθεσία*, 383–389.

68 Τζατζάνης, *Αι κοινωνικαί ασφαλίσεις*, 9.

69 Υπουργείον Εθνικής Οικονομίας/Επιθεώρησις Εργασίας, *Περίληψις εκθέσεων του προσωπικού Επιθεωρήσεως Εργασίας επί της εφαρμογής των εργατικών νόμων δια τα έτη 1927–1928* (Athens: Εθνικόν Τυπογραφείον, 1930), 53.

Interest in occupational health has undoubtedly a long and noteworthy history in Europe, as has already been pointed out. Interest in these matters in Greece has remained exceptionally limited and usually goes no further than a simple citation of the statutory framework. It is worth noting, however, that occupational hazards (accidents at work, occupational illnesses) in Greece was an object of discussions between the liberal government and the ILO in the late 1920s. The ILO estimated that Greek social legislation was on a lower level than in neighboring countries, that social expenditures in Greece was exceptionally low—lower than in countries like Bulgaria and Yugoslavia—and that labor legislation as a whole was not implemented. Within this framework, the ILO argued that in a country with unusually low wages like Greece, social welfare was more necessary than in countries with higher levels of income. The ILO's proposal for a unified system of social security (to include insurance covering illness, maternity, disability, old age, surviving dependants and accidents at work and occupational illness) finally prevailed over the initial planning of the Venizelos government for a separate compulsory insurance to cover occupational hazards. If for the ILO the creation of a unified system of social security in Greece was to function as a serious counterbalance to the generally low standard of living and poor working conditions, for the liberals the social security system would operate as a counterweight to the restriction of civil liberties and to the objectives of the communists.[70] The international economic crisis and the reactions of many sides of the social power alliance delayed the implementation of the unified social security system until 1937.[71] Even after the Agency for Social Insurance (IKA) was founded, occupational hazard insurance continued to be a tripartite funded system in which state, employers and employees contributed equally. This means that, as opposed to other countries like, e.g. Italy, in Greece no distinct occupational hazard insurance existed that was paid for by the employer exclusively. Employers were thus relieved of the responsibility and the financial cost of covering acci-

---

70 Λιάκος, *Εργασία και πολιτική*, 453–454, 460–472. For the discussion on the necessity of a social security system among Greek physicians, see also Λήδα Παπαστεφανάκη, "Από την 'υγιεινή των επιτηδευμάτων' στην 'ηυξημένην νοσηρότητα της εργατικής τάξεως': η Επαγγελματική Υγεία στην Ελλάδα, 1870–1940," Proceedings of the Conference *Δημόσια υγεία και κοινωνική πολιτική. Ο Ελευθέριος Βενιζέλος και η εποχή του* (Athens: Εθνική Σχολή Δημόσιας Υγείας - Εθνικό Ίδρυμα Ερευνών και Μελετών 'Ελευθέριος Κ. Βενιζέλος', Παπαζήση, 2008), 265–288.

71 Λιάκος, *Εργασία και πολιτική*, 508–539.

dents at work and occupational illnesses, the relevant costs being passed on to the IKA. According to some assessments, this led to a "defective system of insurance coverage of occupational risk" in Greece, which to a large degree has survived to the present day.[72]

Even if the principle of occupational risk was initially accepted in Europe and in Greece to cover industries involving very serious hazards in the performance of tasks (mining, shipping, railways, explosives), protection was gradually expanded to cover not only work-related accidents, but an increasing range of occupational illnesses as well. By the mid-1930s, occupational illnesses were covered in almost all European countries while the technological advances in medicine allowed precise definitions of an increasing number of labor-related diseases. By the end of the 1930s, France had registered four occupational illnesses, while Britain had registered 35—including poisoning, pneumonoconiasis and silicosis—and Germany had recognized 22.[73] In Greece only three occupational diseases justifying workplace insurance were recognized, namely lead and mercury poisoning and infection with anthrax.[74] Nonetheless, the interwar years were marked by significant technical progress in preventing accidents at work. In the Greek technical press of this period, for instance, news of technical progress in Europe and the United States in the combating of occupational risks was reported systematically.[75] These technical developments and a (somewhat fleeting) interest in the "scientific management" of labor gave, gradually, an enhanced role to Greek engineers, who were added (after physicians, jurists and labor inspectors) to the list of specialists with working conditions, hygiene and safety in the workplace as a central object of concern.

A new normative discourse thus developed with the assistance of the engineers. A characteristic example of this "specialist" discourse

---

72 Γιώργος Ραχιώτης and Σπύρος Δρίβας, "Συμβολή στη μελέτη του θεσμικού πλαισίου για την υγεία και ασφάλεια στην εργασία στην Ελλάδα (1911–1941) σε συνάρτηση με την ευρωπαϊκή εμπειρία," Proceedings of the Conference *Δημόσια υγεία και κοινωνική πολιτική. Ο Ελευθέριος Βενιζέλος και η εποχή του* (Athens: Εθνική Σχολή Δημόσιας Υγείας – Εθνικό Ίδρυμα Ερευνών και Μελετών 'Ελευθέριος Κ. Βενιζέλος', Παπαζήση, 2008), 289–306.

73 Catherine Omnès, "Jalons pour une histoire comparée de la santé au travail en Europe," *Histoire et sociétés* 23 (2007): 15.

74 Γεώργιος Παπαδόπουλος, *Αι κοινωνικαί ασφαλίσεις και η δημοσία υγεία* (Athens: Εκ του Τυπογραφείου Αλ Βιτσικουνάκη, 1931), 47.

75 For example see Anon. "Αι τεχνικαί επιστήμαι και η πρόληψις των ατυχημάτων της εργασίας. Από την 11ην Διεθνή Διάσκεψιν Εργασίας," *Έργα* 72 (1928): 644–648.

is revealed in the 1935 book, *The Prevention of Accidents at Work* [*Η πρόληψις των ατυχημάτων εργασίας*] published at the expense of the "Association of Employers for Mutual Accident Insurance" [*Αλληλασφαλιστικός Συνεταιρισμός Εργοδοτών και Ατυχημάτων*] in which many important Greek industrialists were members. This publication, written by a former Labor Inspector, the engineer Ioannis Anapliotis, contains practical advice and recommendations on occupational health, accompanied by illustrations with the purpose of better informing workers. Moreover, the tone is didactic and authoritarian: "If you follow our recommendations, your physical integrity, on which your happiness and that of the members of your family depends, will be ensured."[76] And in order to avoid the forms of civil and criminal liability stemming from an accident at work, the attempt to shift responsibility for the accident to the employee is clear enough. It is, in fact, expressly admitted in Anapliotis' foreword: "Accidents at work result from some error, more rarely from a chance happening. The fault is either the poor state of the material that you are using or your own—in most cases the latter is the main cause. You can avoid this cause if you are careful: if each of your actions during your time at work is directed by your reason, experience and care, accidents can be avoided."[77] Instructions are given to foremen, overseers and laborers for the correct performance of tasks at work that best avoid accidents. Among the qualifications that workers should have to be engaged in the business (occupational experience, technical training and so on), they should be "healthy, not suffering from short-sightedness, deafness, neurasthenia, vertigo, rupture, alcoholism, epilepsy, syphilis, tuberculosis or eye complaints."[78] The aim of hygienists in early twentieth century Greece when dealing with mortality, healthy development and an increase in the population's capacity for work was thus turned by an alliance of Labor Inspectors and employers into the social exclusion of 'unhealthy' sections of the population from the labor process.

---

76 Ιωάννης Αναπλιώτης, *Η πρόληψις των ατυχημάτων εργασίας εις βιομηχανικάς επιχειρήσεις, τεχνικά έργα, φορτώσεις, εκφορτώσεις, μεταφοράς. Πρώται βοήθειαι εις παθόντας. Νόμοι – Διατάγματα* (Athens: Ελευθερουδάκης, Εκδοσις Αλληλασφαλιστικού Συνεταρισμού Εργοδοτών κατά Ατυχημάτων, 1935), 3.

77 Αναπλιώτης, Η πρόληψις των ατυχημάτων, 4.

78 Αναπλιώτης, Η πρόληψις των ατυχημάτων, 7.

## Conclusions

The shaping of occupational health in Greece in the first four decades of the twentieth century was intimately linked to the highly acute problems of public health, the fight against tuberculosis and the necessity for social welfare. More precisely, tuberculosis became the nodal point in which policies on public health met with occupational hygiene and labor policy.

The concept of occupational risk, first introduced to Greece in the interwar years, therefore presupposes the development of a certain form of scientific knowledge, the presence of "specialists" (physicians, jurists, labor inspectors, safety experts) and the elaboration of rules and norms to define risk in the workplace. The professionalization of "experts," whose assessment made risk visible, shaped a "technique" in the proper the definition of an occupational risk. By means of this process, both accidents at work and diseases caused by the working environment were progressively dissociated from the content of wage labor, thereby converting them into technical and medical problems requiring technocratic handling and imposing the development of preventive techniques and a variety of initiatives for social discipline.

Thereafter, medical and technical knowledge related to occupational risk crystallized in public institutions in a system of power relations, facets of which were employers, workers and an increasingly interventionist state. The specialists' socially-produced knowledge was connected with and shaped by cultural and political struggles over the meaning and content of health and, more specifically, occupational health. The knowledge of these specialists mediated in the relations between workers, employers and the state, not to mention international organizations like the League of Nations' International Labour Office. Thus, the imputation of responsibility to employees for accidents at work or for illnesses on the grounds of an 'immoral' lifestyle including alcohol abuse, for instance, or for refusing to use protective measures that were part of a medical discourse which, on the pretext of "scientific observation" or of "objective recording," reinforced dominant power relations. Perhaps as a result, workers' campaigns for occupational health remained to a large extent unexplored and are today often regarded as a footnote in the history of the labor movement on the international scene. Yet the demands of workers' organizations for occupational health did not go unnoticed; they made their appearance sporadically or were incorporated into other demands, such as over wages and working hours. The reservations workers had against employing individual means of protection or the imposition of a uniform appearance gave expression to their unspoken resistance to a technocratic 'civilization' process from above, even if it did not always succeed in finding political expression.

The definition of health and diseases by 'specialist' scientists is, in the end, a political issue, one which may be determined in historical terms. Scientific hygiene served not only as a tool in the political economy of health but as a means of debating the 'social question' through reforms and increasing intervention in the everyday life, work and leisure of the lower social classes. This chapter has shown that for Greek physicians and hygienists in the first four decades of the twentieth century, the content of health was inseparable from the new social values of the middle class, the family and social mores of domesticity. Thus the family was considered a 'natural' place for women, contributing to the moral regulation of the social relationships that had arisen with the emergence of industrial capitalism. In this way the views expressed by the hygienists in Greece promoted new social and gender values, while renegotiating longstanding social relations of power.

The history of occupational health in Greece raises questions connected to the policies and practices later followed by every historical subject involved. However, a more comprehensive analysis would require a working knowledge of the policies advanced (institutions, legal committees, the formation of "specialists" and professionalization, demands of labor organizations and so on). It would also require an analysis of the working conditions, which were probably affected by the existing laws; that is, not only of the views but the practices of physicians, as well as of employees and employers. An examination of the organization of labor in various workplaces and further studies of social epidemiology would permit a closer approach to the question of how occupational health was shaped in early twentieth century Greece, not only in theory but also in practice.

# "LIKE YEAST IN FERMENTATION"

## PUBLIC HEALTH IN INTERWAR YUGOSLAVIA*

*Željko Dugac*

To explore public health in interwar Yugoslavia means, first, to consider the enthusiasm of the group of experts headed by Dr. Andrija Štampar (1888–1958) and their resultant ideas and achievements. It is a story of the recruitment of collaborators to realize a radical idea that medicine should be accessible to all social strata—but also an account of the opposition they faced. Yet this tale about advancing a novel idea, of the conflict of interests and power, of progress and the reasons why it was hindered, is neither new nor entirely original. The rise of the young discipline of preventive social medicine to a status equal to the "old" disciplines treating health problems with the scalpel and microscope was no easy process anywhere in interwar Europe. This chapter shows, moreover, that it was especially difficult in the newly-founded Kingdom of Serbs, Croats and Slovenes (renamed Yugoslavia in 1929), which suffered from severe social and political problems.

That said, the present chapter focuses on the brighter side of the story, on the public health programs realized in the Kingdom of Serbs, Croats and Slovenes during the 1920s,[1] for this was the 'Golden Era,' when the Kingdom's public

* For comments on earlier drafts of this paper and for assistance with the English translation, I am grateful to Dr. Tatjana Buklijas and Dr. Marius Turda.

[1] Andrija Štampar, *Deset godina unapređenja narodnog zdravlja* (Zagreb: Narodne Novine, 1934), 22; all translations from the original are my own unless otherwise cited. The chapter is based on the archival sources and printed primary texts in South Slavic languages kept in the Department of History of Medicine of the Croatian Academy of Sciences and Arts, School of Public Health "Andrija Štampar" and in the Croatian State Archives in Zagreb. I have also used material from the Rockefeller Archival Centre in New York and the Archives of the League of Nations Health Organization in Geneva. This research is part of a larger project on *Public Health and Medicine in Croatia: Identity and International Collaboration in the Twentieth Century*, funded by the Ministry of Science, Education and Sports of the Republic of Croatia. I am grateful to the Rockefeller

health system served as an example to other national systems. This was before the turbulent end of the decade, when Štampar lost his position, King Aleksandar Karađorđević (1888–1934) became a dictator and the world was shaken by economic and political crises. An important aspect of this success was the founding of the Ministry of Public Health and its socio-medical institutions and the ideology of the social medicine movement in Yugoslavia. Indeed, the activities of such key institutions and the formation of professions engaged in health education and sanitary engineering were methods of great social impact within society. There is a rich literature on the broader frame of interwar public health and social medicine in Europe and beyond and a growing interest in comparative national studies on these topics.[2] The present chapter contributes to this field by offering an empirical description of the construction of public health, its ideological orientation and the programs to which it gave rise.

## The Foundation of the Ministry of Public Health

Although Croatia and Slovenia had more experience in organizing public health systems because they were longer under the influence of Habsburg sanitary regime than any other Balkan countries, it was in the Kingdom

Archival Centre in New York for financial support and permission to use their archives. I also want to thank the Dugac family for financial support during the preparation of this chapter.

For a detailed discussion of opposition to the public health programs of Andrija Štampar, initially from private practitioners whose interests were challenged by the socialization of medicine, the intrigues of the Belgrade nationalist physicians and, finally, by the confrontation between the predominantly pro-Serb state and Štampar in the 1930s see: Željko Dugac, "New Public Health for a New State: Interwar Public Health in the Kingdom of Serbs, Croats, and Slovenes and the Rockefeller Foundation," in Iris Borowy and Wolf Gruner, eds., *Facing Illness in Troubled Times: Health in Europe in the Interwar Years* (Frankfurt: Peter Lang, 2005), 277–304. See also Željko Dugac, *Protiv bolesti i neznanja: Rockefellerova fondacija u međuratnoj Jugoslaviji* (Zagreb: Srednja Europa, 2005); "Zbor liječnika, Andrija Štampar i javnozdravstvena politika u Kraljevini Srba Hrvata i Slovenaca-Kraljevini Jugoslaviji," *Liječnički vjesnik*, 127 (2005): 151–157; Željko Dugac and Marko Pećina, eds., *The Diaries of Andrija Štampar 1931-1938* (Zagreb: Srednja Europa, 2008) will additionally illuminate these conflicts.

2 See Paul Weindling, "Introduction: Constructing International Health between the Wars," in Paul Weindling, ed., *International Health Organisations and Movements 1918–1939* (Cambridge: Cambridge University Press, 1995), 2; Paul Weindling, "Philanthropy and World Health: The Rockefeller Foundation and the League of Nations Health Organisation," *Minerva* 35, 3 (1997): 269–281; Carl Prausnitz, *The Teaching of Preventive Medicine in Europe* (London: Oxford University Press, 1933), 121–140; Ilona Löwy, Patrick Zylberman, "Medicine as a Social Instrument: Rockefeller Foundation, 1913–1945," *Studies in History and Philosophy of Biological and Biomedical Sciences* 31, 3 (2000): 365–379; 73; E. R. Kohler, "A Policy for the Advancement of Science: the Rockefeller Foundation 1924–29," *Minerva* 16, 4 (1978): 480–515; Patrick Zylberman, "Fewer Parallels than Antitheses: René Sand and Andrija Štampar on Social Medicine, 1919–1955," *Social History of Medicine* 17, 1 (2004): 77–92; and Susan Gross Solomon, Lion Murard, Patrick Zylberman, eds., *Shifting Boundaries of Public Health: Europe in the Twentieth Century* (New York: University of Rochester Press; 2008).

of Serbia that an original and important public health development took place.[3] In 1879 Dr. Vladan Đorđević (1844–1930), then in charge of the Kingdom's health sector and one of the first physicians to work on public health within the state administration, obtained parliamentary support for a law on the establishment of a national health fund. By securing financing for the health service, Đorđević not only ensured its creation but also received official recognition for hygiene being a public good that should be advanced by means of state contributions. Importantly, he became aware of the public health work of Milan Jovanović Batut (1847–1940) and subsequently sent the young physician on an overseas study trip between 1882 and 1885. Batut first traveled to the *Reichgesundheitsamt* in Berlin, where he worked in Robert Koch's department, then to Munich, studying practical hygiene at the Institute of Hygiene under the famous Max Pettenkofer, and finally to Louis Pasteur in Paris. Batut thus had the opportunity to study in Germany, then the leading nation in modern medicine, as well as with the extremely active French bacteriologists. Upon returning to Serbia, Batut unfortunately did not obtain the opportunity to establish a laboratory. Nonetheless, he was appointed professor of hygiene and forensic medicine at the so-called Great School, predecessor of the first Serbian universities. Batut became the leading force of the hygiene movement in Serbia, as well as the chief organizer of the health service in the Duchy of Montenegro. Yet the working environment in that country was made difficult by the lack of financial resources and trained personnel, as Batut wrote: "When I arrived in Montenegro in 1880 to organize the health service, there was not a single university-trained physician in the entire country; and when the government approved the budget to hire ten district physicians, only two candidates applied, one of whom soon left his position and the country."[4] Moreover, Batut was not only in line with modern public health developments across Europe, he was also among the first to link the problems of micro-organisms, lack of education and poverty with the national organization of health.

3 As the first medical school in the South Slav region was only founded in 1917 in Zagreb, physicians before this date had been almost exclusively trained in Vienna. There, they acquired medical knowledge and assimilated Western European methods of approaching public health. See Tatjana Buklijaš, Emese Lafferton, "Introduction" to the special section on "Science, medicine and nationalism in the Habsburg Empire from the 1840s to 1918," *Studies in History and Philosophy of Biological and Biomedical Sciences* 38, 4 (2007): 679–686.

4 Andrija Štampar, *Deset godina unapređivanja narodnog zdravlja* (Zagreb: Narodne Novine, 1934), 7.

Failing to appoint Batut to a leading position within the state health sector was, in a way, corrected in 1918 when the newly founded Kingdom of Serbs, Croats and Slovenes established its health administration. The much-respected and influential Professor Batut now had a chance to employ a young physician in the newly founded Ministry of Public Health, one whose ideas he had noticed earlier. Batut later described his first meeting with Andrija Štampar in the following words: "The depth of conviction, the enthusiastic words he used to express his beliefs, the interest with which this young physician discussed with me various issues in hygiene and public health led me to suspect that this young physician could soon become an excellent worker in the areas of health on which we were working."[5] Štampar initially obtained a position in the Ministry of Public Health on 28 May 1919. From the start, both men understood and complemented each other perfectly: Batut had experience and background, Štampar élan and youth. Many years later, Štampar reflected that: "Professor Milan Jovanović-Batut, the greatest ideologue of health not just in the southern Slav lands but worldwide, built sound foundations for health policy long before the now famous great names of medical science."[6] Indeed, it was entirely through Batut's actions that the Kingdom founded not only a health department in the Ministry of the Interior, as did most other countries, but also created an independent Ministry of Public Health, the first of its kind in Continental Europe.[7]

The latter Ministry was organized into departments dealing with civilian health issues, military health, racial, public and social hygiene and pharmacies. The first Minister of Public Health was Uroš Krulj, a Serb physician from Mostar, famous for his two books: the 1920 *Importance of Hygiene for the State and Nation* [*Važnost higijene za državu i naciju*] and the 1925 *Politics and Race* [*Politika i rasa*].[8] Krulj's interest in eugenics was likely the reason for the establishment of the Department of Racial, Public and

5 Štampar, *Deset godina*, 10.

6 Štampar, *Deset godina*, 17.

7 At the time the Kingdom of SCS was founded, the eastern part of the territory used the Gregorian calendar and the west the Julian calendar. For that reason, the literature mentions two different dates for the appointment of the Ministry of Public Health: 7 December 1918 (Julian calendar) and 20 December 1918 (Gregorian calendar). Ferdo Šišić, *Dokumenti o postanku Kraljevine Srba Hrvata i Slovenaca 1914-1919* (Zagreb: n.p. 1920), 289–291.

8 Uroš Krulj, *Važnost higijene za državu i naciju* (Sarajevo: Zdravstveni odsjek za Bosnu i Hercegovinu, 1920); Uroš Krulj, *Politika i rasa* (Sarajevo: n.p., 1925).

Social Hygiene (usually called the Hygiene Department). This department was charged with public health and health education and was furthermore intended to promote eugenics; it covered the largest scope of activity within the Ministry itself, and was directed by Andrija Štampar. The department's specifications were to promote the health of pregnant women, to prevent voluntary abortions, to monitor workplaces where the health of infants could be endangered, to provide health care to mothers, infants and orphans, to conduct physical examinations of nursing mothers and to manage outpatient clinics for mothers and workers. The Hygiene Department was additionally tasked with promoting healthier dietary habits in the population, inspecting food and grocers and organizing training for food inspectors. Its remit also extended to the supply with clean water, the protection of regional water sources from contamination, the hygiene of residential housing, sewage, the cleanliness of courtyards and public spaces and deforestation. Štampar's department was also in charge of public baths, parks, playgrounds, institutions for the care of sick children and of promoting physical education, gymnastics, sports and games. Furthermore, the department organized care for abandoned children, health care for the youth during periods of education and work, hygiene in domestic life and in the workplace, including worker's insurance against injury, illness and disability. It ran institutions for the injured and disabled whilst supplying the disabled with protheses. It managed health on public roads, including emergency medical care in accidents. It was in charge of prison hygiene. It studied physiological and pathological characteristics by collecting anthropological, biometrical and demographic data. Finally, Štampar's department performed eugenic duties, like promoting marriages between people of similar age.[9]

The statutes founding the Ministry of Public Health emphasized educational health programs, such as medical courses for physicians or physiology and hygiene education in schools. Most women's schools taught the basic concepts and nature of infectious diseases in addition to techniques

---

[9] The history of eugenics and the history of reproductive politics in Yugoslavia have yet to be written. Indeed, the histories of these controversial topics in Southeastern Europe have only just begun to be explored. See Marius Turda and Paul Weindling, eds., *Blood and Homeland. Eugenics and Racial Nationalism in Central and Southeast Europe, 1940-1944* (Budapest: Central European University Press, 2007).

of care for new mothers, newborns and the sick. Holiday courses trained teachers in basic first aid and the prevention of infectious diseases. Sexual hygiene and ethics were part of the secondary school curriculum. The Ministry also collected materials for the museum of social hygiene, while popularizing the tenets of basic hygiene through public lectures, articles in daily newspapers and specialized magazines as well as pamphlets, models, pictures, exhibitions and traveling shows. Moreover, the Ministry worked with associations pursuing similar methods for informing people about health issues.[10]

In 1921 the Ministry of Public Health in Yugoslavia was reorganized into five departments: administrative, hygiene health services, research and education and pharmaceuticals. The Health Board was later added as an advisory body to the Ministry.[11] But on 31 August 1923, both the departments for research and education for pharmaceuticals were abolished, with their areas of activity transferred to the Institute of Social Medicine.[12]

The reformation of the Ministry was accompanied by the preparation and enactment of sweeping health legislation. The Serbian Medical Association had prepared drafts for a general health law in 1919 and sent it to medical associations around the country; the final draft was strongly criticized by the Croatian Medical Association in particular. They stressed that any health law should pay more attention to social issues. Indeed, Štampar himself described the first draft of the Public Health Law as deficient. The meeting of the Croatian Medical Association in Zagreb on 3–4 June 1919 saw a heated discussion on the issues, concluding that the proposed law might have sufficed for the Kingdom of Serbia before the war, but now it was "incomplete and obsolete."[13] The ensuing, fierce argument between Croat and Serb physicians had a strong patriotic undercurrent: it took place during the elections for the constitutional assembly as well as the preparation of the new Constitution. The government employed a range of maneuvers—such as the prohibition of the popular Communist Party, bribery

10 (Anon.), "Uredba o ustrojstvu Ministarstva narodnog zdravlja od 7 lipnja 1919," *Narodne novine* 85 (1919): 1.

11 Adolf Spitzer, ed., *Uredba o ustrojstvu ministarstva narodnog zdravlja od 25. XI. 1921. Zbirka zakona i naredaba tičućih se zdravstva i zdravstvene službe, dodatak XI* (Zagreb: Merkur, 1926), 15–18.

12 (Anon.), "Ukidanje odjeljenja," *Glasnik Ministarstva narodnog zdravlja* 4 (1923): 539.

13 (Anon.), "Izvanredna glavna skupština Zbora liječnika kralj. Hrvatske i Slavonije u Zagrebu, 3. i 4. lipnja 1919," *Liječnički vjesnik* 41 (1919): 429–440.

of smaller political groups and the evasion of the prewar agreements that united the Slovene and Croat lands with the Serb Kingdom—to secure a majority to enact a Constitution cementing Belgrade's control over the newly-formed country: the Vidovdan Constitution of June 1921. The constitution facilitated Yugoslav domination by Serb elites under King Aleksandar. Further centralization of power aroused protests in other parts of the country, especially Croatia (although here, coordination and unification of health seems to have produced better results).[14]

Another draft of the law followed these heated debates. Much of this one was produced by Andrija Štampar in his new position in the Ministry of Public Health. The new draft paid greater attention to health and social medicine. For example, the first chapter, entitled *Health Education,* detailed the way in which the complex mechanism of health education should be put into action. This is followed by chapters on eugenics, protection of mothers and children, school children and youth health care, occupational healthcare, hygienic housing and the protection of public health, measures of sanitation (building the basic infrastructure for hygiene) and, finally, punishments for infringements of the law.[15] Along with these provisions, the *Draft of the Law on the Regulation of the Health Profession* defined the new organization of national healthcare, including the duties of institutions and staff. This law was followed by a number of others: *Draft of the Law on Health Institutions, Draft of the Law on the Prevention and Control of Infectious Diseases, Draft of the Law on the Control of Chronic Infectious Diseases, Draft of the Law for the Combat of Alcoholism, Draft of the Law on Spas, Mineral and Hot Water Springs and Climate Therapy Locations.*[16]

Although a general law for the protection of public health was never fully enacted, these drafts formed the legal basis for further parliamentary acts. For instance, *The Ordinance on the Health Education of the People,* enacted by the Ministry of Public Health in 1928, contained the most important segments of Štampar's aforementioned draft. Critically, this law placed the recently founded Department of Hygiene in charge of all socio-

14 Hrvoje Matković, *Povijest Jugoslavije* (Zagreb: Naklada Pavičić, 2003), 90–100.

15 Andrija Štampar, ed., *Nacrt zakona o čuvanju narodnog zdravlja* (Belgrade: Ministarstvo narodnog zdravlja, 1921), 4–16.

16 *Nacrti sanitetskih zakona* (Zagreb: Zbor liječnika Hrvatske, 1921).

medical work in the territory under its jurisdiction.[17] This greatly contributed to the decentralization of tasks, enabling professionals in the field to assess local needs and priorities.

New legislation also reorganized the structures of health administration in the country. Until 1918 the Kingdom had encompassed regions that had well-organized health institutes as well as those where such structures were more or less inadequate. In 1924, to even out the differences in health organization across the much larger interwar state of Yugoslavia, the Ministry of Public Health established state inspectorates. These inspectorates replaced diverse and unevenly distributed administrative bodies scattered regionally with a national authority for the organization of Yugoslav healthcare.[18] Yet the change proved difficult and health administration had to be further reorganized in 1927 and 1928, when the inspectorates were abolished due to objections over governmental centralization by regions opposed to this form of organization. In this way much of health care system was once again transferred to local governments in the early 1930s.[19]

The remit of each medical department was regulated by legal acts, in the first place by *The Ordinance on the Organization and Development of Hygiene Institutes, Polyclinics and Health Stations*, enacted by the Ministry of Health Care and Social Policy. The task of the hygiene institutes located in the capital of each territorial unit was to manage the health protection of mothers and children, school polyclinics, outpatient clinics for tuberculosis, outpatient clinics for venereal and skin diseases as well as museums of hygiene with traveling hygiene exhibitions. Institutes for the health protection of mothers and children were intended to include a clinic for pregnant women, a clinic for mothers with small children and a section for children's hygiene—staffed by an itinerant teacher and equipped with an exhibition able to accompany traveling health shows. Polyclinics also needed a hygiene exhibition as did health stations in charge of local hygiene promotion. Lastly, this groundbreaking statute prescribed the exact remit of

17 Adolf Spitzer, ed., *Uredba o zdravstvenom prosvjećivanju naroda Ministarstva Narodnog Zdravlja od 3. III. 1928. Zbirka zakona i naredaba tičućih se zdravstva i zdravstvene službe, dodatak XV* (Zagreb: Neva, 1928), 251–254.

18 Adolf Spitzer, ed., *Ustrojstvo i djelokrug inspektorata od 28. III. 1924. Zbirka zakona i naredaba tičućih se zdravstva i zdravstvene službe, dodatak XI* (Zagreb: Merkur, 1926), 72–73.

19 Adolf Spitzer, ed., *Uvodnik. Zbirka zakona i naredaba tičućih se zdravstva i zdravstvene službe, dodatak XV* (Zagreb: Neva, 1928), 3.

outpatient clinics for the control of skin and venereal diseases, trachoma, tuberculosis and other diseases.[20]

## New Social Hygiene Institutions

Andrija Štampar believed in the idea of statist medicine,[21] one whose expenditures could be included in the national state budget. Medicine should follow the example of state-funded and managed education, claimed Štampar, for such an organization would provide security for physicians and adequate care for patients. He thus advocated a broadly based social security system that would incorporate as many people as possible, in the hope, ideally, that one day it would expand to cover the entire world. Furthermore, Štampar thought that legislation could solve the problem of organizing healthcare, but was powerless in another, equally important problem: securing sufficient numbers of cooperating physicians.[22] The aim of this worldview, which Štampar called "our ideology," was to create a system in which medical care would be accessible to as much of the population as possible, not just the wealthy. Influential to this view of medical problems was the idea of society as an organism, where diseases had a social aetiology and pathology, and are treated with the corresponding methods of social therapy. Health education and promotion were key methods in preparing the ground for a particular community to understand health issues in the right way.

The challenges of public health and its promotion were, according to Štampar, not only the purview of the physician but of a wider group of peo-

20 (Anon.), "Uredba o organizaciji i delokrugu Higijenskih zavoda, Domova narodnog zdravlja i Zdravstvenih stanica," *Narodne Novine* 46 (1930), 1–2; Adolf Spitzer, ed., *Naredba o osnivanju ambulatorija za kožne i spolne bolesti od 10. VII. 1920. Zbirka zakona i naredaba tičućih se zdravstva i zdravstvene službe, dodatak XIII* (Zagreb: Tiskara Merkur, 1926), 305–306; Adolf Spitzer, ed., *Pravilnik o radu antituberkuloznih dispanzera od 21. II. 1922. Zbirka zakona i naredaba tičućih se zdravstva i zdravstvene službe, dodatak XIII* (Zagreb: Tiskara Merkur, 1926), 366; Adolf Spitzer, ed., *Pravilnik o radu ambulatorija za suzbijanje trahoma od 12. XI. 1925. Zbirka zakona i naredaba tičućih se zdravstva i zdravstvene službe, dodatak XIII* (Zagreb: Tiskara Merkur, 1926), 412.

21 Štampar, *Deset godina*, 16.

22 A major problem in the health system was the lack of physicians and their concentration in urban areas. The former Habsburg lands, Croatia and Slovenia, had more physicians than the rest of the country and some of them had practiced earlier as state employees in designated localities (districts, counties). But their 'state' position, which involved providing free treatment to the poor and managing public health problems, was in reality secondary to their private practice. Štampar expected little cooperation from these physicians.

ple from different professions—engineers, social workers, teachers, writers, artists—indeed the entire national community. The physician was conceived first and foremost as a social worker using social therapy because individual therapy had failed to produce the desired outcome.[23]

Following World War I and especially during the 1920s, many social hygiene institutions were established and many physicians were trained to work in them. These processes were of key importance, both in the development of public health and the achievement of good results. The distribution of the institutions was like a spider's web, spreading from the center to the periphery. The center of the network was occupied by the main institutions, the School of Public Health and various hygiene institutes in the capital cities. From there, this web extended to smaller cities and towns now furnished with polyclinics and still further out to rural health stations. These latter units acted as the bases for nurses and physicians traveling to the smallest settlements.

On 1 January 1931, the year of Štampar's retirement, the Kingdom of Yugoslavia had established the following public health institutions: one School of Public Health, ten hygiene institutes, 44 polyclinics, 44 bacteriological stations, 52 anti-rabies stations, 81 school polyclinics, 28 infant outpatient clinics, 35 outpatient clinics for the control of tuberculosis, 66 outpatient clinics for the control of skin and venereal diseases, 17 outpatient clinics for the control of trachoma, ten socio-medical departments, 12 sanitary engineering departments, 19 chemistry departments, two biological-immunological departments, five parasitological departments, 16 antimalarial departments, one department of veterinary medicine, one general medicine department, two hospital departments, 111 health stations, ten public baths, 19 accessory stations, and 20 sanatoria and resorts. There were 606 institutions in total.[24] The difference between 1918 and 1931 is vividly communicated by a map showing the distribution of this web of social medicine institutions across Yugoslavia (see fig. 1).

All hygiene institutes were located in large cities: the Central Hygiene Institute was in Belgrade, with the others in Ljubljana, Split (beside the Institute for the Control of Malaria in Trogir), Osijek, Novi Sad, Banja Luka,

23 Andrija Štampar, *Pet godina socijalno-medicinskog rada u Kraljevini Srba, Hrvata i Slovenaca 1920–1925* (Zagreb: Institut za socijalnu medicinu Ministarstva narodnog zdravlja, 1926).

24 Štampar, *Deset godina*, 31.

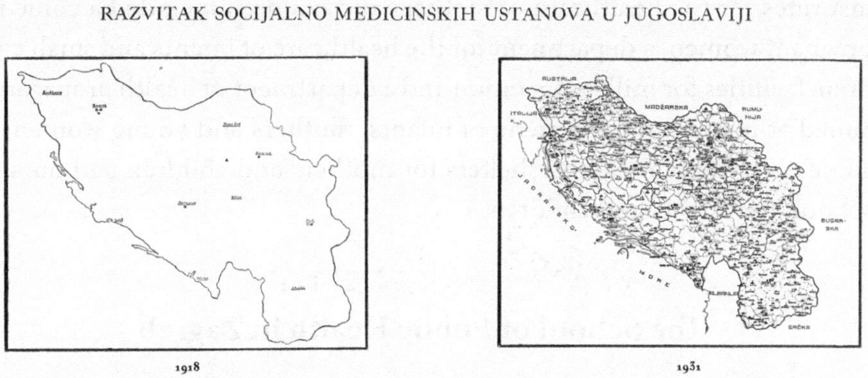


**Fig. 1. Increasing of the social-medicine institutions in the Kingdom of Yugoslavia from 1918–1931. (© Division for the History of Medicine, Croatian Academy of Sciences and Arts)**

Cetinje, Niš, Skopje (where the Hygiene Institute was incorporated into the Institute for Tropical Diseases) and finally in Zagreb, where the Hygiene Institute was attached to the School of Public Health. These urban hygiene institutes each had a bacteriological-epidemiological department, a chemistry department for food control, a sanitary engineering department for the planning of human and animal housing, a socio-medical department for the study and education of the people and a hospital department for the prevention of malaria and other infectious diseases. Some hygiene departments also had additional divisions, for example for parasitology, rabies prevention or statistics. Polyclinics usually comprised outpatient clinics for mothers and children, school polyclinics, a bacteriological-epidemiological department, an anti-tuberculosis outpatient clinic, a hygiene exhibition and, where needed, outpatient units for venereal diseases, trachoma, malaria and the like. These clinics were scattered in towns, small and large, as well as villages.

Institutions smaller in scale than polyclinics were called "health stations." They consisted of a general practice outpatient unit, specialized units for malaria, syphilis or trachoma and the ubiquitous section for hygiene education. Sometimes they also included public baths. School polyclinics were specialized institutions comprising general practice departments for children, ear, nose and throat departments, tooth and mouth departments and a department for health education. They often comprised children's public baths and, sometimes, included a school kitchen and cafeteria. Finally, the institutes for the healthcare of mothers and children included a clinic for pregnant women, a department for the healthcare of infants and small children, facilities for milk preparation and a department of health propaganda aimed at bettering the hygiene of infants, mothers and young women. In some cases these included shelters for mothers and children and nursery schools for infants and children.

## The School of Public Health in Zagreb

The establishment of the School of Public Health in Zagreb was part of the international program that the Rockefeller Foundation undertook across Europe after World War I. The program gave rise to a network of institutions of this type, especially in Central and Southeastern Europe. The Zagreb School was founded in 1926, although the last building was only officially opened on 3 October 1927.[25] The 1926 *Regulations on the Foundation, Organization and Activity of the Hygiene Institute with the School of Public Health in Zagreb* stated that the purpose of the school is to study, prevent, combat and treat infectious diseases and to manage the problems of hygiene and social medicine more generally. The opening statute furthermore stated: "The key task of the school is to educate the people theoretically and practically, especially with respect to the hygiene needs of the rural areas."[26] Zagreb's Hygiene Institute and School of Public Health consisted of the following units: administration, smallpox vaccine production, hospital (an inpatient institution for the treatment of infectious diseases), parasitology, technical services and social hygiene. From the health education viewpoint, two departments were of particular importance: first, the Department of Social Medicine, further divided into the Division for Hygiene Promotion, the Division for the Research of Public Pathology and

25 (Anon.), "Otvaranje higijenskog zavoda i škole narodnog zdravlja u Zagrebu," *Liječnički vjesnik - Staleški glasnik* 49 (1927), 346; Dugac, *Protiv bolesti*, 104. A meeting of the International Health Board from the Rockefeller Foundation on 6 November 1924 decided that the School of Public Health would receive a contribution in the amount of $145,000. The first installments were paid between May 1925 and 1926, when an additional sum of $34,000 was added to compensate for the variation in dollar exchange rates between 1924 and 1926. See Minutes, International Health Board, June 23, 1927. RG.710, S.1.1, F.30. RAC. New York.

26 (Anon.), "Pravilnik o osnivanju, organizaciji i radu Higijenskog zavoda sa školom narodnog zdravlja u Zagrebu," *Narodne Novine* 42 (1926): 1–3.

Statistics and the Division for Public Education, and second, the Technical Services Department, which was divided into the Division for Rural Sanitation and the Division for Sanitary Engineering. The Department of Social Medicine supervised all socio-medical institutions falling within the territory of the Institute's mandate. The divisions in charge of health education had the following tasks:

- The Division for Hygiene Promotion organized popular lectures, published books, pamphlets, leaflets and posters, produced films and directed a film workshop, ran the central library and the printing department of the Institute, managed the hygiene museum and organized the temporary hygiene exhibition;
- The Division for Public Education organized permanent and occasional courses at the Institute, produced curricula for schools training the lower level health personnel in the country, organized and supervised instruction in the schools for nurses, laboratory technicians, food inspectors and midwives, organized courses and schools for housewives, set up and monitored the education of school physicians, supported all private physical education initiatives for the public, collaborated with all other national and international institutions sharing its goals and supervised both technical publications and the production of educational tools.

Situated within the Technical Services Department, the Division for Rural Sanitation produced plans for houses, stalls, outhouses, latrines, dunghills and wells. It organized courses and practical exercises in the schools as well as in the field. For its part, the Division for Sanitary Engineering provided expert consultation on hygiene matters, produced sanitation plans and conducted practical work. In addition to basic sanitary engineering tasks, this division also organized local education and participated in health promotion by producing and distributing plans, models and replicas of sanitary objects, wells and information on how to bore them, water tanks, outhouses, dunghills and other facilities of this kind. The Division housed a museum collection that included 7,967 photographs and 2,335 slides, an extensive collection of supporting materials for courses and lectures, a collection of materials from the field of hygiene and water sanitation and a small library of foreign sanitation plans and publications.[27]

Finally, the photo and film laboratory in the Division for Public Education was in charge of the production of photographs and slides used in public health lectures and exhibitions, the production of educational films and the translation and editing of imported films. It also created and purchased other educational tools like pictures, models, moulages, preparations and so on. The interwar era saw the names of both units and the school's overall organizational structure change several times, even though its remit of activity remained largely the same.[28] Until World War II the School of Public Health maintained a key role in education, research, sanitary engineering and collaboration, the latter extending to the exchange and training of stipend recipients with international institutions. An especially intense cooperation developed with other schools of this type in Europe and the United States, and with the League of Nations Health Organization, the institution most visible in promoting the internationalization of public health. As the foremost institution for health education in the country, the School's work was initially planned to encompass the entire territory of Yugoslavia, but political conditions and local resistance limited its remit to today's Croatia and Bosnia and Herzegovina.

## The Central Hygiene Institute in Belgrade

Negotiations for the establishment of a central institution in the Kingdom's capital to research infectious diseases and produce vaccines commenced immediately after World War I. Yet the lack of consensus on its scope, overly ambitious expectations and a general inability to find both the right place and sufficient financial support prevented the immediate realization of this plan.[29] This protracted discussion about a future institute lasted until

27 Izvještaj za 1936. godinu, Tehničko odjeljenje, 539 T.O.7. K. 1937. Opći spisi - personalije, predračuni za nadničarske liste, dopisi upravnih vlasti. Dokumentacija Škole narodnog zdravlja "Andrija Štampar". Zagreb.

28 (Anon.), "Uredba o organizaciji i delokrugu Škole narodnog zdravlja u Zagrebu," *Narodne Novine* 46,1 (1930): 1–2.

29 For the chronology of the long and clumsy negotiations see Bogdan Konstantinović, "Historijat Centralnog higijenskog zavoda," *Socijalno-medicinski pogledi* 8 (1936): 1–18.

1924 when, on Štampar's invitation, the Rockefeller Foundation entered the debate. The future Central Hygiene Institute was the recipient of the second largest Rockefeller grant to be made to Yugoslavia.[30] The grant was seen to be a diplomatic move of a sort, an attempt to show that Belgrade would not be outdone by Zagreb. The foundation proved to have made a clever decision since in doing so, Štampar also gained an opportunity to take matters into his own hands, bringing the exhausting business involved in founding the institution to an end. It also ensured that balance was maintained in the persistently tense relationship between Zagreb and Belgrade.

The records of the International Health Department of the Foundation show that the institution was intended to coordinate all epidemiological institutes and bacteriological laboratories in the country for the purpose of infectious disease control. Its task was, furthermore, to provide laboratory support to public health work, to produce sera and vaccines, and to control all such product use in the country. The plans show that the Central Hygiene Institute was also expected to educate public health professionals. Before approving the grant, the Rockefeller Foundation requested from Slavko Miletić (Minister of Public Health) a confirmation that the Institute would indeed undertake these tasks. Miletić replied to Selskar Gunn[31] that he agreed with the proposed tasks and that any financial support provided by the Foundation would be used for these purposes.[32] In 1925, following a discussion with Štampar and Miletić, the International Health Board of the Foundation approved 2,000,000 dinars ($30,000) for equipment for the Central Hygiene Institute in Belgrade.[33] This grant, paid through the Belgrade branch of the Yugoslav Bank, was accompanied with the following text: "at the disposal of the President of the committee, Dr. Andrija Štampar," thus ensuring all financial transactions would have to be overseen by the man the Foundation trusted, Štampar.[34] Štampar, Gunn and Russell were furthermore preoccupied with the problem of appointing

30 The largest grant was given to the School of Public Health in Zagreb.

31 Selskar Gunn (1883–1944) was the charismatic regional director of the Rockefeller Foundation European office (with headquarters in Paris) and the vice-president of the Foundation. He was also a close friend of Andrija Štampar. See Dugac, *Protiv bolesti*, 51-57.

32 Minutes, Executive Board, International Health Division, 1 April 1925. RG.710, S.1.1, B.1, F.1. RAC. New York.

33 Memo, November, 1924. RG.710, S.1.1, B.1, F.3. RAC. New York.

34 Letter Gunn to Russell, 1 May 1925. RG.710, S.1.1, B.1, F.1. RAC. New York.

a future director of the Institute. Štampar asked the Foundation's advice, in order to find a qualified person overseas, since it was initially deemed that there were no suitable candidates in Yugoslavia.[35] In spite of lively discussion, and even a preliminary list of potential candidates, a local functionary, the well-connected and strongly Serb nationalist-oriented Stevan Ivanić, was finally appointed. Ivanić would later become a leading opponent of Štampar and his public health programs.

The Central Hygiene Institute began work in 1926. It assimilated three institutions founded in 1922/3: the Permanent Bacteriological Station, Institute of Social Medicine and the Institute for Tropical Diseases. It was organized into the following departments: bacteriological-epidemiological, chemical, parasitological, sanitary engineering and socio-medical. It also encompassed several institutions located in Belgrade and its surroundings: the Hospital for Infectious Diseases, the Museum of Hygiene, the Institute for the Protection of Mothers and Children, the School Polyclinic, the Institute for the Research and Control of Cancer as well as other socio-medical institutions in the city.[36]

## Hygiene Institutes in Other Towns

One of the chief socio-medical institutions in the country beside the School of Public Health in Zagreb and the Central Hygiene Institute in Belgrade was the Institute for Tropical Diseases in Skopje. The Institute was formally a department of the Central Hygiene Institute in Belgrade but in reality it worked independently. It managed all institutions whose task was the study and control of tropical diseases. This extremely active institution—whose scope of activity was not limited to the study of tropical diseases (mostly malaria)—was of great interest to the Rockefeller Foundation in the second half of the 1920s. To further develop the network of socio-medical

35 Letter Štampar to Gunn, 8 August 1925. RG.710, S.1.1, B.1, F.1. RAC. New York.
Many important candidates with experience in public health or previous sojourns in Serbia, for example as American Red Cross activists, were suggested for the position of Central Hygiene Institute Director, including Andrew W. Sellards, Carl E. Buck, W. Lloyd Aycock and others (Memo, 14 November 1925. RG.1.1, S.710, B.1, F.3, RAC. New York.)

36 Vojislav Milovanović, ed., *Medicinski godišnjak Kraljevine Jugoslavije* (Belgrade: Jugoreklam, 1933), 33-44.

institutions in Macedonia (under the supervision of the Skopje Institute), Štampar applied to the Foundation to support the creation of small health stations staffed with a nurse, sanitary inspector and an assistant in impoverished Macedonian villages that had hitherto lacked any health service. Štampar reported to the Foundation that 23 such institutions had already been established and that he needed support for an additional 24. In larger Macedonian towns Štampar hoped to establish school polyclinics. The International Health Board approved his request and granted him $14,000 toward the proposed projects.[37]

With respect to Bosnia and Herzegovina, Štampar applied to the Foundation in 1927 for $20,000 to establish polyclinics in Banja Luka, Bihać, Mostar, Pljevlja and Travnik. Gunn supported his request and the grant was approved. This money was used to purchase furniture and equipment for the newly founded institutions in the aforementioned towns.[38] In 1924 and 1925 Štampar also applied to Gunn to help establish a public health institution for Montenegro. He estimated the cost of building such an institution to be $10,000. The Ministry would supply half the amount needed as well as the initial equipment and maintenance costs. Štampar's application for $5,000 was supported by Gunn, who stated that such help would be especially welcome for the improvement of public health conditions "in this very poor corner of Yugoslavia." In spite of his support, on this occasion the Foundation declined to provide financial help.[39]

## New Professions

As new institutions were established one by one, the professional members of staff were also trained. The foundation of the Medical School in Zagreb in 1917 was followed, in 1921, by the appointment of Emil Prašek as the full Professor and Head of Department of Hygiene and Bacteriology. In 1922/23 he gave a first practical course in this subject. The same year,

37 Minutes, International Health Division, 11 November 1926, RG.710, S.1.1, B.5, F.44. RAC. New York.

38 Minutes, International Health Division, 29 November 1927. RG.710, S.1.1, B.5, F.50. RAC. New York.

39 Letter Gunn to Russell, 7 September 1925. RG.710, S.1.1, B.5, F.50. RAC. New York; Report, Mitchell to Gunn, 2 September 1926. RG.710, S.1.1, B.5, F.50. RAC. New York; Leland Mitchell, Public health activities in the West Montenegro, 16 April 1926, RG.710, S.1.1, B.5, F.46. RAC. New York.

*Privatdozent* A. Štampar offered a seminar in *Social Hygiene, Social Diseases: Main Chapters in Social Medicine and Health Policy,* which he repeated in 1923/24.[40] In Belgrade, Milan Jovanović Batut taught hygiene at the Medical School founded in 1920.

Students of medicine learnt about health education at the Medical School in conjunction with compulsory fieldwork.[41] Young physicians who obtained positions at newly founded institutions saw health education as an important component of socio-medical work. Some of these doctors received scholarships for further training in Western Europe and the United States. At international social medical institutions they acquired knowledge of the most recent advances in social medicine and public health, all of which could be utilized once they returned to their homeland.[42] The School of Public Health organized courses for the training of health professionals. In 1930 the "Regulations of the Educational Courses for Physicians" stipulated that all physicians employed in state administration, hygiene institutes or, more generally, in public health had to attend a six-month course where they would be taught, among other things, public health, social medicine, history of medicine and so on.[43]

Nurses had to be educated in social medicine as well. A nursing school was first established in Zagreb in 1921 and then in Belgrade and Skopje. The Rockefeller Foundation generously supported the equipment of the school and provided scholarships for nurse training overseas.[44] At the meeting of the directors of European hygiene schools in Paris in 1930, Štampar emphasized the importance of education for nurses and the results achieved by the School of Public Health in this area.[45]

Many other professionals worked with physicians and nurses in social medical institutions: sanitary and civil engineers, film directors, editors and cameramen, painters and illustrators, writers, especially of short-sto-

---

40 Biserka Belicza, "Kronologija najvažnijih događaja do 1944. godine," in Anton Švajger, ed., *Medicinski fakultet u Zagrebu 1917–1997* (Zagreb: Sveučilište u Zagrebu Medicinski fakultet, 1997), 12–15.

41 Minutes. Conference of Directors of Schools of Hygiene, Paris 20-23.V.1930. Health general R5874/8A. League of Nations Archive. Geneva.

42 Dugac, *Protiv bolesti*, 89–102.

43 (Anon.), "Pravilnik o kursevima za lekare," *Narodne Novine* 46, 3 (1930): 1.

44 Vladimir Ćepulić, "Državna škola za sestre pomoćnice u Zagrebu," *Liječnički vjesnik-Staleški glasnik* 48 (1926), 75–87; Dugac, *Protiv bolesti*, 121–123.

45 Minutes. Conference of Directors of Schools of Hygiene, Paris 20-23.V.1930. Health general R5874/8A. League of Nations Archive. Geneva.

ries and many others. Along with those who worked full time or on an occasional basis for social medical institutions, others, such as teachers and priests, took part in the advancement of public health as well. Teachers were expected to recognize a sick child, give first aid and, if necessary, to send them to a physician for further tests and treatment. This was especially important for the teachers without access to a school physician.[46] Furthermore, teachers and priests were entrusted with teaching the population basic principles of hygiene and disease protection, for instance the importance of prophylactic vaccinations against smallpox.[47] It was of particular benefit to physicians to have priests and teachers lecture on health since it saved them from traveling themselves to distant places without medical practices. A 1921 memorandum sent to the county authorities in Vukovar proposed that teachers and priests read extracts and articles from the health education magazine *Zdravlje* [Health] to their classes and congregations.[48]

## Popular Health Education

Popular health education was viewed as a key activity for public health institutions. The main tasks of the School of Public Health were to produce health education materials, educate personnel and plan and coordinate action in the field. As mentioned previously, all new social medical institutions had to incorporate health education into the core of their activities. The Health Department in Zagreb, with an organizational structure inherited from Habsburg times, included in the early 1920s a Division for Health Promotion, Research and Public Education and Social Hygiene. The Division had been very active prior to the 1924 foundation of the Institute of Social Medicine in Zagreb. This institute had health education as one of its main tenets and for that purpose established a photography and film laboratory for the production of slides and films. In 1926 the Zagreb Insti-

46 (Anon.), "Pravilnik za suzbijanje zaraznih bolesti u školama," *Liječnički vjesnik - Staleški glasnik* 44, 2 (1922): 299–303.

47 (Anon.), "Poziv učiteljstvu i svećenstvu, da pouče narod o važnosti i koristi cijepljenja protiv boginja (Razne vijesti)," *Liječnički vjesnik* 42 (1920): 615.

48 Dopis, Zdravstvena propaganda 6126. F.134 Zdravstveni odsjek za Hrvatsku, Slavoniju i Međimurje 1919-1924. Opći spisi. Hrvatski državni arhiv. Zagreb.

tute of Social Medicine was incorporated into the School of Public Health. The laboratory became the site of the largest and most important production of health education films, slides, pictures and printed material: posters, books, booklets, magazines and so on.[49]

Health education greatly developed in interwar Yugoslavia. It adapted to the social circumstances that included widespread illiteracy, traditional culture in rural areas, deeply rooted customs and prejudices. Health educators tailored their activities to suit the needs of the local population while respecting cultural specificities.[50] In addition to well-known international approaches, they also developed their own original approaches. Booklets and pamphlets were viewed as highly suitable materials for health education, primarily because they were more concise than books. For a rural population that was largely illiterate or without any developed reading culture, such succinct forms presented more suitable media. Booklets were distributed on occasions such as lectures in schools or in various work places. In the Split inspectorate area, pamphlets proved very popular, so much so that the local inspectorate printed the following publications in editions of several thousand copies: *How to Protect Ourselves from Diseases, How to Protect Ourselves from Spotted Typhus, Pestilential Diseases, Rabies, Foot care, Scarlet fever and Measles, On Trachoma* and *On Scarlet Fever*. Furthermore, it was reported that 3,120 copies of other booklets were distributed.[51] Pamphlets were another popular printed educational form. Soon after World War I, in 1921, thousands of pamphlets titled *Spotted Typhus* were distributed in Dalmatia alone.[52] Pamphlets were also visibly placed in public spaces to warn against widespread and unsafe habits such as spitting on the floor.[53]

Health education slogans were used independently or incorporated into other content. When used independently they were usually printed on paper or written on other surfaces such as walls, wooden boards and cloth. They

49 (Anon.), "Sekcija za zdravstvenu propaganda," *Liječnički vjesnik-Staleški glasnik* 44 (1922): 339; (Anon.), "Otvorenje instituta za socijalnu medicinu u Zagrebu (Razne vijesti)," *Liječnički vjesnik-Staleški glasnik* 46 (1924): 337; and Mirna Zebec, Silvije Vuletić, Antun Budak, eds., *70 godina rada na promicanju zdravlja hrvatskog puka* (Zagreb: Gandalf, 1997).

50 See Aida Brenko, Željko Dugac, Mirjana Randić, *Folk Medicine* (Zagreb: EMZ, 2001).

51 (Anon.), "Zdravstvene propaganda (Razne vijesti)," *Glasnik ministarstva narodnog zdravlja* 5 (1924): 472–473.

52 Ivan Petković, "Zdravstvene propaganda u Dalmaciji," *Glasnik ministarstva narodnog zdravlja* 3 (1921): 313–317.

53 Josip Hribar, "Narodno zdravlje Savske banovine," *Liječnički vjesnik - Staleški glasnik* 56 (1934): 17–22.

were placed on the walls of rooms in which people tended to congregate or pass through and exhibited in health education exhibitions and public places during promotional campaigns. In addition to such short forms, books and especially readers were highly popular. They discussed topics of general health importance such as the structure of the human body, personal, household and public hygiene and individual diseases. Some readers focused on specific segments of health education or were targeted at specific audiences, for instance children. The most popular readers were the *Peasants' Reader on Health* (1924) and *A Popular Reader on Health: Part One* (1930, 1933). These books organized material on the human body and diseases not in relation to organ systems but according to how these diseases entered into the body.[54]

The book used most extensively by the School of Public Health in its public education work was *A Popular Reader on Health: Part One.* Its first edition (1930) was printed in 20,000 and the second (1933) in 37,000 copies. Some of the texts incorporated within the reader were composed by Milan Jovanović-Batut and Andrija Štampar.[55] For children the School of Public Health published a picture book titled *Children's Reader on Health* (1927). Children learned the basics of health through simple and easy to remember terms such as cleanliness, sun, fresh air, play and freedom. This book illustrated a healthy child's bedroom that was bright, airy and "well-swept." Furthermore, this room was contrasted with a room in the house where "ignorance ruled." The words associated with this house were suffering, pain and poverty. Such a house where sun, light and fresh air were lacking was represented as a center for tuberculosis. The picture book also introduced the subjects of alcoholism and smoking, topics rarely present in contemporary health education campaigns and literature.[56]

The publication of health education materials was accompanied by the creation of a cultural milieu encouraging their use. An intense campaign to increase interest in books and learning supported the establishment of public libraries and reading rooms. This campaign was linked with the foundation and strengthening of associations such as gymnastics and singing societies as well as voluntary charitable organizations.

54 Ivan Haslinger, *Seljačka čitanka o zdravlju* (Karlovac: Zakladna tiskara Narodnih Novina, 1924).
55 *Narodna čitanka o zdravlju. Knjiga I* (Zagreb: Škola narodnog zdravlja u Zagrebu, 1930).
56 *Dječja čitanka o zdravlju* (Zagreb: Higijenski zavod sa Školom narodnog zdravlja u Zagrebu, 1927).

Health education magazines had been published in the South Slavic regions, especially Serbia, since the late nineteenth century. For instance, Batut published and edited the popular magazine *Zdravlje* in Sombor and Cetinje (1880–1882), then *Narodno zdravlje* [Public health] (as an addition to *Srbski arhiv* [Serbian archives]) until 1895 and *Zdravlje* in Belgrade (1906–1911). After the First World War the magazine *Zdravlje* was revived in a much larger print edition: 15,000 copies in Cyrillic script (Serbian), 10,000 in Latin script (Croatian) and 5,000 in Slovenian. The magazine was published by the Society for the Protection of Public Health in Belgrade and edited, again, by Batut.[57] Archival documents testify that articles from *Zdravlje* were recommended for use in health education lectures to rural communities.[58]

The newly founded School of Public Health launched another health education publication in 1927, first named *Obavijesti* [Notices] and then *Narodni napredak* [Public progress]. It published adapted scientific texts on various health problems but also included articles on farm management as well as instructive stories and texts authored by students from the Peasants' University and other literate peasants more generally. *Glasnik ministarstva narodnog zdravlja* [Herald of the Ministry of Public Health], published in Belgrade between 1919 and 1929 and edited by Štampar, had a twofold role of health education and informing the public of current affairs in the Ministry's area of authority. Finally, there were many specialized journals devoted to specific topics such as, for instance, the struggle against alcoholism.

## Lectures and Courses on Health

In the interwar era, the lecture—a highly important health education tool—no longer relied simply on the speaker but also introduced other media. New technologies such as the portable projector enabled the screening of films outside of cinemas. Lectures were, furthermore, often accompanied by other forms of visual presentation such as slides. In 1921, *Glasnik Mini-*

57 Novković, "Zdravlje," *Liječnički vjesnik* 42 (1920): 80.

58 Korespondencija Kraljevska kotarska oblast u Irigu - Kraljevska županijskoj oblasti u Vukovaru 12. IX. 1921, 590/21. F.134 Zdravstveni odsjek za Hrvatsku, Slavoniju i Međimurje 1919-1924, Opći spisi. Hrvatski državni arhiv. Zagreb.

*starstva narodnog zdravlja* reported on the production of slides for use in lectures on alcoholism, tuberculosis and venereal diseases. These slides were distributed throughout the country.[59] The overall changes in the lecture format may be observed in the reports of the Department of Social Medicine of the School of Public Health for January 1927: of 42 lectures, only three were delivered without visual aids. [60]

Lectures were such a widespread form of health promotion that it is impossible to find a campaign where they were not applied. In some cases, lectures stood independently, but they also frequently accompanied film screenings, exhibitions and other educational forms. They formed part of large campaigns advocating cleanliness and temperance or promoting measures against tuberculosis. Another media form used in health education was the radio lecture, an area in which Radio Zagreb was especially active.[61]

A variety of bodies—including the Ministry of Education, voluntary societies and political organizations—organized courses, but the leading role in this sphere was played by the newly founded socio-medical institutions such as the School of Public Health in Zagreb, the Central Hygiene Institute in Belgrade and other hygiene institutes (especially in Ljubljana and Sarajevo) as well as numerous polyclinics and outpatient clinics throughout the country.[62] Two types of course were of particular importance. The first were courses that took place in the field, for instance home economics courses. The second were institutional courses, of which the most important and original were held in the School of Public Health in what was called the Peasants' University. Most participants were peasants, but some courses were organized in cities for workers, students and other elements of the urban population. Courses were organized for volunteer activists in social hygiene work, particularly those responsible for the care of children. These included, for instance, teachers, priests and municipal officials.

---

59 Vladimir Majcen, *Obrazovni film, Pregled povijesti hrvatskog obrazovnog filma* (Zagreb: Hrvatski Državni Arhiv – Hrvatska Konoteka, 2001); (Anon.), "Fotografska radionica," *Glasnik ministarstva narodnog zdravlja* 3 (1921): 144.

60 Izvještaj o radu od 7.2.1927, 247/1927. K.1. F.517 Higijenski zavod sa Školom narodnog zdravlja. Hrvatski državni arhiv. Zagreb.

61 (Anon.), "Popularna informativna predavanja o zdravstvu putem radija (Razne vijesti)," *Liječnički vjesnik* 51 (1929): 47.

62 (Anon.), "Uredba o organizaciji i delokrugu Higijenskih zavoda, Domova narodnog zdravlja i Zdravstvenih stanica," *Narodne novine* 46 (1930): 1–2.

In between the short courses and the longer, more complex Peasants' University were courses provided by Schools of Health Education. These had no designated location and took place in different cities. They offered health education lectures, usually accompanied by films and slides, together with literacy courses and the provision of promotional material such as leaflets, booklets and posters for the general population.[63]

Most home economics courses were organized by the School of Public Health between 1927 and 1937. In total, 262 courses of this type were organized for 4,135 female students.[64] Their program included everything considered necessary for a woman to become a good housekeeper, mother and wife. Lectures and practical classes included food science, cooking, growing vegetables, preserving food, dairy farming, good manners, food hygiene, anatomy and physiology, bacteriology, protection from infectious diseases, the most common diseases (fever, dysentery, scabs, scabies, smallpox and vaccination), first aid (wounds, hot water burns, caring for the sick at home and home pharmacy), the most important medicines and bandages at home, childbirth, child care and, finally, propaganda warning against rural 'quack' healers. Theoretical instruction was accompanied by practical experience, especially in cooking, preserving food and cleaning dishes. Students were taught to make hygienic baby clothes and bed linens. They were trained to correctly clean the house and make beds, especially for children and the sick, wash laundry and so on.

Sometimes courses were planned ad hoc in response to the needs assessed in the field, but usually they followed highly structured plans. Some of the larger courses followed guidelines prepared by national or even international institutions. For instance, the program implemented by the School for Mothers in Sarajevo in 1927 was prepared based on a curriculum obtained from Warsaw that in turn was modeled on Danish courses for peasants.[65]

---

63 (Anon.), "Zdravstvena propaganda na teritoriju ljubljanskog, splitskog i beogradskog inspektorata u godini 1924," *Glasnik ministarstva narodnog zdravlja* 6 (1925): 159–164.

64 Živko Prebeg, *Higijensko-domaćinski i tkalački tečajevi u selima Škole narodnog zdravlja, Izvještaj o dvanaestogodišnjem radu* (Zagreb: Škola narodnog zdravlja, 1938), 4–5.

65 Program rada "Materinske škole" u Sarajevu, 218/1927. K.1. F.517 Higijenski zavod sa Školom narodnog zdravlja. Hrvatski državni arhiv. Zagreb.

The largest and most important of these courses was the aforementioned Peasants' University, which held its first session in 1927/28.[66] Each year, courses for men were organized between 15 October and 15 March, and for women between 1 April and 1 June. These courses lasted five and two months respectively and all prospective students had to be both healthy and literate. The aim of the Peasants' University was to educate the rural population and prepare them to take an active role in various health education and sanitation campaigns. The goal was, furthermore, to teach peasants scientific knowledge useful for the rationalization of farm labor as well as teach them how to live a hygienic life. All students—usually organized into groups of 30–40—learned about health and methods of health improvement, efficient agriculture, rural husbandry, construction and sanitary technology. They also studied Croatian, mathematics and general culture. Women's courses paid particular attention to children's hygiene, care and education, care for the sick and housekeeping and cooking. The Peasants' University brought together two kinds of education, that of hygiene and farming and in addition, selected topics from chemistry, physics, biology, history, political economy, mathematics, grammar, visual arts, music and poetry. Great attention was paid to human anatomy and physiology. There were also lectures in veterinary medicine, agriculture, vegetable and fruit growing and animal husbandry.[67] Such courses also offered a route for promoting abstinence since participants had to abstain from alcohol consumption.[68] The course and Peasants' University participants were recruited to promote hygiene in their villages in the absence of nurses and physicians and to assist the staff of socio-medical institutions in their field campaigns.[69]

## Health Education Posters and Hygiene Exhibitions

Social medicine also encouraged different approaches to health education, including the introduction of new methodologies, including counseling and home visits. It no longer sufficed for a physician to give patients

66 The first course lasted from 15 October 1928 until 15 March 1929; see the inscription on a photograph in the Photo Archive of the School of Public Health "Andrija Štampar" in Zagreb.

67 Poslovnik o radu u Seljačkom Sveučilištu. Record Group 6.1. Series 1.1. Box 29. Folder 345. RAC. New York.

68 Zakletva o apstinenciji od 26. II.1927, 379-27. K.1. F.517 Higijenski zavod sa Školom narodnog zdravlja. Hrvatski državni arhiv. Zagreb.

69 Hribar, Josip. Narodno zdravlje Savske banovine. *Liječnički vjesnik - Staleški glasnik* 56 (1934), 17–22.

advice; new counseling offices were organized within socio-medical institutions. Furthermore, house calls after World War I acquired an entirely different meaning. They were no longer a commercial visit by the physician to the patient but an important part of prevention work usually performed by nurses.

With the formation of a network of socio-medical institutions, the duties and roles of physicians and other medical personnel were legally regulated, especially with respect to the role of counseling, counseling centers and house calls. Regulations relating to the work of polyclinics stipulated counseling individuals and groups as well as pregnant women and mothers on the care for children.[70] Mothers' counseling centers were expected to organize frequent consultations, courses and house calls where they could assess the requirements of the family and extend help when needed, especially after childbirth.[71]

The visual form of presenting health education was an extremely important methodology, especially as the majority of the target audience was illiterate. The Ministry of Public Health played an active role in the publication of posters by helping other private and state institutions and organizations.[72] After the foundation of the Institute of Social Medicine and later the School of Public Health, these institutions took over the publication of health education posters almost entirely. Indeed, the School of Public Health became the main center of this activity,[73] with posters that were either designed in the school or translated from originals obtained from abroad, usually from the German Hygiene Museum in Dresden.

Posters were mainly used for exhibitions but were also employed to advertise such events. Furthermore, they were exhibited at public events, from charity concerts to public health campaigns for cleanliness or against

70 (Anon.), "Uredba o organizaciji i delokrugu Higijenskih zavoda, Domova narodnog zdravlja i Zdravstvenih stanica," *Narodne novine* 46 (1930): 1–2.

71 Adolf Spitzer, ed., *Pravilnik o radu dispanzera za majke, dojenčad i malu djecu od 12. XII. 1925. Zbirka zakona i naredaba tičućih se zdravstva i zdravstvene službe, dodatak XIII* (Zagreb: Tiskara Merkur, 1926), 413–414; Štefanija Vinter, "Rad i organizacija dječjeg ambulatorija od god. 1908. do 1928," *Liječnički vjesnik* 50 (1928): 758–778; Amalija Šimec, "Sestra u službi javnog zdravstva," *Liječnički vjesnik* 50 (1928): 728–729.

72 (Anon.), "Odluka o pomaganju društva koja rade na narodnom zdravlju i o nagrađivanju literarnih spisa," *Liječnički vjesnik-Staleški glasnik* 43 (1921): 267, and (Anon.), "Antialkoholni dan u školama," *Liječnički vjesnik-Staleški glasnik* 44 (1922): 108–109.

73 (Anon.), "Uredba o organizaciji i delokrugu Škole narodnog zdravlja u Zagrebu," *Narodne novine* 43, 1 (1927): 1; and (Anon.), "Uredba o organizaciji i delokrugu Škole narodnog zdravlja u Zagrebu," *Narodne novine* 46 (1930): 1–2.

Fig. 2. Health promotion poster: *If you whant to save your health,* interwar period School of Public Health, Zagreb. (© School of Public Health Andrija Štampar)

alcohol.[74] They were placed in various institutions and public spaces: schools, train stations, prisons, offices and squares in order to ensure their visibility.[75] Štampar reported that in the five years between 1920 and 1925, 1,500,000 units of various promotional material, including posters, were distributed[76] (see fig. 2).

Hygiene exhibitions, similarly, had a central place in the new health legislation. For instance, the aforementioned ordinance on the organization of the Ministry of Public Health (1919) stipulated that the Ministry's task was to collect materials for a social hygiene museum and to ensure the popu-

74 Lada Kavurić, *Hrvatski plakat do 1940* (Zagreb: Institut za povijest umjetnosti, 1999), 112; (Anon.), "Kongres trezvenosti na Cetinju i proslava Trezvenosti," *Liječnički vjesnik-Staleški glasnik* 54 (1932): 129; (Anon.), "Proslava trezvenosti," *Glasnik ministarstva narodnog zdravlja* 4 (1923): 280–281.

75 Ivan Stalio, "Venerija i njeno suzbijanje u Dalmaciji," *Glasnik ministarstva narodnog zdravlja* 3 (1921): 113–126; Pismo od 1. V 1922, 1060/1922. F.134 Zdravstveni odsjek za Hrvatsku, Slavoniju i Međimurje 1919–1924. Opći spisi. Hrvatski državni arhiv. Zagreb.

76 Andrija Štampar, "Pet godina rada na higijeni," *Glasnik ministarstva narodnog zdravlja* 6, 12 (1925): 289–291.

Fig. 3. Health exhibition, model of the health child room, interwar period, School of Public Health, Zagreb. (© School of Public Health Andrija Štampar)

larization of hygiene principles using models, pictures, traveling museums, exhibitions and so on.[77]

In the mid-1920s the Zagreb and Novi Sad inspectorates of the Ministry of Public Health were particularly active. In 1923 the Novi Sad inspectorate organized a childcare exhibition to travel across the country. In 1925 this inspectorate also put on a hygiene exhibition in Vukovar. In the same time, the Zagreb inspectorate organized an exhibition entitled *The Mother and Child*, which was also presented in many parts of the country[78] (see fig. 3).

The various medical and hygiene institutions were charged with organizing health education exhibitions.[79] The founding ordinance of the School of Public Health specified that this institution should set up and manage a hygiene exhibition and traveling show.[80] The School of Public Health estab-

77 (Anon.), "Uredba o ustrojstvu Ministarstva narodnog zdravlja od 7 lipnja 1919," *Narodne novine* 85 (1919): 1.

78 (Anon.), "Higijenske izložbe na teritoriju novosadskog inspektorata," *Glasnik Ministarstva narodnog zdravlja* 6 (1925): 491-496; Žiga Švarc, "Higijenska izložba Mati i dijete," *Liječnički vjesnik* 46 (1924): 647-650.

79 (Anon.), "Uredba o organizaciji i delokrugu Higijenskih zavoda, Domova narodnog zdravlja i Zdravstvenih stanica," *Narodne novine* 46 (1930): 1-2.

80 (Anon.), "Uredba o organizaciji i delokrugu Škole narodnog zdravlja u Zagrebu," *Narodne novine* 43 (1927): 1.

lished a museum on its premises collecting various hygiene exhibits and materials such as pictures, posters, statistical maps and so on. The museum was visited by the School's students and the general public.[81] Train carriages were also used to house exhibits, for instance an agricultural show in the early 1930s.[82]

## Health Education Films

With the end of World War I, discussions in medical circles began on the role of film in health education. In a presentation on the "Care for Children in the Kingdom of Serbs, Croats and Slovenes" at a conference on social hygiene in Paris in 1919, Štampar argued that, in principle, young people under the age of 16 should not be admitted to cinemas, except to view films explicitly allowed by the authorities. In doing so he emphasized the important role of film in the development of the youth.[83] The 1920 budget of the Ministry of Public Health shows purchases of health education films.[84] In 1923 *Glasnik Ministarstva narodnog zdravlja* wrote about the use of film for health education purposes at the Yugoslav Temperance Association's celebration of temperance, where *In the Grip of Passion* and *Children of Satan* were shown.[85] This news report was accompanied by an article discussing in depth the purpose of film in health promotion together with the criteria that such a film should fulfill. The article argued that film was an art form and that a film director should not only be an artis, but also a scientist well acquainted with the topic of the film.[86] For Štampar, film was a highly successful and persuasive health education tool.[87]

81 Pismo od 18. prosinca 1927, 11570/927. K.2. F.17 Higijenski zavod sa Školom narodnog zdravlja. Hrvatski državni arhiv. Zagreb. Pismo 14993/927. K. 2. F.517 Higijenski zavod sa Školom narodnog zdravlja. Hrvatski državni arhiv. Zagreb.

82 Mladen Širola, "Pokretna poljoprivredna izložba," *Narodni napredak* 5 (1933): 149–151.

83 Andrija Štampar, "Zaštita djece u kraljevstvu Srba, Hrvata i Slovenaca," *Liječnički vjesnik* 41 (1919): 381–383.

84 *Budžet rashoda ministarstva narodnog zdravlja i ministarstva prostornih dela* (Sarajevo: Zemaljska štamparija, 1920).

85 (Anon.), "Proslava trezvenosti," *Glasnik ministarstva narodnog zdravlja* 4 (1923): 280–281.

86 T. Fodor, "Higijensko prosvjećivanje i film," *Glasnik Ministarstva narodnog zdravlja* 6 (1925): 477–481.

87 Andrija Štampar, *Socijalne medicina uz sardnju jugoslavenskih socijalnih lekara, prvi deo* (Zagreb: Izdanje Instituta za socijalnu medicinu u Zagrebu, 1925).

From its inception in 1924, Zagreb's Institute of Social Medicine took over the acquisition, distribution and production of health education films. The institute provided the necessary facilities for both the production and editing of films.[88] Its film laboratory was equipped with the complete inventory, including the machinery, of the recently liquidated Zagreb film company *Jugoslavija ltd.* [Yugoslavia]. Initially the laboratory simply provided subtitles for imported films. However, in 1926 it came under the management of the School of Public Health, which allowed the laboratory to further develop its substantial film production. The first films translated from English, German and French and adapted to local conditions included: *Childcare, Child Play, Child Joy, Professor Pirquet's Clinic in Vienna, Wash Your Hands, Dust is Dangerous, When Disease Hunts Us, Bad Habits, Open Your Eyes, Malaria, Rabies, Infections, Raising Children in America* and so on. The first films to be produced locally had the same titles but used local actors, usually in folk costumes, and were filmed in Croatia. They explored everyday, well-known problems and were frequently filmed on location in the model village of Mraclin near Zagreb. Its homes, people and folk costumes became an inseparable part of many films. In the early films, sequences from foreign films were supplemented with those made in Yugoslavia. The first health education film produced in its entirety at the School of Public Health was *Sanatorium Topolšica* [*Topolsica Sanatorium*], in 1927. A series of films followed, including the 1929 titles *Rural Sanitation* and *In Nature*. The same year also saw the making of the first children's and thematic health education films, and in the early 1930s the first films with sound were made. In addition to documentary films, feature films played an important role as their stories communicated the health education message in a convincing, interesting and vivid way. Good examples are *The Pub, Rescue at the Right Moment* and *Sinners*.[89]

These public events aimed at communicating the principles of hygiene were not purely an interwar novelty. Events that engaged with health issues had been organized before, but new forms were more intense, frequent and diverse. These new activities took place in public spaces so as to attract

88 Elaborat o filmskom poslovanju Škole narodnog zdravlja u Zagrebu, 6. lipnja 1934, 11964. K.3. F.517. Higijenski zavod sa Školom narodnog zdravlja. Hrvatski državni arhiv – Hrvatska kinoteka. Zagreb.

89 Vjekoslav Majcen, *Filmska djelatnost Škole Narodnog Zdravlja «Andrija Štampar» (1926–1960)* (Zagreb: Hrvatski državni arhiv, 1995).

as many people as possible and in order to influence their everyday life in interesting and unusual ways. During such health education events there were lectures, exhibitions and parades, posters were exhibited, leaflets distributed and preventive physical examinations offered. The goal of these events was to stimulate public interest in health, especially with regard to the most dominant diseases and hygiene, and also to encourage the public to take up athletics, exercise, reading, participation in singing societies and so on.[90]

The so-called children's exhibitions were particularly interesting. In 1922 a highly successful exhibition of babies and toddlers took place in Lazarevac, Serbia. Its goal was to promote childcare issues. The success of this exhibition encouraged the organization of similar events elsewhere. The best groomed child received a money prize and other children were given small gifts such as clothes, stockings and so on.[91] These events were recorded in the films of the School of Public Health, such as in *The Child Beauty Pageant* from 1928.[92]

## Sanitary Engineering: Demonstrating Practical Hygiene to the Public

Sanitary engineering was the public health method that, along with health education, had the greatest impact in interwar Yugoslavia. The majority of sanitary engineering projects concerned the construction of water supply facilities: wells, water tanks and animal watering stations, hygiene facilities such as public baths, public and private toilets, dunghills, houses, stalls and silos and various public buildings and places such as schools, sanatoria, community halls, cemeteries, parks, playgrounds and dairies. Sanitary engineering was therefore inextricably linked with health education and their symbiosis resulted in an entirely new approach. This may be summarized as "community work toward the solution of health problems" or, so the *Regulations of the Educational Courses for Physicians*, a "demonstration of practical

90 (Anon.), "Program povjerenstva za suzbijanje tuberkuloze," *Liječnički vjesnik* 41 Suppl 10 (1919): 1–11.

91 Rukopis zapisa o izložbi djece, 7442 . F.134 Zdravstveni odsjek za Hrvatsku, Slavoniju i Međimurje 1919-1924. Opći spisi. Hrvatski državni arhiv. Zagreb.

92 Majcen, *Filmska djelatnost,* 36.

hygiene examples to the public."[93] This method represented a move toward more complex ways of working that involved cooperation between participants. It necessitated a wide spectrum of well-trained personnel, such as sanitary engineers and technicians as well as members of local communities.[94]

Community work towards solving health problems may be divided into the following segments: the study of the needs and abilities of the local population, preparation of suitable promotional material, sensitization and education of the population in the field, activation of the local community towards solving the problem, implementation of the planned action and demonstration of its use. Health education thus permeated sanitary engineering at different levels, both during the planning and building of hygiene facilities in the field. 'Preparation of the ground' belonged to the first phase of the project and was conducted with the help of community members connected in some way to the School of Public Health, such as, for instance, former students of the Peasants' University. These individuals were expected to set a good example for their peers in order to help their perception of health and hygiene problems and also open them up to new knowledge. It was important to gain not just compliance but also the interest of the local community, which had to realize the benefit of the proposed project.[95]

Hygiene facilities—latrines, dunghills, wells, sewage and water supply lines—were of particular importance. These projects were used to communicate knowledge about personal and public hygiene, infectious diseases, disease vectors and routes of infection, the importance of hygiene infrastructure for farms and the benefits of rational hygiene. Innovative methods such as lectures, slide shows and films, leaflets and booklets were used. Hygiene facilities were presented using standard floor plans together with three-dimensional models, produced in the Technical Services Department of the School of Public Health in Zagreb.[96] The goal was to show

93 (Anon.), "Pravilnik o kursevima za lekare," *Narodne novine* 46 (1930): 1.

94 Petrik Milivoj, "Sanitarno tehnički rad Škole narodnog zdravlja," *VPS Časopis za vodnu, plinsku, sanitarnu i municipalnu tehniku* 4 (1938): 7–31.

95 Andrija Štampar, "Škola narodnog zdravlja u Zagrebu, njezina povijest i sadašnji položaj," *Zdravstvene novine* 10 (1957): 35–37.

96 Prijedlog za budžet tehničkog odjeljenja Škole narodnog zdravlja za godinu 1933/34, 895/33-T.O. K. 1932, 1933, 1934. Opći spisi tehničko odjeljenje. Dokumentacija Škole narodnog zdravlja «Andrija Štampar». Zagreb.; Prijedlog budžeta III odjeljenja za god. 1936/1937, 790/T.O. K. 1936. Opći spisi tehničko odjeljenje. Dokumentacija Škole narodnog zdravlja «Andrija Štampar». Zagreb.

the audience what a good hygiene facility should look like. These floor plans and models were then used for the actual construction work. It was believed that using the pedagogic method of copying a model would result in best practice. The staff of the school not only distributed floor plans and built facilities according to them, they also encouraged local builders to use them in their projects.[97] Outhouses were, for instance, built based on the *Kentucky Model* promoted and used by the Rockefeller Foundation in their programs worldwide and principally in the United States.[98] After the completion of the project, knowledge of its use and utility could be put to the practical test.

Importantly, the School tried to expand the scope of its activities in rural communities, even when the staff of the School was not present in the community. That is, it tried initiate campaigns for increasing the standard of hygiene even in those areas where public health experts were not active. The School encouraged peasants to build hygiene facilities themselves, in the first place outhouses, dunghills and baths. The Ordinances of the School envisaged financial awards for peasants who built hygiene facilities independently according to the instructions of the School. These awards were paid from the School budget every year.[99]

Štampar emphasized the importance of the close contact between health education workers and the rural population and of their joint work on specific practical problems. In the article "Observation on the advancement of health in rural areas," a version of the lecture delivered to Harvard students in 1938, he wrote:

> The best way to conduct health education in rural areas is to work on something from which peasants can directly benefit. That is how it is done in Yugoslavia. The School of Public Health and Hygiene Institutes are the centers of health work in villages. They advise peasants in matters of sanitary technology,

97 Kolarić Kišur, "Deset godina asanacije sela," *VPS Časopis za vodnu, plinsku, sanitarnu i municipalnu tehniku* 4 (1938): 32–82.

98 Korespodencija o sudjelovanje na zemaljskoj higijenskoj izložbi, 1933, 113-Z-I-1933, 1591/33T.O, 4378/II, 2. XII 1933, Dopis iz Odjeljenja za propagandu upućen u odjeljenje za sanitarnu tehniku, 2.XII 1933, 4378/II, Dopis o nužnosti uređenja zbirke u Tehničkom odjeljenju, 23. XII 1933, D3326TO, Dokumentacija Škole narodnog zdravlja "Andrija Štampar". Zagreb.

99 (Anon.), "Pravilnik Škole narodnog zdravlja u Zagrebu o podjeljivanju nagrada za higijenske nužnike i đubrišta," *Narodni napredak* 3 (1931): 28–31; Dopis od 25. V. 1934, 1079/T.O. K. 1932, 1933, 1934. Opći spisi tehničko odjeljenje. Dokumentacija Škole narodnog zdravlja «Andrija Štampar». Zagreb.

> work with them, build wells, latrines and dunghills and fix houses. Their work, naturally, incites interest among the rural population and inspires them to achieve better health. In winter, farmers take courses in general education and especially in agriculture, veterinary medicine, animal husbandry and health. This method achieves better results than exhibitions and lectures alone.[100]

Putting socio-medical programs into action required close knowledge of the state of affairs in the field; the perceptive abilities of the population, literacy levels and the general lack of education that made the understanding of basic scientific and medical concepts difficult. In addition to the adherence to traditional customs, nationalism presented an issue that had to be taken into account. For instance, some health-related topics were never discussed within traditional culture, such as venereal diseases or genital hygiene.[101] Lectures on venereal diseases thus required a great deal of delicacy. They were rarely advertised as "Lectures on Venereal Diseases," as such titles attracted small audiences and, especially, few women.[102] Most of these lectures were organized separately for women and men.

The content of health education materials occasionally failed to reach their intended audiences. For instance, although posters had initially been seen as promising, easily comprehensible forms of health promotion, they were proven to be unsuccessful since illiterate peasants failed to decode their pictorial messages. *Glasnik Ministarstva narodnog zdravlja* wrote: "In contrast to our urban population, our farmers still lack the mental maturity for looking at posters." Furthermore, it was argued that peasants could not grasp the deeper meaning of the poster: "The peasant sees the poster, nods and does not stop to think about it, they only see a "snake," a "skull," a "woman with a child," "some insects" and nothing more. This is how they explain it!" The author of this text, furthermore, wrote that peasants could conclude nothing from images and that they distorted and misunderstood simple messages. For instance, a poster slogan such as "Brandy kills," could be understood by peasants as excluding other beverages, which could therefore be drunk with impunity.

100 Andrija Štampar, "Opažanja jednog radnika na unapređenju zdravlja na selu," *Liječnički vjesnik* 61 (1939): 1–5.
101 Jovan Barać, "Borba protiv veneričnih bolesti," *Liječnički vjesnik* 41 (1919): 419–422.
102 Ervin Meixner, "Nešto o suzbijanju spolnih bolesti," *Liječnički vjesnik* 42 (1920): 319–322.

Problems related to the cognitive abilities of the health education message audience were augmented by other obstacles encountered by health education workers. Their activities were often badly received by local authorities who were supposed to help public health projects. Priests, for instance, saw these campaigns as the "wicked activities of liberal gentlemen."[103] Temperance campaigns conflicted with local interests. The report on the celebration of temperance organized by the Yugoslav Temperance Association in 1923 stated that "It was impossible to organize a temperance event in public spaces. The population of Dalmatia is starving and has no money. The temperance movement is seen with suspicion because it might stop the export of wine, one of the few sources of income in Dalmatia."[104]

Health educators frequently encountered problems caused by local interests. A 1927 letter from a field nurse to the School of Public Health speaks vividly of the need to build trust as a prerequisite of successful public health work: "I was ordered to organize a hygiene course for women. But as the law does not stipulate attendance of the course and as the rural population does not understand its importance; thus it was very hard to do. By working day-by-day in the village, I have acquired the peasants' trust and they have allowed their daughters to enroll in the course."[105] Poverty was a major problem in health education projects. For instance, the students of home economics courses had to bring food to use in practical work, presenting a substantial expense for rural households. For that reason, cooking classes used only those foods that were easily available, such as flour, potatoes, beans and cabbage. Teaching women to cook more complicated dishes was avoided as poverty made the ingredients inaccessible.[106]

Attending courses, and especially of the Peasants' University in Zagreb, was beyond the financial means of many peasants. Their studies were in some cases paid for by associations such as the "League for the Control of Tuberculosis." This association financed several girls who took courses at

---

103 Ivan Petković, "Zdravstvene propaganda u Dalmaciji," *Glasnik ministarstva narodnog zdravlja* 3 (1921): 313–317.

104 (Anon.), "Proslava trezvenosti," *Glasnik ministarstva narodnog zdravlja* 4 (1923): 280–281.

105 Pismo od 25. I. 1927, 239/27. K.1. F.517. Higijenski zavod sa Školom narodnog zdravlja. Hrvatski državni arhiv. Zagreb.

106 Pismo od 6. II. 1927, 246/27. K1. F517 Higijenski zavod sa Školom narodnog zdravlja. Hrvatski državni arhiv. Zagreb.; Program rada "Materinske škole" u Sarajevu, 218/1927. K.1. F.517 Higijenski zavod sa Školom narodnog zdravlja. Hrvatski državni arhiv. Zagreb.

the School of Public Health.[107] Lecture advertisements had to stress that there was no fee as the fear of such a fee would have prevented many from attending.[108]

An evaluation of the success of home economics courses revealed that of those women who had attended courses in 1927, 17 showed no improvement in the management of their kitchens and households. Furthermore, six women could not reform their households because of the opposition of older female household members. Thus, for a total of 23 students a lack of betterment was recorded, while the remaining 41 were more or less successful.[109] Popular magazines frequently wrote about the opposition of other household members to new ways of managing households and to hygiene at home. Women were encouraged to disregard the opposition in their households, especially that of older family members.[110]

## A Conflict of Ideas

Andrija Štampar and his colleagues Berislav Borčić, Ivo Kuhn, Milivoj Petrik, Milivoj Rankov and Ivo Pirc created and managed the described public health programs (see fig. 4). Štampar attentively observed everything that took place in the field, made personal rounds of all the new institutions, never stayed in his office and was permanently in the field. He phoned his employees' offices regularly to check if they were at their work places. The staff of new institutions was positively indoctrinated with the new ideology, condensed in the aforementioned manifesto. Like a prophet, Štampar preached "our ideology," allowing his deep belief to shape all of his attitudes; he expected, almost fanatically, the same from his team. On one hand he was a dictator who created rules his colleagues had to obey, on the other hand he gave them freedom to do whatever they thought best

107 Hribar Josip, "Narodno zdravlje Savske banovine," *Liječnički vjesnik- Staleški glasnik* 56 (1934): 17–22.

108 Pismo od 16.XII. 1927, 1117/27. K2. F517 Higijenski zavod sa Školom narodnog zdravlja. Hrvatski državni arhiv. Zagreb.

109 Drago Chloupek, "O poboljšanju narodne ishrane," *Sestrinska riječ* 1 (1933): 16–21.

110 About popular health education see more in Željko Dugac, *Kako biti čist i zdrav: Zdravstveno prosvejćivanje u međuratnoj Hrvatskoj*, (Zagreb: Srednja Europa, 2010), currently in press.; and Dugac Željko, "Popular Health Education and Venereal Diseases in Croatia between Two Worl Wars," *Croatian Medical Journal* 45 (2004): 490–498.

Fig. 4. Andrija Štampar and Berislav Borčić, cc 1927. (© Division for the History of Medicine, Croatian Academy of Sciences and Arts)

within the prescribed frame. Thus, giving him all the credit for the positive results is fair because Štampar ran the entire project. From his colleagues he expected, as was explained by his student and junior colleague Branko Cvjetanović, "imagination and courage; imagination to solve the encountered problems and courage not to flee from unexpected difficulties."[111]

Public health in interwar Yugoslavia experienced many ups and downs. Štampar's time was certainly a time of greatest success. Problems and conflicts within the medical profession influenced the course of public health throughout this period. Nonetheless, it was not the medical profession but political instability and growing Serb domination over other ethnic groups in Yugoslavia that had the greatest impact on public health in this artificial conglomerate state. Politics intervened in all segments of social life, including public health. Its impact must be taken into account when analyzing public health in interwar Yugoslavia.

This region saw a collision of the idea of national self-sufficiency as promoted by the regime and the idea of broader international collaboration advocated by international public health institutions. Both sides

111 Cvjetanović, Branko. Štampar i svijet, njegovo međuratno djelovanje, rukopis, Dokumentacija Škole narodnog zdravlja "Andrija Štampar." Zagreb.

argued for unification, yet each differently. The regime wanted unification and centralization under the national symbols of the ruling Serb apparatus that negated the distinctiveness of other ethnic groups and aimed at integrating them all into a greater Yugoslav—in reality, Serb—nation. The ideas promoted by Andrija Štampar favored unification and centralization within medicine alone, namely the centralized supervision of activities that were autonomous in the field and readily adaptable to local circumstances. National boundaries were to be removed not by erasing ethnic groups, but by removing obstacles between them and by establishing cooperation without prejudice or distance. These ideas were supported by the League of Nations Health Organization, whose renowned activities to stop the spread of typhus from the Soviet Union in the 1920s mobilized many national governments and initiated cooperation and information exchange across state borders. Štampar's ideology of 'broad unification' appealed to the Belgrade regime in the 1920s because, in a way, it corresponded to their mode of activity aimed at an increasing level of unification. Nevertheless, it soon became apparent that these two ideologies had very different goals. Consequently, the good results achieved by new public health programs in the field—which in the beginning had contributed to the increased authority of the new state—no longer appealed to the regime. The opening of the School of Public Health in Zagreb instead of Belgrade, the conflict between Andrija Štampar and the Ministry of Agriculture,[112] the conflict between Štampar and health cooperatives [*Zdravstvene Zadruge*] in Serbia[113] and, finally, the attempt of the powerful Belgrade medical clan headed by Ste-

[112] Andrija Štampar came in to conflict with the Ministry of Agriculture because he believed that the Ministry should do some of the work that was relevant to public health, for instance the replacement of traditional wooden ploughs with those made of steel and the organization of field veterinary services. The Ministry of Agriculture was fairly inert, so Štampar criticized it to a great extent. His activities aimed at improving rural farms caused displeasure at the Ministry of Agriculture.

[113] Štampar believed that these cooperatives could offer more and better health services and that other health institutions, such as those in other parts of the country and within the remit of the relevant Ministry, could co-exist with them. The cooperatives, in contrast, argued that they were neglected by the Ministry and that Štampar intentionally ignored their needs. The crux of the conflict were the local strongmen who saw the new institutions organized by Štampar as competition. They furthermore feared Štampar's authority over their domains and were possibly also worried about financial losses since the cooperatives were financed by overseas foundations such as the *Milbank Memorial Fund* and *Serbian Child Welfare Association of America*. The cooperatives and the conflict with powerful Serbian physicians such as Stevan Ivanić—who after Štampar's departure became the most powerful man in the national public health system and who was close to the cooperatives—were the main reasons for Štampar's dismissal. On the conflict with the cooperatives, see Dugac, *Protiv bolesti* and Dugac, Pećina, ed., *The Diaries of Andrija Štampar*.

van Ivanić to occupy Štampar's position resulted in Štampar's dismissal and, consequently, the end of a highly productive era in South Slavic public health.

## Conclusion

The building of a public health infrastructure in the Kingdom of Serbs, Croats and Slovenes (later Yugoslavia) started after World War I under the leadership of Andrija Štampar with the goal of developing a network of new socio-medical institutions. Their aim, in turn, was preventative medicine, protection of certain more sensitive population groups and the systematic fighting of disease. The process of institutionalization was paralleled by a process of professionalization in which people were trained in previously unknown public health disciplines.

These two processes were tightly linked with a process of intense international cooperation. Štampar and his colleagues worked closely with the Rockefeller Foundation, which made possible the building and equipping of many important institutions, such as the Zagreb School of Public Health, the Belgrade Central Hygiene Institute and others. They also helped start programs for the study of diseases, the collection of vital statistics, the laboratory production of medicines and vaccines, and above all, to encourage and advance public health instruction at universities, in nursing schools or through scholarships for training abroad.

In addition to the Rockefeller Foundation, Štampar developed a useful cooperation with the League of Nations Health Organization, focused in the first place on the knowledge exchange that included an exchange of scholars. Another aspect of cooperation was participation in the meetings and conferences of the League of Nations Health Organization, for instance meetings of directors of the European schools of public health, as well as joint work on projects like international conferences, such as the one on rural hygiene.

The methods used to solve public health problems were diverse, but popular health education and sanitary engineering were the most popular. Both methods were of particular importance in campaigns against acute infectious and social diseases that in the interwar period presented the greatest health problems in poverty-stricken, rural Yugoslavia. These methods produced good results, particularly as they evolved further with the persistent work of public health professionals. Their success together with their acceptance by rural communities were based on the adaptation of the work to the needs and perceptive abilities of the local users. Inhabitants of rural regions evolved from perceived objects to active participants in projects the purpose and benefit of which they could understand. Thus, one-dimensional health information was transformed into multi-dimensional health education that not only imparted knowledge but also shaped habits, attitudes and patterns of behavior.

The ideology of the new health campaigns was acceptable to the local, mostly poor population because it argued that the benefits of science and medicine should be accessible to everyone, not just the rich. The provision of free health care, advice and help when building hygiene facilities was something new, hitherto never experienced by the poor rural and suburban people who now received attention from the state. It is perhaps no wonder that Štampar's ideas and of his colleagues took such deep root and continued to blossom, even when, in the 1930s, the Ministry changed its politics and dismissed Štampar.

# PART III
# EUGENICS AND REPRODUCTION

# MARITAL HEALTH AND EUGENICS IN BULGARIA
## 1878–1940

*Gergana Mircheva*

"Crisis" might be the best word to describe the cultural situation in Bulgaria between the two World Wars. Experienced as a collective identity crisis questioning peacetime values of social, economic and political discourses, it was conceptualized mainly in nationalistic terms. However, in interwar cultural reflections the negativity of economic and social decay was treated as an incentive for recovery and creative identity reinforcement. Public visions of a wholesome national regeneration were formulated in utopian constructions "populated" with organic images of *Volk* and State. In the 1920s and especially the 1930s, eugenic ideas gathered strength in Bulgaria, based on the widely shared conviction that "the Balkan states must justify the aim of their independent existence by ensuring the peoples' health and the advancement of their spiritual culture."[1] In other words, "[i]n order to survive, our small people surrounded by bigger neighbors must not only grow in number by means of reasonable reproduction, but should be comprised of healthy, strong and resistant individuals."[2]

The complex background of the eugenics discourse in Bulgaria was produced within processes of adoption and emancipation of Western scientific models[3] from the late nineteenth until the mid-twentieth

1 Konstantin Pashev, "Borbata protiv slepotata i natsionalniya solidarizam," *Oftalmologichen pregled* 2, 3 (1934): 46; all translations from the original are my own.

2 Stefan Danadzhiev, *Printsipite na psihicheskata higiena* (Sofia: Pechatnitsa Hudozhnik, 1935), 213.

3 The emergence of eugenics in Bulgaria was facilitated by certain organic utopias within Bulgarian culture. The nineteenth century's culture of the so-called "Bulgarian national revival" used the metaphorical representa-

century.[4] Projects for their institutionalization could be perceived as a radical element within the system of "biopower"[5] that the modern state exercized in politics and techniques for the "naturalization" of socio-historical phenomena. In a paradoxical way, the discourse of eugenics blended conservative anti-liberal attitudes with modern aspirations of social engineering.[6] The imaginary transformation of the "nation" into an "organism" together with its eugenic "transfiguration" were laid out in the utopian horizon of "the Bulgarian right-wing project" after World War I.[7] Its substance was provided by the concept of "native" [*rodno*] as taken from the pre-modern collective value system and successfully adapted within the early modern period to be re-integrated into the cultural political framework of the interwar years. In the 1920s and 1930s the value of "native" as "national" expressed in organic images inspired critical social moods, a readiness to violate human rights and anti-liberal attitudes hardened in a teleological stance toward the mastering of a national future.[8] As an expression of this will for social catharsis, eugenic projects represented a striving toward a "normalization" of the crisis. Where its symptom, "degeneration", was defined as an "acute social disease,"[9] and health as the most precious

tions of the body as a central rhetorical instrument, as a "generator of figurative speech" to describe the rise of nation-building processes and the construction of modern collective identity. See Ina Peleva, *Botev. Tyaloto na natsionalizma* (Sofia: izdatelstvo "Kralitsa Mab," 1998). This nationalistic dispositive, the "body of nationalism" as Peleva called it, was adopted and further developed by the Bulgarian modernist culture at the beginning of the twentieth century. See Boyko Penchev, *Tagite na kraevekovieto* (Sofia: Literaturen vestnik, 1998), 107–122.

4 The chronological framework of my study encompasses the period between the establishment of the modern Bulgarian state between the end of the Russo-Turkish war in 1878 and the start of World War Two. The first, more considerable attempts to introduce eugenic concepts and practices in the public sphere took place at the beginning of the twentieth century. It was in the 1920s and 1930s, however, that the eugenic discourse in Bulgaria became more socially influential.

5 Michel Foucault, *The History of Sexuality*, vol. 1: *The Will to Knowledge* (London: Penguin, 1998), 140.

6 Paul Weindling discusses the inherent cultural contradictions of the Weimar period in Germany with regard to eugenics: "Whereas conservatives idealized the remnants of the traditional social order, the artists and social critics attacked bourgeois conventions as pathological and diseased. Experts in social hygiene and eugenics attempted to reconcile these antitheses on the bases of social health." See Paul Weindling, *Health, Race, and German Politics between National Unification and Nazism* (Cambridge: Cambridge University Press, 1993), 400–401.

7 See Ivan Elenkov, *Rodno i dyasno. Prinos kam istoriyata na nesbadnatiya "desen proekt" v Balgariya ot vremeto mezhdu dvete svetovni voyni* (Sofia: LIK, 1998). On the interwar right-wing cultural visions see also Ivan Elenkov and Rumen Daskalov, eds., *Zashto sme takiva? V tarsene na balgarskata kulturna identichnost*, (Sofia: Svetlostruy, 1994).

8 Elenkov, *Rodno i dyasno*, 58.

9 Petar Monev, "Izrazhdane, kultura i obrazovanie," *Mediko-pedagogichesko spisanie za zdravno-sotsialni i vazpitatelni grizhi za deteto i yunoshata* 5, 1 (1938): 26.

capital of the organic people-state,[10] the victory of national regeneration through a bio-political restoration of the social environment to health was envisaged. Degeneration could be reversed, marking a utopian return to the primary condition of the Bulgarian race.[11] As Zygmunt Bauman argued "[g]ardening and medicine supplied the archetypes of constructive stance, while normality, health, or sanitation offered the archmetaphors for human tasks and strategies in the management of human affairs."[12] The concepts of "degeneration," "heredity" and "social danger" were misused as interrelated formulae for coping with individuals who were deemed "unfit" for the ideal type of "national organism."

In 1926, on the initiative of the Bulgarian zoologist Stefan Konsulov (1885–1954), the "Society for the Study of Racial Hygiene" [*Obshtestvo za rasovo-higienichni prouchvaniya*] was established in Sofia.[13] This informal intellectual circle was transformed into the "Bulgarian Society for Racial Hygiene" [*Balgarsko druzhestvo za rasova higiena*] in 1928.[14] There is almost no historical evidence on the activities of these two short-lived doctrinal societies apart from data on a small series of lectures and publications. This is to a certain extent true for the reformed "Bulgarian Society for Eugenics (Racial Hygiene)" [*Balgarsko druzhestvo za evgenika (rasova higiena)*], established in 1934/5 as well. It, however, had more members,[15]

10 Petar Tsonev, "Zdravno vazpitanie," *Mediko-pedagogichesko spisanie* 3, 1 (1936): 13–17.

11 The concept of race was used by most Bulgarian eugenicists in the wider meaning of a "biological community." The racial anthropological theory of the prominent Bulgarian biologist Metodi Popov (1881-1954) was widely shared by them. Popov denied the existence of pure races and claimed that all European nations had originated from the mixing of five main races. The specific ratio of the different racial elements formed the European national bodies and determined their national physiognomies. See Metodi Popov, *Nasledstvenost, rasa i narod. Rasova prinadlezhnost na balgarite* (Sofia: Pridvorna pechatnitsa, 1938). In eugenic usage the concept of race was more or less fluid, for it functioned as a synonym for "people" or "nation." "Race" signified the biological substrate or quintessence of the Bulgarian people.

12 Zygmunt Bauman, "Modernity, Racism, Extermination," in Les Back and John Solomos, eds., *Theories of Race and Racism: A Reader* (London: Routledge, 2000), 221.

13 It included the Civil Servants Insurance Company's physician Damyan Ivanov, the psychiatrist Lyubomir Rusev, the economist Petar Penchev, who was head of the Economic Department at the Central Statistics Office, and the sociologist Ivan Kinkel, associate professor at the Faculty of Law, Sofia University.

14 Its members also included the prominent Bulgarian histologist and cytologist Asen Hadzhiolov.

15 Members of the society included the anthropologist and pediatrician Stefan Vatev, Nikola Saranov, professor in criminal proceedings, and the famous ophthalmologist Konstantin Pashev, academic since 1938, among others. For more details on the history of the Bulgarian eugenics societies see: N. N., "Obshtestvo za rasovo-higienichni prouchvania," *Priroda* 25, 9 (1926): 142; N. N. "Bulgarsko D-stvo za evgenika," *Narod i potomstvo* 1, 1 (1935): 3; Christian Promitzer, "Taking Care of the National Body: Eugenic Visions in Interwar Bulgaria (1905–1944)," in Marius Turda and Paul Weindling, eds., *Blood and Homeland: Eugenics and Racial Nationalism in Central and Southeast Europe, 1900–1940* (Budapest: Central European University Press, 2007), 223–252.

and a newspaper of its own, "People and Progeny" [*Narod i potomstvo*]. Although eugenics in Bulgaria was not grounded in influential scientific societies, some of its advocates held eminent and authoritative positions within the state institutional system.

There also existed followers of eugenic ideas, who were not members of eugenic societies. One of these was Stefan Danadzhiev (1866–1943), one of the founders of Bulgarian psychoneurology, who had studied medicine in Vienna (1893) to later specialize in psychiatry there (1893–1894). From 1901–24 Danadzhiev was head of the neuropsychiatric ward at Sofia's Alexandrovska Hospital.[16] Some of the participants in the eugenic debates were members of an organization named the "Society for Hygiene and Preventive Medicine" [*Druzhestvo za higiena i predpazna meditsina*], which was active in the 1930s and 1940s and was established and presided over by the founder of the Hygiene Department at the Bulgarian Medical Faculty, Dr. Toshko Petrov (1872–1942).[17] The members of the society were not self-claimed "eugenicists," but in a moderate way they actually supported ideas and practices relevant to eugenics where it overlapped with social hygiene and preventive medicine.

A basic methodological premise of this text is that a thorough study of eugenics in the Bulgarian context requires the reconstruction of both direct and indirect contributions to its discursive potential, in the way it was accumulated into a shared field of strategic options. This field was structured not solely through the ideological self-determination of experts as "eugenicists" or "racial hygienists," but also through implicit mutual references

Promitzer's chapter is the first profound study on the history of Bulgarian eugenics societies published to date. It examines the ideas of some of the most distinguished advocates of eugenics and racial science in Bulgaria, the dispute about racism at the end of 1930s and the 1943 "Law on Families with Many Children."

16 In 1921–22 and 1935–36 Danadzhiev was a secretary at the Natural Sciences and Mathematics Section of the Bulgarian Academy of Sciences and its chairman in 1928–29. He was also a member of the "International League against Epilepsy." See Ivan Petrovski, *Savremennoto Balgarsko zdraveopazvane ot 1877 g. do nashi dni* (Sofia: IK "DNK", 2001), 299–300. His sympathy for the right-wing was declared in his book *Printsipite na psihicheskata higiena* [The principles of psychic hygiene], where he wrote about the "praiseworthy initiatives" of the new political regime since 1934. See Stefan Danadzhiev, *Printsipite na psihicheskata higiena*, 37.

17 Toshko Petrov was also a founder of the Hygiene Institute at the Central Directorate of National Health and its director from 1910 to 1919. He graduated in medicine from Geneva (1894). In 1897 he received his PhD in medicine at the medical institute of Lyon (France) and in 1905–1936 he specialized in hygiene in Toulouse and Paris. He was a member of the French Société de Médicine Publique and of the Société d'Hygiéne Publique, Industrielle et Sociale. See Mihail Arnaudov, ed., *Almanah na Sofiyskiya universitet Sv. Kliment Ohridski. Zhivotopisni i knigopisni svedeniya za prepodavatelite. Vtoro izdanie. Za petdesetgodishninata na universiteta 1888–1939* (Sofia: Pridvorna pechatnitsa, 1940), 467.

between conflicting positions and through the inclusion of common topics in different discursive settings. Thus, the eugenics discourse in Bulgaria was represented explicitly—within the projects of its proponents, as well as "in partnership"—at its points of intersection with social hygiene, preventive medicine and physical, criminal and racial anthropology.

Here I trace some aspects of Bulgarian "(pre)marriage hygiene," with a special focus on eugenic projects for the "hygienization" and qualitative improvement of the Bulgarian population.[18] The recommendations to introduce sterilization of hereditarily "inferior" individuals formed the "hard core" of Bulgarian eugenic discourse from the middle of the 1920s onwards.[19] These radical recommendations, however, did not gain public support and were not implemented. Instead, marriage restrictions, prenuptial certificates and marital health propaganda were promoted by most eugenicists as "softer" alternatives to sterilization. They were suggested as more urgent measures that were also more adaptable to the Bulgarian cultural situation. Indeed, marriage hygiene based on pre-marriage medical examinations was also regarded as difficult to put into practice because of deficits in eugenic knowledge among the general public. And yet, certain eugenic measures were formulated into legislative initiatives and some were given initial legal consideration.

In order to outline the characteristics of marriage hygiene in Bulgaria, first, it is necessary to reconstruct the legal regulation of illnesses as impediments to marriage and, respectively, as a reason for its termination under Church law and court practice. In the second part of the chapter I discuss some projects for amending this legal regime of "marital health" that were proposed by eugenicists and social hygienists. Furthermore, I examine the forms of preventive medical knowledge that were institutionalized by the 1929 "Law on National Health" and the 1936 "Decree-Law on Fighting Venereal Diseases." A central aim of this chapter is to outline the cultural patterns on which the discussed legal acts were based. The analysis focuses on how

18 The present chapter refers to some of the problems raised by the above-cited text by Christian Promitzer but focuses mostly on the influence exerted by the eugenics discourse on the cultural and legal framework of marital health in Bulgaria.

19 The Bulgarian discussions over the prospective sterilization of individuals with supposed hereditary defects will not be subject of detailed reconstruction here since this topic requires a separate study. On the issue of sterilization debates in Bulgaria see Gergana Mircheva, "Bulgarskiyat evgenichen proekt ot 20-te i 30-te godini na minaliya vek i normativniat kod na 'rodnoto'," *Kritika i humanizam* 17, 1 (2004): 207–221.

these patterns were negotiated in a process of competition, harmonization and (im)possible compromise between Church doctrine, state policies and expert medical (including eugenic) arguments. Last, but not least, the specifics of the eugenic rhetoric of national "degeneration" and "regeneration" are interpreted here as providing links between the concepts of collective health and the (re)construction of modern Bulgarian national identity.

My research interest in the legal state of eugenic initiatives is conditioned by one more factor—the status of the law in the framework of modernization in Bulgaria. Bulgarian law after 1878 was legitimated not as a product of civil consensus laying the limits of personal freedom, but as an absolute and unconditional prescription of state power. Since the Bulgarian state functioned as the central agent of modernization, law was charged with projective utopian energy, produced in the absence of the classical liberal subject as an agent of legislative reasoning.[20] That is why the legal constructions of modern life often had no practical value and many laws "suffered" from insufficient implementation. In this gap between state law and socio-cultural reality, eugenicists took the position of self-proclaimed mediators between the nation and the state. The nation, they argued, had to be rejuvenated and brought to its normative status through national eugenic enlightenment. The state, as a proprietor of the legislative discourse, had to support this transformative action. Moreover, state intervention was called to impose by law the imperatives of the eugenic mission on those individuals who would not voluntarily accept them. Thus the nation's health, progress and happiness were envisioned as a goal to be achieved in the synthesis of technocratic practices, national integration theory and the instruments of state biopower.

20 The reasons lie in the specifics of Bulgarian liberalism, which developed as a populist ideology instead of having the individual as its ideological centre. Here I share Ivan Elenkov's ideas, *Rodno i dyasno*, 13–30, that during the parliamentary debate on the new Bulgarian Constitution in 1879, two strategies of legitimizing modern political power crystallized. In the concepts of the Conservative Party, which drafted the constitution, the law was promoted as the central instrument for constructing a modern public sphere. However, from this perspective, the authority of law did not derive from the mutual consent of the individuals. Law was evaluated as a final act with binding force, that is, as a "self-sufficient" source of modern life. Hence the state, which enacted the laws, was invested with ultimate power to initiate modernization initiatives. In the argumentation of the Bulgarian Liberal Party, on the other hand, it was not the state but the people (*narodat*) that was conceptualized as the supreme authority. From this point of view, the concept of the people functioned as the most valuable ethno-national sign of collective identity. In this constitutional debate, individuals were interpreted in communal and egalitarian terms. As Elenkov argued, the rhetoric of this discussion was adopted and further developed in later projects aimed at the creation of modern society in Bulgaria.

## The Church, the Law and the Eugenic Discourse on the Issues of Healthy Marriage: The Regime of Illness under the 'Statute of the Exarchate'

During the whole period under review here, the regime of conclusion and termination of marriage lay within the competence of the Church. In 1883, Parliament adopted the "Statute of the Exarchate Adapted for the Principality"[21] elaborated on ecclesiastical grounds together with an Annex[22] regulating family law matters. The matrimonial jurisdiction of the Bulgarian Exarchate, which was divided into *eparchies*,[23] was distributed between priests, who concluded marriages, the Eparchy Ecclesiastical Councils (*eparchiyski duhovni saveti*), which acted as a first instance in cases of marriage termination, and the Holy Synod as a final appellate.[24]

The last of the listed "requirements that should be kept in order to conclude a lawful marriage" under the "Statute of the Exarchate" was that "the young man and woman must not suffer from insanity or epilepsy" (Art. 1, para d). "If one or both of the spouses have suffered from incurable insanity or epilepsy before marriage, it may be dissolved and those who are to be blamed for this shall be prosecuted" (Art. 6). If an impediment on the grounds of mental illness appeared after the marriage was concluded and "it is proven that the illness has not preceded the marriage, it shall not be dissolved, except in extreme cases" (Art. 7). "Dissolution" of marriage on the grounds of Art. 6 was possible only if the "healthy husband" claimed it. The 1883 "Statute of the Exarchate" did not make a strict legal differentiation between termination of marriage on grounds of its illegal conclusion (annulment) or by reason of divorce.[25] The dissolution of marriage, even

21 See "Ekzarhiyskiy ustav, prisposoben v Kniyazhestvoto," *Darzhaven vestnik* 5, 19 (1883): 155–156; *Darzhaven vestnik* 5, 20 (1883): 158–159; and *Darzhaven vestnik* 5, 21 (1883): 166–171.

22 See "Dopalnenie kam Ekzarhiyskiy ustav, prisposoben v Knyazhestvoto," *Darzhaven vestnik* 5, 22 (1883): 174–178.

23 *Eparchies* were the administrative-territorial units of the *Exarchate*.

24 The Synod was the supreme canonical authority in the Principality and comprised the bishops under the presidency of the Patriarch.

25 The common sanction of "dissolution" covered cases of illegal (voidable) marriage due to pre-marital "spiritual incapacity" as well as divorce cases due to insanity or epilepsy appearing after the consecration of marriage. In the latter case a legal union could be canceled and the effects of the termination would take place in the future. In the former case, annulment is a stronger sanction having retroactive action.

in the first case, was only a non-binding option. Moreover, it was no longer possible if the insanity or epilepsy was healed, which "healed" the marriage itself. Therefore, at the end of the nineteenth century, Christian values of matrimony, as a sacred union, prevailed over any other considerations of "family health."

In 1895 "feeble-mindedness (idiotism)" was added to the impediments to matrimony,[26] and in 1897 the 1895 "Statute of the Exarchate" was supplemented with a Part VI regulating marriage[27] and introduced syphilis to the ecclesiastical regime of marital health (Art. 186, para 5). On grounds of the strict differentiation that the "Statute of the Exarchate" now made between annulment of voidable marriages on the one hand and divorce on the other, the sanctions for the violation of the impediments to marriage were strengthened: it was no longer only subject to *optional* termination;[28] instead, a marriage *should* be rendered null and void if any of the four illnesses were proved to have existed before its conclusion (Art. 202, para 4). In the case of post-marriage insanity, epilepsy, idiocy or syphilis, divorce could be granted only "in the last resort", "after all possible curative measures had been tried" (Art. 187, para 7). Thus the ruling of a marriage as intolerable was provided with an additional medical aspect, which strengthened the position of the court medical expert.

The ecclesiastical doctrine pointed out that the right to define impediments to marriage belonged to the Church in conjunction with the civil authority. According to "Orthodox Canon Law," written by the prominent canonist and Bishop of the Austrian eparchy of Dalmatia, Nikodim Milash (1845–1915), the reasons for the termination of matrimony could be classified into three groups. The first included those introduced by the Church authority, supported by arguments in canon law while the second group embraced impediments established by the secular authority, which were adopted by the Church post factum. The third one comprised arguments arranged in state laws that were "not adopted by the Church either essen-

26 In 1895 a new "Statute of the Exarchate" was enacted, but Part VI, which was dedicated to marriage law, did not enter into force until 1897. In that time the regime of marriage was regulated by "Interim Provisions" (*Vremenni polozheniya*). See Aleksandar Belev, " Pregled na sydebnata praktika," *Yuridicheski arhiv* 6, 4–5 (1934): 470.

27 See "Zakon za pribavlenie chast shesta kam Ekzarhiyskiya Ustav, prisposoben v Knyazhestvoto, kakto i za dopalnenie i izmenenie nyakoi chlenove ot sashtiya ustav," *Darzhaven vestnik* 19, 85 (1897): 1–5.

28 I mean the wording "could be dissolved" used in Art. 6 of the 1883 "Statute of the Exarchate."

tially or formally."[29] Insanity was mentioned in some Byzantine source as a reason for divorce, but was not provided for by the canon law of all Orthodox Churches.[30] Insanity and mental derangement affected the validity of matrimonial consent, i.e. they caused defects in it, which was the initial basis for their legal recognition by the "Statute of the Exarchate." As for syphilis, Milash's argument that infectious diseases were introduced as impediments to marriage not by Orthodox canon law, but by state law should be taken into account.[31] The legal effect of syphilis was underlined in later interpretations of the "Statute of the Exarchate" from the 1930s onwards: "The social, eugenic character of the conditions regulated by Art. 186, para 4 is further stressed by the adding of syphilis to the listed illnesses, as well as through the inadmissibility of marriage on grounds of the spouse's healing in the course of marriage."[32]

Neither statute regulated procedures for ascertaining health requirements for a proper marriage. In the absence of special instructions, the general rules of "pre-marital examinations" carried out by priests were to be considered.[33] There is scant and contradictory data on the existence of cases of medical examinations and pre-nuptial health certificates required by some eparchies under the framework of the period studied. In a 1925 article, Stefan Vatev (1866–1946) noted the case of a physician who was sentenced under the Criminal Code for having issued a false pre-nuptial certificate. In the same text, Vatev wrote that, as far as he knew, only the Dorostol-Cherven Eparchy required submission of pre-nuptial certificates, but without giving further information about the period in which this requirement had been put into practice.[34] In a 1932 publication, Toshko Petrov wrote: "In our country, 30 years ago, much earlier than elsewhere, the Ruse Eparchy required the bridegroom to submit a pre-nuptial certificate stating that he is healthy in order to be given a marriage license."

29 Nikodim Milash, *Pravoslavno tsarkovno pravo. Sastaveno po obshtite tsarkovno-yuridicheski iztochnitsi i chastnite zakoni, deystvuyushti v avtokefalnite tsarkvi* (Sofia: pechatnitsa Sv. Sofia: 1904), 584.

30 Milash, *Pravoslavno tsarkovno pravo*, 636.

31 Milash, *Pravoslavno tsarkovno pravo*, 636.

32 See Belev, "Pregled na sadebnata praktika," 469.

33 See Milash, *Pravoslavno tsarkovno pravo*, 577.

34 Stefan Vatev, "Otgovornost na lekarite po chl. 38 i chl. 367 ot Nakazatelnia zakon za prestaplenia protiv chuzhdi tayni i lazhlivi lekarski svidetelstva," in *Godishnik na Sofiyskiaya universitet. Meditsinski fakultet* 4, (1924/1925): 16–48.

Later, according to Petrov, the same practice was introduced in the Vratsa Eparchy.[35]

However, in spite of the psychiatrist Lybomir Rusev's[36] statement that "the Holy Synod is not aware of a single case of marriage prohibition under this article,"[37] a single precedent can be found in the existing collections of the Synod's court documents.[38] According to the Holy Synod, the refusal of the Eparchy Council to give marriage permission to a feeble-minded person, (which was appealed in this particular case), did not have a peremptory legal effect. Therefore, the appellant could always initiate medical examination and renew his request to enter into marriage on grounds of medically certified "new data" about his mental state. Among other things, this precedent indicated the lack of synchronization between Church and expert medical opinion. From the position of psychiatric knowledge, the clinical presentation of feeble-mindedness was "resistant" to temporal factors, meaning that medically relevant "new data" about the course of the illness in terms of perspective improvement was not possible.

Initially, the legal effects of pre-marital illness were limited to the private sphere of both spouses. Only the healthy spouse could lodge a claim and open proceedings in order to nullify the marriage; neither the clergy nor any other third party could do so. The Church, therefore, did not see a reason to activate religious sanctions towards the ailing "family organism," despite the risks that other organizations saw for the "national organism." However, in the late 1920s the Holy Synod's position began to change. In 1927 the Holy Synod declared in Decision No. 205 that the legislative goal of divorce by reason of illness to be "above all to prevent the creation of a generation burdened by serious diseases."[39] This new strategy was fur-

35 Toshko Petrov, "Higienata pri braka i predbrachnoto osvidetelstvuvane," *Letopisi na Balgarskiya lekarski sayuz* 7 (1932): 5.

36 Lyubomir Ivanov Rusev, b. 1898, studied medicine in Vienna and Prague and graduated from the Medical Faculty of Sofia University in 1924. He specialized in neural diseases and psychiatry in Paris, Munich and Vienna. In the period 1924 to 1926 he was an assistant professor at the Institute of Hygiene and in 1927 he worked as an assistant professor at the neuro-psychiatric university clinics for six months (see: Arnaudov, *Almanah na Sofiyskiya universitet,* 531. Rusev was one of the staunchest supporters of eugenics in Bulgaria. He expressed his ideas in a number of publications, including *Osnovni printsipi na evgenikata ('rasova higiena')* (Sofia: pechatnitsa Poligrafiya, 1934).

37 Rusev meant Art. 202, para 4. See Rusev, *Osnovni printsipi na evgenikata,* 69.

38 See Decision No. 277/1932 in Ivan Dzhamdzhiev, ed., *Sadebna praktika na Svetiya Sinod po brakorazvodni i duhovno-nakazatelni dela,* vol. 1 (Sofia: pechatnitsa Kozhuharov 1943), 143.

39 See Dzhamdzhiev, *Sadebna praktika na Svetiya Sinod,* vol. 2, 458.

ther developed by court decision No. 253 in 1933, which reinterpreted the meaning of pre-marital insanity. Another consideration was added to its initial relevance as the incapacity to valid matrimonial consent: "the Statute of the Exarchate takes into account illnesses that make the conclusion of marriage socially undesirable and inadmissible. By insisting on the personal health of future spouses, and indirectly on that of their offspring, the ecclesiastic legislator has intended to prevent the degeneration of the family. The aim of Art. 202, para 4, is eugenic, social and cannot, therefore, be placed in relation to issues of matrimonial consent and shortcomings in this consent."[40]

The above-cited court decisions demonstrate the opposite points of heterogeneous strategies for the protection of the family health, both ecclesiastic and state. This move by the Church to reconsider some of its court practice in compliance with eugenic rhetoric manifested in the 1920s and 1930s was rooted in the ambivalence of Bulgarian marriage law, i.e. it was open to both Orthodox and secular reasoning to a certain extent. Hence the comparison that the Church historian Stefan Tsankov (1881–1965) made in 1924 between the larger number of illnesses listed as marriage impediments under Bulgarian legislation on the one hand, and the other European legal regimes on the other: "in comparison to all European countries we have gone so far that we take the first, unmatched position. Whether this is good or bad depends on many social conditions, depends on our culture and morality."[41]

## Projecting Marital Hygiene in Eugenic Terms

As early as 1900, in an article titled "Illnesses in Reference to Marriage and Divorce: A Forensic Medicine Study,"[42] S. Danadzhiev expressed his vision of a prospective amendment of the "Statute of the Exarchate," generally regarded by him as a tool of social ethics and hygiene. The Statute should

40 See Dzhamdzhiev, *Sadebna praktika na Svetiya Sinod*, vol. 2, 688.

41 Stefan Tsankov, "Bolestite v balgarskoto brachno pravo," in Georgi Todorov Danailov et al., *Lektsii varhu semeystvoto, darzhani prez mesetsite fevruari i mart 1925 g.* (Sofia: Balgarsko dermatologichno druzhestvo, 1924), 10.

42 Stefan Danadzhiev, "Bolestite po otnoshenie na braka i razvoda: sadebno-meditsinska studiya," in *Periodichesko spisanie na Balgarskoto knizhovno druzhestvo v Sofia* 12, 1 (1900): 34.

make additional requirements for future spouses; they "must have reached full physical development," to be provided in a pre-nuptial certificate.[43] In 1905 the ophthalmologist Konstantin Pashev (1873–1961) published his monograph "Hygiene of the eye" [*Higiena na okoto*], which touched upon the issues of heredity. For the purposes of health prevention, Pashev stated that individuals suffering from tuberculosis, syphilis, extreme myopia, eye defects and blindness should not enter into marriage.[44]

However, it was not until the 1920s and 1930s that the medico-psychiatric tendency to link marital hygiene with projects for improving the Bulgarian race became tangible. The care for "the family from a biological point of view" was conditioned by the awareness that "the foundation of the nation, the commitment to its prosperity, for its living through the centuries, these are the families of which it is comprised."[45] The welfare of the community and the fight against degeneration called for the imposition of marriage restrictions on people who were "below the average norm of society," argued the zoologist Stefan Konsulov (1854–1954).[46] Arguments con-

43 In 1904 the physician N. Stankov expressed the same idea as Danadzhiev in an article published in the popular science journal "Medical Talk" [*Meditsinka beseda*]. Moreover, Stankov warned against marriages between individuals suffering from "illnesses with evident degenerative effects on the offspring." These illnesses included not only somatic illnesses but also epilepsy, tuberculosis and syphilis. See N. Stankov, "Rannoto ojenvane i vredata ot nego," *Meditsinska beseda* 8, 10 (1904): 582. On the pages of the journal "New Time" [*Novo vreme*], the organ of the Workers' Social Democratic Party, Strashimir Dochkov suggested from a socialist perspective that degeneration should be prevented via socio-economic reforms. However, he also considered marriages between healthy and vital individuals as a form of biological selection that would weaken the social effects of various hereditary illnesses. See Str[ashimir] Dochkov, "Nastledvennostta u choveka," *Novo vreme* 5, 12 (1901): 1232. I am grateful to Christian Promitzer for making this reference available to me.

44 Konstantin Pashev, *Higiena na okoto* (Sofia: Pechatnitsa "Sveta Sofiya," 1904), 43. Like Toshko Petrov, Konstantin Pashev also graduated in medicine from Lyon (France) and received a doctoral degree there (1899). In the period 1898–1906 he specialized in ophthalmology in Lyon, Paris, Berlin and London. Pashev was a regular member of the French, German and Oxford Ophthalmic Societies and a corresponding member of the German Academy in Munich (see Arnaudov, *Almanah na Sofiyskiya universitet Sv. Kliment Ohridski*, 442–443). In the 1930s he stated his sympathy for German racial hygiene in a number of publications, especially those in the journal *Oftalmologichen pregled. Meditsinsko-obshtestveno spisanie za borba protiv slepotata* [Ophthalmic review: medico-social journal for combating blindness], whose editor-in-chief he was. See Konstantin Pashev, "Grizhi na maykata za ochite na deteto," *Oftalmologichen pregled* 2, 3 (1934); Konstantin Pashev, "Borbata protiv slepotata i Natsionalniyat solidarizam," *Oftalmologichen pregled* 2, 2 (1934): 46–47; Konstantin Pashev, "Evgenichnoto ochno dvizhenie v Germaniya," *Oftalmologichen pregled* 2, 1 (1933): 17–19; and Konstantin Pashev, "Nasledstvenata slepota i germanskiya zakon za evgenichnoto obezplodyavane," *Oftalmologichen pregled* 2, 3 (1934): 69–81.

45 Stefan Konsulov, "Semeystvoto ot biologichno gledishte," in Danailov et al., *Lektsii varhu semeystvoto*, 1.

46 Konsulov, "Semeystvoto ot biologichno gledishte," 7–8. Prof. Stefan Konsulov was one of the founders and most distinguished promoters of racial hygiene in Bulgaria with publications in the field of social biology, anthropology, pedagogy and philosophy of natural science.

cerning mental hygiene also "toughened" the interwar eugenic discourse. "In stormy and unrestful times like the recent ones and those in which we are living now," stated Danadzhiev, "when the nervous system of the masses is weakened due to the adversities caused by the war and where the ethic and religious feelings are loosened, psychic infection has reached its peak."[47] The normalization of mental life—individual, family and national—was seen by Danadzhiev as the sacred mission of the medical expert, a hygienist and healer of "the common soul of the people."[48]

In this context of fighting the consequences of the postwar cultural crisis, two proposals for change in the regulation of illness were addressed to the Bishops Assembly of the Synod, during the ongoing discussions on a new "Statute of the Exarchate" in 1933. One of these was made by Dr. Toshko Petrov in his capacity as chairman of the "Society for Hygiene and Preventive Medicine." The other came from the University Ophthalmic Clinic headed by Dr. Konstantin Pashev, who was also chairman of the "Bulgarian Ophthalmic Society." The contents of both proposals may be interpreted as an expression of comparable, but not identical strategies. The identical answers provided by the Holy Synod reveal the attitude of the Bulgarian Orthodox Church to the regime of "social" and "national" diseases in the medical projects of the 1930s. According to the "Society for Hygiene and Preventive Medicine," marriage should be prohibited in cases of "1) incapability of leading matrimonial life; 2) insanity, feeble-mindedness, epilepsy or any other incurable disease; 3) uncured venereal disease, pulmonary tuberculosis and tuberculosis of the throat."[49]

As a reaction to the above proposed health norms in Petrov's project, the Bishops Assembly pointed out that the diseases under Paragraph 2 had already been envisaged by the revised statute. As regards to the rest of them, they "could hardly be ascertained when a marriage is being concluded, for this requires pre-nuptial certificates, which could not yet be introduced in Bulgaria." With no comment on the justification of the medical claim, the Assembly declined it. The argument was that "the National Health Depart-

47 Stefan Danadzhiev, "Dushevnite bolesti i semeystvoto," in Danailov et al., *Lektsii varhu semeystvoto,* 11.

48 Danadzhiev, *Printsipite na dushevnata higiena,* 165.

49 See the 1933 Records of Proceedings of the Bishops Assembly, Central State Archive, fund 791k, inventory 1, archival unit 51, record No. 18 of 8.12.1933, 42.

ment should be empowered to do what is necessary in order to prevent marriages of people with such disabilities."[50] This answer was in line with the position of the Synod declared as early as 1907. It pointed to the lack of regulation for pre-nuptial certificates under the "Statute of the Exarchate" and stated that "things being so, it [the Synod] considers itself incompetent to solve the matter. If the state, which is entrusted with the care of [its] citizens' health, considers such a measure necessary, it can duly legalize it."[51] In both acts the supreme Church authority referred to the legislative will of the state as alone competent and interested in the promotion of measures for national health care control. However, the reaction in 1933 was somewhat inconsistent: on the one hand it was admitted that pre-marital medical examinations were not suitable for the cultural situation in Bulgaria; on the other it was left to the executive to do "what is necessary" in this respect.

The second project submitted to the Exarchate authority insisted on the "prohibition of marriage of blind-born individuals and the preventive ophthalmic examination of future spouses."[52] This proposal by the group around Konstantin Pashev was motivated by extreme eugenic considerations: all blind-born people bore indications of a hereditary defect and therefore, any procreation should be denied. The prohibition of marriage in Pashev's publications was discussed as one among other measures of similar character, such as a prospective "law on compulsory extraction of the blind eye" or sterilization of "eye invalids."[53] While the proposal of the "Society for Hygiene and Preventive Medicine" was initiated from more moderate positions, its eugenic grounds were part of current scientific trends for the prevention of dangerous diseases considered of national importance and far from a radical version of racial hygiene. Hence, the cure that Dr. Petrov prescribed against hereditary diseases was not surgical sterilization but a social treatment, namely marriage prohibition. Depriving the individual of their right to decide about their own reproduction was considered sufficient to ensure collective health.[54] At the same time, the common topics and refer-

50 See the 1933 Records of Proceedings of the Bishops Assembly, 43

51 See Decision of the Bishops Assembly of 24.05.2007, in Hristo Tanchev, ed., *Ekzarhiyskiy ustav s talkuvaniyata i naredbite na Svetiya Sinod, Ministerstvata, Varhovniy kasatsionen sad i saotvetstvenite zakonopolozhenia* (Sofia: pechatnitsa St. Atanasov, 1932), vol. 2, 20.

52 See the 1933 Records of Proceedings of the Bishops Assembly, record No. 20 of 11.12.1933, 45.

53 Pashev, "Grizhi na maykata za ochite na deteto," 407–408.

54 Toshko Petrov, "Higienata pri braka," 3.

ences implied by both societies and their projects also represented a tendency to merge social hygiene, preventive medicine and eugenics as allied strategies of public health "security."

Consequently, it was no coincidence that the reaction of the clergy was one and the same as regards both projects. On the occasion of the second:

> The Assembly established that this disability [blindness at birth] could not be considered by the draft Statute as an impediment to marriage either. The Church possesses neither the resources nor the power to impose sanctions and take measures affecting the health of the people. This should be done by the National Health Department by law, binding all religious groups in the kingdom, not only the Orthodox Church.[55]

In this instance the proposal was declined but not condemned. The argument, evasive in a sense, was the statement of a passive position. The decision did not make it clear whether it stood for the refusal the Church to get involved in suspicious scientific undertakings or if it only acknowledged the Church's limited capacity to enforce them while giving implicit approval or at least having no general objection against them.

Hence the contrary opinions about the relevance or irrelevance of the "Statute of the Exarchate" to eugenics. Pashev criticized the retrospective nature of the regulation, because "even while in Germany, laws on artificial and compulsory sterilization are being enacted, in our country the Statute of the Exarchate [...] does not mention defective heredity at all,"[56] while Danadzhiev considered that in the health requirements of the "Statute of the Exarchate" "the legislator had meant to lay down the principle of natural selection based on the common biological law of heredity and thus avert family degeneration."[57] The type of argumentation used by Danadzhiev reflects the strategies proposed by American eugenicists regarding the legislation of marriage laws in the United States.[58]

Furthermore, according to Danadzhiev, the legal experience of the Church in the matter of restrictions suggested that it "would not object to

---

55 See the 1933 Records of Proceedings of the Bishops Assembly, record No. 20 of 11.12.1933, p. 45.

56 Konstantin Pashev, "Nadzorat varhu ochite v uchilishtata ni," *Oftalmologichen pregled* 2, 1 (1934): 5–6.

57 Danadzhiev, "Dushevnite bolesti i semeystvoto," 17.

58 See the discussion in Daniel Kevles, *In the Name of Eugenics: Genetics and the Uses of Human Heredity* (Cambridge, MA: Harvard University Press, 1995).

the application of certain eugenic measures in line with the aspirations of the Statute of the Exarchate precisely for the bearing of healthy children and future generations."[59] The principles of "mental hygiene," as a branch of racial hygiene as defended by Danadzhiev, demanded that "hereditarily burdened psychopaths, or better, psychopaths in general, must not have children and members of mentally healthy stock must not marry them and procreate."[60] He further insisted that family hygiene should not only effect people with "heavy" defects but "mild cases" as well because they were considered most active in terms of reproduction.[61]

The draft Statute of the Exarchate elaborated by the Holy Synod itself in 1933 envisaged the introduction of leprosy to the impediments to marriage.[62] This suggestion was heavily criticized by Ivan Moskov, expert in forensic medicine and a member of the Society for Hygiene and Preventive Medicine.[63] Moskov argued that the register of pathological grounds for restricting marriage should be extended in other directions instead of sanctioning "an extremely rare disease in our country, the danger of which, considering our climate and other conditions, is doubtful." Citing examples of Austrian and Italian marriage laws and especially the 1933 German "Law on the Prevention of Hereditarily Diseased Offspring," Moskov insisted that the "Statute of the Exarchate" should regulate "all chronic, contagious, hereditary and [other] serious diseases and infirmities, which were undoubted and constant obstacles to the purpose of marriage, family life and the capability of conceiving and bringing-up children." Moskov also remarked that a modern-day state should be interested not only in animal husbandry but in human breeding as well. The care for the resulting offspring must also be provided by the contemporary state not only with regard to the race of domestic animals.[64]

59 Danadzhiev, *Printsipite na psihicheskata higiena*, 133.

60 Danadzhiev, *Printsipite na psihicheskata higiena*, 16.

61 Danadzhiev, *Printsipite na psihicheskata higiena*, 118–119.

62 See Central State Archive, fund 791k, inventory 1, archival unit 49, p. 407; as well as Ivan Moskov, "Sadebno-meditsinski saobrazhenia za reformirane na brachnoto pravosadie u nas," *Yuridicheski arhiv* 7, 4–5 (1935): 386.

63 Ivan Moskov studied law at Sofia University and medicine in Florence. He then specialized in forensic medicine in Vienna and Paris. He was head of the Forensic Medicine Department at the Medical Institute from 1939 to 1952 and is considered the founder of medical deontology in Bulgaria. See Arnaudov, *Almanah na Sofiyskiya universitet*, 392.

64 Moskov, "Sadebno-meditsinski saobrazheniya," 385.

## The Hygiene of Marriage from the State Civil Perspective: Toward a "Nationalization" of Family Health

In line with the Church's position to devolve to the state, public concerns over pre-marital hygiene, Andrey Lyapchev (1866–1933)[65] submitted to parliament a draft "Law on National Health" in 1928.[66] One of the main contributors to its creation was Dr. Ivan Kesyakov (1871–1956), Director of the Central Directorate of Public Health.[67] The official motivation of the law underlined the need for the adoption of "the contemporary science of national health care, which goes beyond the framework of personal medical care for the ill and extends to the purpose of a reasonable health policy in view of the entire organism of the nation and its individual separate groups."[68] The role of this "contemporary creative work of health care" was stipulated as follows: "to explore the reasons for those common defects that undermine the health and strength of the nation and to eliminate them in order to provide a healthy generation capable of work, to prevent the curable contagious and national diseases with all contemporary protective measures and thus to restrict untimely aging and death so as to preserve the national capacity for creative labor."[69]

Under Article 69 of the draft law, "only healthy persons may get married; such persons shall receive printed instructions on the importance of health and the harmful influence of diseases on the spouses and the next generation" (Para 1). "For that purpose the municipal authorities may open special counseling bureaus, with physicians in charge of them" (Para 2). "Epileptics, the mentally ill and those suffering from contagious venereal diseases shall be forbidden to marry" (Remark to Art. 55, para 2).

65 Andrey Lyapchev was Bulgarian Prime Minister in three cabinets, from 1926 to 1931. In 1923 he joined the "Democratic Accord" (*Demokraticheski sgovor*), a political formation with a right-wing orientation that embraced several political parties, and became the leader of its moderate wing.

66 The draft "Law on National Health" was published as an Annex No. 42 to the 1927 records of proceedings of the National Assembly. See *Prilozheniya kam stenografskite dnevnitsi na XXII-to obiknoveno narodno sabranie, I redovna sesia, otkrita na 28 oktomvriy 1927 g.*, vol. 1: *Zakonoproekti i predlozheniya* (Sofia: Darzhavna pechatnitsa, 1927), 66–86.

67 Petrovski, *Savremennoto Balgarsko zdraveopazvane*, 306.

68 See *Prilozheniya kam stenografskite dnevnitsi*, 69.

69 See *Prilozheniya kam stenografskite dnevnitsi*, 69.

In the final text of the enacted law[70] these provisions were united in one rule with a more narrow focus. The imperative "only healthy persons may get married" was cut, as was the enumeration of particular marriage prohibitions. However, the regulation for counseling bureaus and printed instructions entered into force. The "Care for the Offspring" under the Law's Section VI of the same name, was paid special attention to. Moreover, this was done with awareness of a recognized biological risk. According to official arguments, the drafting of the law was necessary because "[t]he nation and the state suffer considerable material and spiritual damage from hereditary defects resulting from marriages concluded between physically or mentally ill persons [...] To impede such degeneration of the race through sanitary control of marriage is a task for health care legislation in all civilized countries."[71]

The legislative initiatives in the field of national health care provoked a parliamentary debate that finally limited the project horizon of the draft. The arguments were grouped around two major opposing lines: "public versus private" (were the reasons for the state to intervene in the intimate sphere acceptable?) and "native culture versus culture of civilized countries" (did the envisioned medical interventions have local relevance or were they mere imitation?).

The concept for national health for the "Democratic Accord" was defended by deputy, Nikola Pydarev, among others. With clear enthusiasm for both a new state and its scientific mission, he declared:

> If life in general is making great strides, medical science is making even greater progress, hence the necessity of state intervention in private lives since public health care concerns is getting stronger and stronger. This intervention cannot otherwise be performed except by means of legislative action, by laws which provide for the state's right to infringe in private interests and individuals' rights in view of its great, supreme care for national protection.[72]

Petko Staynov, a deputy from the same parliamentary group, also supported the role of the state as a "guardian of citizens' health." However, he

70 See "Zakon za narodnoto zdrave," *Darzhaven vestnik* 50, 277 ( 1929): 1–25.

71 *Prilozheniya kam stenografskite dnevnitsi*, 85.

72 *Stenografski dnevnitsi na XXII Narodno sybranie, I redovna sesiya*, 668.

warned that state interference in private relations was "a dangerous issue" needing a delicate approach. The ambitions of the state to manage a regime of marriage health were treated as though the state were the "natural" successor of Church authority and with legal continuity: "We know that for centuries, religion has intervened in the private lives of people [ … ]. This is an intervention of a certain authority, no matter what type of authority [ … ]." There were further warnings that if state authoritarianism was not precisely measured and defined, "religious conflicts" and "the indignation of public opinion" would most probably ensue.[73]

The same aspiration of a balanced regulation of public/private interests in terms of marriage restrictions was also presented by Nikola Tyrkalanov from the "Democratic Accord": "It is beyond doubt that this is a very intimate sphere and difficult to deal with, but I believe that if this is done in countries, much more advanced than ours, like Sweden, Holland, Norway, etc., we should do what is possible without going to extremes."[74] This approach was contested by Georgi Petrov, a representative of the "National-Liberal Party" [*Natsionalliberalna partiya*]. Despite his declared sympathy with preventive medicine and pre-nuptial certificates, which, according to him, were adopted by the "Law on National Health" under American influence, Petrov was convinced that this regime was inadequate for the Bulgarian economic and cultural situation. Stressing the fact that marriage in Bulgaria was an "ethical-religious institution," not a civil one, he asked:

> In what way do you mean, then, to subsume this institution, which touches the most delicate fibers of the human soul, under the norms of the draft law and thus affecting the enduring religious traditions of our people? In my last visit to the countryside I told our peasants that pre-nuptial certificates might be introduced by law. They answered: 'That won't do!'—'Why?'—They answered: 'That is a shame' [ … ] Now you see what is the content of these words—this is the whole philosophy of our people.[75]

In the parliamentary debate it was concluded that pre-nuptial certificates would have crossed an integral cultural border. The "consulting stations"

73 *Stenografski dnevnitsi na XXII Narodno sybranie*, 686.
74 *Stenografski dnevnitsi na XXII Narodno sybranie*, 687.
75 *Stenografski dnevnitsi na XXII Narodno sybranie*, 773.

were instead motivated as "a proper measure that without violating the personal and religious feelings of the people could still protect them."[76] However, even this conciliatory regulation was only of a legally-instructive and advisory character. Apart from traditional morality and the skeptical attitudes toward the medical profession there was another obstacle for the institutionalization of compulsory medical controls prior to marriage. This was the limited potential of cadres and the poorly structured administrative network of medical care.

Nevertheless, the references made to the legal regimes of marital hygiene in Germany, the USA and the Scandinavian countries were indicative of a more general legislation pattern. According to Harry Bruinius, by 1914, at least thirty states had passed some type of marriage restriction.[77] By 1912, thirty-four American states had adopted some type of marriage restriction. In Connecticut, Washington, Utah, Michigan and Colorado, pre-nuptial examinations were required in order to prevent the spread of venereal diseases. Most marriage laws in the United States also prohibited the insane and the feeble-minded from marrying. In nine states epilepsy, and in two "drunkenness" were considered marriage impediments.[78] By 1941, fourteen states in the United States had introduced obligatory pre-nuptial certificates of fitness for marriage.[79] The Nordic countries are another early example of enforcing marriage restrictions and pre-nuptial health certificates. In Sweden, the marriage laws of 1915 listed three impediments to marriage: mental retardation, mental illness and epilepsy.[80] The 1918 *Lex Veneris* restricted marriage of the infected with venereal diseases and in 1920 pre-nuptial health certificates were required.[81] In Norway, too, compulsory pre-nuptial examinations were required since 1918 to certify the lack of impediments to marriage, particularly advanced mental illness and

76 *Stenografski dnevnitsi na XXII Narodno sybranie*, 686.

77 Harry Bruinius, *Better for All the World. The Secret History of Forced Sterilization and America's Quest for Racial Purity* (New York: Alfred Knopf, 2006), 147.

78 Ruth Clifford Engs, *The Eugenics Movement: An Encyclopedia* (Westport, CT: Greenwood Press, 2005), 53.

79 S. P. Cupcea, "Examenul prenupţial," *Buletin eugenic si biopolitic* 12, 7–8 (1941): 119.

80 Gunnard Broberg and Mattias Tydén, "Eugenics in Sweden: Efficient Care," in Gunnard Broberg and Nils Roll-Hansen, ed., *Eugenics and the Welfare State: Sterilization Policy in Denmark, Sweden, Norway, and Finland*, (East Lansing: Michigan State University Press, 1996), 100, 118.

81 Peter Baldwin, *Contagion and the State in Europe, 1830–1930* (Cambridge: Cambridge University Press, 1999), 450.

syphilis,[82] while in Denmark, compulsory pre-nuptial certificates were introduced in 1922 to rule out venereal diseases.[83] In Weimar Germany the state encouraged voluntary pre-nuptial tests.[84] It was the Nazi "Law for the Protection of the Genetic Health of the German Nation" [*Gesetz zum Schutze der Erbgesundheit des Deutsches Volkes*] passed in 1935, which demanded from couples that they undergo medical examination prior to marriage in order to exclude potential "racial damage."[85] However, even the Nuremberg marital health law did not stipulate a compulsory "test of fitness for marriage" [*Ehetauglichkeitszeugnis*] for all citizens.[86] According to Paul Weindling, registry officials only referred those people suspected of ill-health to the Reich Health Offices.[87] The officials examined the future spouses and their family members for any "negative registration." In case that relevant data was found, the couple had to pass pre-marital examinations.[88]

The ways in which the concept of pre-nuptial health certificates was developed in the above discussed countries reveal several trends. The introduction of such certificates merged efforts for the social prophylaxis of venereal diseases, population policies of birth control, and eugenic values of hereditarianism. The groups of experts and policy makers who suggested pre-nuptial certificates varied, but they shared a common ground. That was

82 Petrov, "Higienata pri braka," 10–11. In the case of certified epilepsy, leprosy or gonorrhea, marriage was allowed, but only after both partners had informed each other of their health status. See also S. P. Cupcea, "Examenul prenuptial," 119.

83 Petrov, "Higienata pri braka," 11. If the medical examination had proved epilepsy or a contagious disease, the obligation of mutual information prior to marriage applied to both partners. See Cupcea, "Examenul prenupțial," 119.

84 See Michelle Moutton, *From Nurturing the Nation to Purifying the Volk: Weimar and Nazi Family Policy, 1918-1945* (Cambridge: Cambridge University Press, 2007), 36.

85 Robert N. Proctor, *Racial Hygiene: Medicine under the Nazis* (Cambridge, MA: Harvard University Press, 1988), 132. The 1935 Nuremberg Marital Health Law prohibited marriage between people suffering from venereal disease, feeble-mindedness, epilepsy or other "genetic infirmities" specified under the 1933 Sterilization Law. See Proctor, *Racial Hygiene*, 132. See also Sylvia Kesper-Biermann, "'Ehegesundheit' als bevölkerungspolitisches Problem. Internationale Dimensionen von Diskussion und Gesetzgebung in der Weimarer Republik," in Ursula Ferdinand and Josef Ehmer eds., *Herausforderung Bevölkerung. Zu Entwicklungen des modernen Denken über die Bevölkerung vor, im und nach dem, Dritten Reich'* (Wiesbaden: VS-Verlag für Sozialwissenschaften, 2007), 123–132.

86 See Richard Bessel, *Fascist Italy and Nazi Germany: Comparisons and Contrasts* (Cambridge: Cambridge University Press, 1996) 101.

87 Weindling, *Health, Race, and German Politics*, 531. Bessel suggests, however, that registry officials were obliged to refer all pending marriages to the Reich Health Offices.

88 Bessel, *Fascist Italy and Nazi Germany*, 101–102. The examination included blood tests, lung X-rays and other clinical and neurological tests, after which certificates valid up to six months following the examinations were issued. See Proctor, *Racial Hygiene*, 139.

the justification of state and professional intervention into the sphere of individual marital choice and reproduction. The prevention of collective health (social, national or racial) was performed at the expense of the liberal humanitarian values of universal medical care. However, despite the fact that pre-nuptial health certificates were generally accepted as a less radical measure of negative eugenics as compared to sterilization,[89] the laws that regulated them were not always strongly enforced.

In the late 1920s and in the 1930s, Bulgarian adepts of marital hygiene were aware that pre-nuptial health certificates had not been introduced on a regular basis even in "more advanced countries."[90] By 1929, when the Bulgarian "Law on National Health" became effective, such measures had not been legalized in any of the neighboring countries either. In Greece, for instance, a law on pre-nuptial certification had not been passed, despite the lively discussions on the topic in the interwar period.[91] In Romania, local campaigns for pre-marital examinations were carried out in the 1930s. These pro-eugenic practices on the part of village physicians exceeded the scope of national marriage law.[92] In Turkey pre-marital examinations were required under the 1930 Public Hygiene Act, which prohibited citizens with venereal diseases, leprosy, mental illnesses and tuberculosis from getting married.[93]

The advocates of pre-nuptial health certificates in Bulgaria realized that this instrument of social hygiene and negative eugenics was not yet appli-

89 As Nancy Stephan has argued: "Many Latin Americans viewed such [pre-nuptial] tests as defining the special form of negative eugenics in Catholic countries" because this form provided "control of birth without birth control." See Nancy Stephan, *The Hour of Eugenics: Race, Gender, and Nation in Latin America* (Ithaca, NY: Cornell University Press, 1991), 122. In contrast to Bulgaria, however, many Latin American states recognized civil marriage. This fact facilitated the efforts of eugenicists and physicians to institutionalize pre-nuptial health certificates. In Chile (1925), Mexico (1928), Brazil (1934) and Argentina (1937) some type of health certificates prior to marriage was legalized, mainly certificates of venereal health. See Asuncíon Lavrin, *Women, Feminism, and Social Change in Argentina, Chile, and Uruguay, 1890–1940* (Lincoln, NE: University of Nebraska Press, 1998), 169, and Stephan, *The Hour of Eugenics*, 124–128.

90 In Britain, Galton's idea of "eugenic certificates" did not become law. In France, compulsory pre-nuptial health certificates were not legally introduced before the revision of Art. 63 of the Civil Code in 1942, during the Vichy regime. See Baldwin, *Contagion and the State in Europe*, 450.

91 See Sevasti Trubeta, "Eugenic Birth Control and Prenuptial Health Certification in Interwar Greece," in this volume.

92 Bucur, *Eugenics and Modernization*, 204. For how Romanian physicians and eugenicists regarded pre-nuptial health certificates and for how these debates related to eugenic sterilization, see Marius Turda, "'To End the Degeneration of the Nation': Debates on Eugenic Sterilization in Interwar Romania," *Medical History*, 53, 1 (2009): 77–104, especially p. 79, 95, 99.

93 See the general discussion in Marius Turda, *Modernism and Eugenics* (Basingstoke: Palgrave, 2010).

cable to the Bulgarian cultural situation. Therefore it was hoped that the institutionalization of "pre-marital counseling" could mediate more intensive future interventions:[94] "The counseling stations [...] will have the opportunity to gather data and gain experience relevant for a future eugenics law, and they will become an intermediary body with regard to its practical application."[95]

The parliamentary and media debate over (pre-)marital hygiene exemplified the tendency to express the problems of the medical discourse on the broader context of moral, social, "folk-psychological" and everyday problems. In the case of the "Law on National Health," medicalization was to a certain extent facilitated by legal strategy itself: in compliance with its own legislative motives, the law contained "only instructions and counseling," for those who wish to get married.[96] By declining to introduce pre-nuptial health certificates, the legislator deprived the medical establishment of administrative power, yet it did open a new institutional horizon for medical knowledge in its capacity of "instructing," "counseling" and other similar forms of moral impact. Although they did not get the institutional support promised by the law (no marriage counseling bureaus were established until 1944),[97] in the 1930s medical and eugenic projects displayed their propaganda potential. For them the "Law on National Health" had the effect of a normative "impulse."

An amendment to the 1929 Law was requested by the Hygiene Section at the "Society for National Health and Social Hygiene." Following a public lecture by Toshko Petrov at the Hygiene Section, a special commission was established which included Petrov, Ivan Moskov and Bogomil Beron (1866–1936), among others. Their proposal was adopted by the commission on the amendment of the Law on National Health that was set up by the Managing Board of the "Bulgarian Physicians Union."[98] Being of the opinion that "compulsory pre-nuptial health certification is an ideal to strive for but is unfeasible under the present circumstances," the Hygiene

94 Stefan Konsulov, *Izrazhdaneto na kulturnoto chovechestvo. Borba s izrazhdaneto* (Sofia: Populyarna nauka, 1928), 37.

95 Danadzhiev, *Printsipite na psihicheskata higiena*, 130.

96 *Prilozheniya kam stenografskite dnevnitsi*, 85.

97 Zahari Bochev, *Zdraven spravochnik (azbuchen ukazatel i poyasnitel na zakonite, pravilnitsite i naredbite po zdravnoto delo v Balgaria)* (Sofia: pechatnitsa P. Glushkov, 1943), 308.

98 Petrov, "Higienata pri braka," 13–14.

Section urged the establishment of a special committee in the Main Directorate of Public Health to be charged with "the propaganda and direction of the national diseases' medical treatment." The Section insisted on the necessity to provide future spouses with "printed instructions" and a "compulsory written declaration" that "immediately before marriage and with regard to it they have consulted a physician".[99] Similar measures against venereal diseases were recommended before the enactment of the 1929 "Law on National Health" by the prominent Bulgarian dermatologist and venerologist Bogomil Beron.[100]

In his 1932 article discussing the suggestions of Bulgarian social hygienists, their leader Toshko Petrov underlined the eugenic importance of compulsory pre-nuptial certificates. Only in this way, Petrov argued, could healthy offspring of "good stock" be guaranteed or would it be possible "to preserve and improve the race in eugenic terms."[101] Bulgarian eugenicists, in turn, argued that along with restricting the reproduction of hereditarily inferior individuals, racial hygiene also suggested measures for the prevention of the "idioplasm" from the so called "hereditary" or "racial poisons." The latter included syphilis, alcohol and tuberculosis, among others. In this field of fighting social diseases, Lyubomir Rusev argued that eugenic measures should not differ from those recommended by social hygiene. However, such measures should consider not only the interests of the present generations but also these of the future ones.[102]

Most of the concepts regarding the "improvement of the race" used by Bulgarian eugenicists were based on the alleged primary importance of heredity and biological selection. The importance of environmental change, however, was not to be underestimated from a eugenic point of view. Indeed, the majority of eugenicists in Bulgaria considered the Lamarckian theory of the inheritance of "acquired traits" to be scientifically wrong and outdated. But their constructions of "heredity" and "social environment" were ideologically interwoven in such a way that sometimes it became difficult to differentiate between them. Lyubomir Rusev, for example, referred to

99 Petrov, "Higienata pri braka," 13.

100 Vasil Beron, "Opazvane semeystvoto ot venericheskite bolesti," in Danailov et al., *Lektsii varhu semeystvoto,* 6–14.

101 Petrov, "Higienata pri braka," 9.

102 Lybomir Rusev, "Merkite na rasovata higiena," *Zdravna prosveta* 6, 39 (1927): 3.

the Austrian racial hygienist Max von Gruber (1853–1927) to illustrate the understanding that the improvement of the "external conditions" was a general, though indirect, eugenic measure.[103] Only good hygienic and economical conditions that were equal in all social groups were expected to provide equal chances for hereditary qualities to develop. On the other hand, certain social policies could prevent the emergence of some negative hereditary dispositions. Hence, social reforms could facilitate obtaining optimal results from biological selection. Stefan Konsulov, in turn, considered the mutation theory of the Dutch geneticist Hugo de Vries (1848–1935) and stated that some strong "external influences" could also change hereditary traits. He argued further that none of the evolutionary theories was perfect and that the scientific truth was the golden mean.[104] The same tendency to theoretical hybridization was expressed by Rusev, who used the concept of "temporary heredity" to explain the effect of "social plagues."[105]

During the 1930s, criticism against eugenics in Bulgaria focused on the idea of sterilization of the alleged "degenerates" and "hereditarily inferior."[106] Almost all Bulgarian eugenicists regarded sterilization as a proper measure, at least in certain cases. At the same time almost all of them considered its introduction in Bulgaria premature.[107] Instead, based on the alleged late development of eugenics in Bulgaria, its adherents adopted a more cautious approach. This was a strategy of gradual change. Sterilization could be considered only after other, more urgent eugenic goals had been accomplished. Above all, a need for the preliminary popularization of "the eugenic view of life" and of "the biological way of looking at social changes" was emphasized.[108] The Eugenics Society also favored the introduction of (initially voluntary) pre-marital examinations, the encouragement of the "biologically fit" via positive eugenic policies and the measures to combat

103 Rusev, *Osnovni printsipi na evgenikata*, 63–64.

104 Stefan Konsulov, "Evolyutsionnoto uchenie v dneshnata biologiya," *Priroda i nauka* 1, 5–6 (1930–1931): 65–66.

105 Rusev, *Osnovni printsipi na evgenikata*, 14, 21.

106 See the editorial "Evgenichni obezplodyavaniya," *Narod i potomstvo* 1, 4 (1935): 1. See also N. Karev, "Evgenika i sterilizatsiya," *Mir* (2 February 1935): 4; and Konstantin Kantarov, "Za rasovata higiena," *Zdravna prosveta* 14, 3 (1935): 1.

107 See, for example, Stefan Konsulov, *Zakonite na nasledstvenostta. Nasledstvenost pri choveka* (Sofia: Populyarna nauka, 1938), 94; Lyubomir Rusev, "Zadachi i sredstva na evgenikata," *Zdravna prosveta* 14, 8 (1935): 1; and Danadzhiev, *Printsipite na psihicheskata higiena*, 223.

108 See the editorial "Evgenichnite obezplodyavaniya," *Narod i potomstvo* 1, 4 (1935):1.

the social ills.[109] Eugenics tended to become an all-embracing discourse of improving national health,[110] that is, a multiform hygienic utopia, coexisting and interwoven with other public health discourses.

## Medical Counseling and Moral Hygiene Against Venereal Diseases

On 19 May 1934 the "Military League" [*Voenen sayuz*] together with the right-wing formation The Link [*Zveno*] succeeded in their coup d'état. The new government of Kimon Georgiev (1882–1969) dismissed parliament and banned all political parties. In 1935, King Boris III (1894–1943) removed Georgiev from office. This situation lasted until 1944, when the communists came to power. Throughout this time the political situation in the country was determined by an authoritarian regime. In the absence of a parliament, legislative power was exercised by the executive, which issued decrees with the rank of laws. This was the political background to the 1936 "Decree-Law on Fighting Venereal Diseases,"[111] which may be defined as a further step towards "etatization" of sanitary projects designed to control marriage as well as a radicalization of their regulation. The recommendation for drafting a separate legal act on venereal disease[112] was adopted. The Decree-Law replaced the anti-venereal disease provisions of the 1929 "Law on National Health" as well as other legal rules that would contradict its stipulations (Art. 32). The 1936 Decree-Law established a special "Committee for Fighting Venereal Diseases" in the Central Directorate of National Health (Art. 1). In general, the Decree-Law promoted an overall police-sanitary regime for socio-medical health. Under the "Marriage Hygiene" Section it ruled that: "A person willing to enter into matrimony shall be obliged to submit a medical certificate to the marriage authority

109 See the "Reviews" section [*Knizhnina*] of *Narod i potomstvo* 1, 4 (1935): 1.

110 As Mircho Sirakov, chief of a division at the Main Directorate of Labor, argued later in 1943: "All measures aimed at the preservation of the population's health are recommendable from a eugenic point of view." See Mircho Sirakov, "Nachala na rasovata higiena i organizatsiya na truda kato evgenichen problem," Izvestiya na Glavna direktsiya po truda i instituta po obshtestveno osiguryavane 3, 5–6 (1943): 14.

111 See *Izvestiya na Glavnata direktsiya za narodnoto zdrave* 21 (154–157) (1936): 3278–3283.

112 Vasil Beron, "Po deynostta na godishnoto sabranie na upravitelniya savet na Mezhdunarodniya sayuz protiv venerichnata opasnost (2-10 okt. 1925 – Parij)," *Izvestiya na Glavnata direktsiya za narodnoto zdrave* 11, 1-2 (1926): 17.

or state under oath before the same authority that he/she does not suffer from a venereal disease" (Art. 15, para 1). The Decree-Law repeated in a more particularized way the provisions of the "Law on National Health" for pre-marital medical counseling: "In the regional cities and where the conditions are appropriate for the sanitary authority, bureaus shall be arranged for pre-marital counseling in cooperation with the Bulgarian Dermatological Society" (Art. 16). However, even for this authoritarian regime—which was not dependent on the initiative of the municipalities but was obligated to particular medically-empowered bodies and subjects—putting this into practice remained impossible. One alternative to counseling bureaus was the free of charge "counsel on marriage" that state and municipal hospitals and medical officers were entrusted to provide in dispensaries to people who wished to get married (Art. 18). "The Central Directorate of Public Health shall issue instructions on the nature of venereal diseases and their effect on the family and it shall distribute them free of charge amongst the population, especially among people who want to get married" (Art. 17).

In compliance with the 1936 Decree-Law, the "Committee on Fighting Venereal Diseases" at the Central Directorate of Public Health, published (in the Directorate's Bulletin) official samples of the pre-nuptial certificate and its alternative, the "oath-declaration," together with "Instructions for Those Who Wish to Get Married," "Instructions for Syphilitics" and "Instructions for People Suffering from Gonorrhea (Blenorrhea)."[113] The 1936 Decree-Law evidenced the fact that above all other "national diseases" venereal diseases provided the strongest arguments for the legalization of more intensive, coercive medical measures including those taken from the "arsenal" of eugenic marital hygiene. In addition to the legal regime, "anti-venereal propaganda" was outlined as one of the most effective instruments that the eugenics discourse could provide.[114]

As in other countries, the epidemic character of these ills of "unbridled passion and moral lapse" evoked the highest degree of risk to state bio-politics and motivated the medical reaction against them.[115] Therefore, Art. 30

[113] See *Izvestiya na Glavnata direktsiya za narodnoto zdrave* 21, 154–157 (1936): 3282–3285.

[114] Hristo Ivanov, "Venericheski bolesti i potomstvo," *Narod i potomstvo* 1, 1 (1935): 2.

[115] This is a quote from one of the many articles, brochures and instructive texts published between the 1920s and the 1940s as part of the socio-medical reaction against "national" illnesses. See Georgi Payakov et al., *Zdravo tyalo - zdrav duh* (Sofia: pechatnitsa "Knipegraf" AD, 1943), 25.

bound police and administrative authorities "to assist the sanitary authority in its fight against venereal diseases." A series of punitive measures connected with the spread of such diseases also included a minimum of one year prison sentence or the imposition of a fine for anyone "who knows, or under the given circumstances was able to know that he/she suffers from a venereal disease and enters into marriage" in spite of it (Art. 20). This provision was more restrictive compared to Art. 416 of the 1929 "Law on National Health." The latter had criminalized entering into marriage only in cases where the affected partner *had known* about his/her illness, irrespective of their ability to get informed. This new stipulation of the Decree-Law was compared by Ivan Moskov to the 1927 German law that sanctioned entering into marriage under the same circumstances but had not informed their spouse-to-be of their illness.[116] Moskov argued that "from a social and eugenic point of view, the language of our text is preferable" compared to the "somewhat delicate formulation" of the German law. Marrying an informed partner did not diminish "the fatal consequences of the venereal infection for the diseased party, the family and the generation."[117] This more restrictive penal regulation was expected to compensate for the optional character of the pre-nuptial medical examinations. All things considered, Moskov assessed the institutionalization of an alternative "oath-declaration" as a "capitulation before the great idea for pre-nuptial certification."[118]

While compulsory medical control before marriage was not monopolized by the state through regal rule, pre-nuptial certificates were legally arranged for the first time. This was made possible with the approval of the Holy Synod; in Circular No. 795 of 17 February 1937 all eparchy authorities were ordered to apply the 1936 Decree-Law.[119] The published model pre-nuptial certificate, however, included a stipulation that no marriage

116 Ivan Moskov, "Venericheskite bolesti ot pravna i sadebno-meditsinska gledna tochka," *Pravna misal* 2, 9 (1936): 631.

117 Moskov, "Venericheskite bolesti," 634.

118 Moskov, "Venericheskite bolesti," 635. Again, the case of Romania is comparable to that of Bulgaria regarding the combination of anti-venereal measures and marriage prohibition. In the same year that when the Bulgarian "Decree-law on Fighting Venereal Diseases" was enacted (1936), Art. 377 of the Romanian Criminal Code criminalized the marriage of anyone suffering from venereal or other epidemic diseases. See Bucur, *Eugenics and Modernization*, 203.

119 See Lyuben Dikov, *Kurs po balgarsko grazhdansko pravo*, vol. 2 (Sofia: pechatnitsa Ovcharov, 1937), 65.

impediments "of a venereal nature" existed. As regards the oath, it envisaged a double (religious and juridical) obligation: "I take an oath in the name of God Almighty that I do not suffer from a venereal disease. If I do not say the truth, let me be punished by the just judgment of God and the laws of the country."

The taking of these responsibilities were to be preceded by the acquirement of medical knowledge to which the official instructions contributed:

"You have decided to settle down to married life, to have a family and children; so be careful not to suffer from a venereal disease, otherwise you are at risk of:

1. being crippled, incapable of work and even dying prematurely.
2. infecting your future spouse and exposing them to all the dangers of the disease.
3. having children with the same disease, with poor health, feeble-mindedness, at risk of dying, or not having children at all;.
4. contributing to the degeneration of our people [...]

You can avoid all these calamities if you go for a medical examination and make sure that you are healthy. And in case you suffer from a venereal disease, if you go for medical treatment until you are finally healed. If you keep to this advice, you will be able to get married and hope for a healthy family and healthy children."[120]

It was only considered possible to disrupt the ascending gradation of endangered interests—personal, spouse, extended family and last but not least the interests of "our people"—with the power of common medical sense. Thus, medical expertise was transformed into a reliable guarantee for personal moral comfort and family/national happiness. The instructions stated the aspirations of the medical professionals toward different forms of spiritual and moral patronage. Apart from the warning to avoid "getting married in any way," syphilitics should also consider the instruction "to keep doctor's orders and advice (for example to live a proper life, not to drink alcohol, not to smoke tobacco and to avoid everything which wears out the body and nerves)."[121] Similar requirements were set for those suffering from blenorrhea "to not try to self-cure yourselves or follow other peo-

120 See *Izvestiya na Glavnata direktsia za narodnoto zdrave* 21, 154–157 (1936): 3282–3283.

121 *Izvestiya na Glavnata direktsia za narodnoto zdrave*, 3284.

ple's recommendations, but to address a physician in due course [...]" as well as, "to refrain from tense physical actions (jumping, long walks, dancing, etc.)."[122] These pseudo-scientific rules to live a proper lifestyle derived their meaning from a medicalized moral hygiene. This idea of an "improper" lifestyle with regard to venereal diseases was also denounced by the rules of eugenic mental hygiene. According to Stefan Danadzhiev, for instance, syphilis resulted in progressive paralysis when there was hereditary predisposition to it; individuals with a "disabled brain" (i.e. having "reduced resistance to brain damage") were more "susceptible" to syphilis as a powerful factor of degeneration.[123]

As mentioned, the notion of "degeneration" itself was widely used in the public discourses of the interwar period. It functioned as a discursive matrix, whose content was derived from various uses of the organic code, for the stigmatization of politically "ineligible" cultural phenomena. "Degeneration" could be seen as a reversal of evolutionist-progressivist schemes. Since Benedict Morel (1809–1873) it was conceived as a "hereditary" process of the consecutive regression of humans from a "normal" type. "Environment" became a negative signifier, a symptom of "nature" lost in civilization's interventions.[124] In the Bulgarian cultural context of the late nineteenth to mid-twentieth century, one of the meanings of "degeneration" as a concept was "de-naturalization", or, according to Boris Yotsov, "to deprive a nation of individuality, that is, to eliminate a whole genus in the kingdom of nature."[125] The "native" (national) was regarded as the utmost significant feature of the "natural," i.e. the "birthmark" of human nature. In the eugenic discourse, "degeneration" marked a postwar crisis of collective identity whose early "symptoms," however, dated back to the early twentieth century, when the first negative effects of modernization appeared. Thus, "degeneration" was represented as a crisis of Bulgarian modernization which, in turn, was part of the crisis of modernity in general.

The Bulgarian eugenic discourse had a double bias. It was inscribed in the broader framework of the project of modernity, *thus* providing *crit-*

---

122 *Izvestiya na Glavnata direktsia za narodnoto zdrave*, 3285.

123 Danadzhiev, *Printsipite na psihicheskata higiena*, 156.

124 For a discussion of these aspects, see Michel Foucault, *Madness and Civilization. A History of Insanity in the Age of Reason* (New York: Random House, 1965).

125 Boris Yotsov, "Malkiyat narod," *Otets Paisiy* 7, 2 (1934): 33. Boris Yotsov (1894–1945) was an expert in Slavic studies and minister of education in the period 1942 to 1943.

*ical revisions and projective versions* of modern Bulgarian cultural identity. Eugenics was a critique of modernity, but in modern terms, representing the contradictions and complexity of "multiple modernities." Following Zygmunt Bauman, eugenics should be considered a strictly modern product, unthinkable without the development of modern science, technology and state power.[126] Such a theoretical approach corroborates the idea that eugenics defies easy dichotomist categorization as either "traditionalist" or "modernizing."[127] Therefore it should not be qualified in terms of backwardness and traditionalism. While traditionalism signified a passive reaction towards modern change, eugenics was "mobilizing traditions rather than preserving them."[128] The supreme value of this activist attitude towards nature and the modern world, however, was established by the Enlightenment project with its imperative of progress guaranteed by the "heroism" of scientific rationality. Eugenics challenged the negativities of modernity's crisis by means of all-embracing institutional efforts for reconstructing the public field. Bulgarian eugenic discourses should therefore be regarded as having a certain modernization potential rather than as a completely anti-modern effect of scientific knowledge projected in socio-political action.

This "negative-positive" ambivalence of eugenics in general was highlighted in its Bulgarian versions. Some adherents of eugenics and racial hygiene, like Danadzhiev and Pashev, ascribed degeneration not simply to the modern way of life in general but to the "unnaturalness" of Bulgarian modernization processes in particular.[129] The Bulgarian routes to modernity put the healthiness, toughness and authenticity of the "national organism" to a test. Hence the question asked by Lyubomir Rusev: "[...] is it only the external luster and the negativities of Western culture [...] that we are going to borrow and thus fall into decay and degeneration without being in a position to rise culturally?[130] It was a common belief, though, that

126 Bauman "Modernity, Racism, Extermination," 212–229. For the conception of "racism" as a technology of biopower aimed at social stratification and fragmentation of the biological field, which is inherent to the functioning of all modern states, though under specific circumstances and to a different degree, see Michel Foucault, *"Society Must be Defended." Lectures at the Collège de France, 1975–76,* (New York: Picador, 1997).

127 Maria Bucur's case study of Romanian interwar eugenics may be given as an example of this position. See Maria Bucur, *Eugenics and Modernization,* especially p. 11.

128 Bucur, *Eugenics and Modernization,* 11.

129 The Romanian case was similar, see Bucur, *Eugenics and Modernization.*

130 Rusev, *Osnovni printsipi na evgenikata,* 55.

the Bulgarian nation was still "young," with good racial characteristics.[131] In the eugenic perspective, the rural population—which was predominant in the country—constituted the "healthy core" of the nation.[132] Degeneration was reversible because the "core" of the Bulgarian genus was considered inviolable. That is why the signs of decline were ascribed to the collective identity crisis that overlapped with the crisis of modernity.

In 1935, Danadzhiev wrote:

> After the victory of the Liberation[133] [...] civilization forced its way in all spheres of our life. We were terrified by our self-recognized ignorance and therefore started working on our development with feverish haste so as to catch up with the rest of the civilized peoples as soon as possible [...] Our brain can not adapt so easily to the sudden change of concepts, way of life and thinking (...) In this respect, like in many others we can be compared to a parvenu [...][134]

The social and temporal crisis produced by the challenges of modernity was thus "diagnosed" as too rapid a transition to civilization, a violation of the "natural" laws of gradual evolution manifested in the insufficient adaptive capacity of the collective "psyche." Danadzhiev's explanation of the crisis of Bulgarian modernization implied an interpretation of the Darwinian concept of adaptation to environmental change. He used this concept for his approach to mental hygiene, assuming that the psychological disorders of individuals were often rooted in their maladjustment to the social milieu. But what Danadzhiev did, in fact, was provide this process a collective character, i.e. the nation, as a central character of the Bulgarian eugenic discourse. "Mental hygiene has much work to do," Danadzhiev stated fur-

131 See, for example, Stefan Konsulov, *Prosvetnata politika u nas* (Sofia: pechatnitsa Glushkov, 1938), 3; Toshko Petrov, "Znachenieto na higienata kato faktor za fizichnoto, intelektualno i moralno razvitie na narodite," *Zdravna prosveta* 5, 4 (1922): 2; Metodi Popov, *Nasledstvenost, rasa, narod. Rasova prinadlezhnost na balgarite. 4 lektsii* (Sofia: Pridvorna pechatnitsa, 1938): 147, Petar Penchev, "Estestvenoto dvizhenie na naselenieto u nas," *Zdravna prosveta* 6, 38 (1927): 1. This vision was shared not only by eugenicists but also by many intellectuals, public health activists and politicians.

132 In the words of Stefan Konsulov: "Everyone realized by now that the village is the fundament of everything in our country; it is the spring of the nation." See Stefan Konsulov, *Prosvetnata politika*, 20. For a similar discourse in neighboring Romania, see Bucur, *Modernization and Eugenics* and Turda, "To End the Degeneration of the Nation.'"

133 Danadzhiev meant the establishment of the modern Bulgarian state in 1878.

134 Danadzhiev, *Printsipite na psihicheskata higiena*, 98–103.

ther, "so as to be able to correct through education the shortcomings of our race's character."[135]

Claiming to be in possession of ultimate competence to reform society, physicians, biologists, psychiatrists, social scientists, among others, expressed their ambition to "compensate" the marginalization of expert knowledge in Bulgarian society after 1878. During the whole period under review, the process of breaking the dominance of so-called "folk (traditional) medicine" remained incomplete and rather conflictual. The insufficient number of medical experts, the poor material base and the problematic social legitimization of medical expertise were just some of the difficulties faced in the institutionalization of modern professions within Bulgaria. Yet the interwar crisis provided arguments for the eugenicists' strategy towards the "regeneration of the faith in competence," which became stated as a condition *sine qua non* of national regeneration.[136]

The "new and difficult eugenic task," argued Konstantin Pashev, suggested that the medical expert "was not to look for the invalids among the people like a policeman, but that the people should look to him as their teacher and patron, who has the noble aim of freeing them from the chains of what today is an inexcusable burden of hereditary defects."[137] The two main mediators of this eugenic will were law and national enlightenment. Danadzhiev, for example, believed that eugenic legislation was needed but that "proper education and the proper emphasis on ethical rules in the sexual field necessary for the health of the whole nation, will be more useful than the laws themselves."[138] On the other hand, restrictive laws should inevitably be imposed on "ignorant and unprincipled" people who "perpetuated their hereditary suffering."[139] The ambivalence of this double bind between the compulsory effects of law and the voluntary affiliation to the aims of eugenic engineering and propaganda produced hybrid effects like those implied by the 1936 "Decree-Law on Fighting Venereal Diseases."

---

135 Danadzhiev, *Printsipite na psihicheskata higiena*, 103.

136 Konstantin Pashev, "Borbata protiv slepotata i natsionalniya solidarizam," *Oftalmologichen pregled* 2, 2 (1934): 46.

137 Pashev, "Nasledstvenata slepota i germanskiya zakon," 80.

138 Danadzhiev, *Printsipite na psihicheskata higiena*, 147.

139 Danadzhiev, *Printsipite na psihicheskata higiena*, 71.

## Conclusion

The ambivalence of knowledge and power relevant to eugenic projects in interwar Bulgaria becomes clear when considering the regulation of marriage. First, the regime of matrimonial impediments under the "Statute of the Exarchate" combined Church and state law, a fact that was conditioned by the specific relationship of the Church and the state in Bulgaria. Marital law, in the way it was fixed in state-enforced "Statutes of the Exarchate" and in the decisions of the Holy Synod, led the Church to gradually take national health care reasoning into consideration. Indeed, the lack of a binding procedure for issuing pre-nuptial certificates, and the Church's refusal to extend the register of relevant illnesses, expressed its resistance to public interpretations of family health. However, in the late 1920s and in the 1930s, there began an unfinished process, in which Bulgarian marital hygiene shifted its focus toward social hygiene and eugenics. This tendency was represented in the framework of the Church court practice on the one hand and in the fact that the state enacted laws provided for certain national (bio-)policies of pre-marital hygiene on the other. The fact that the "Law on National Health" did not introduce more radical eugenic measures may be regarded as an impetus for the projects to amend the "Statute of the Exarchate" in 1933: after the state did not provide for marriage restrictions and pre-nuptial certificates, eugenicists turned to the Church. Its refusal to carry out the proposed eugenic undertakings was, in turn, natural in view of canon doctrine. Yet the Church did not strictly oppose them. On the contrary, it referred to the responsibility of the state as a legitimate source of power and implied its willingness to tolerate the strategies and instruments of state bio-politics. This regime with its double regulation of marriage health based on a somewhat parallel form of legal competence was one of the possible reasons why the Synod approved voluntary pre-nuptial certification under the 1936 "Decree-Law on Fighting Venereal Diseases." However, the balance of state and Church arguments and the mutual recognition of authority was delicate. The supremacy of the Church regarding all issues of marriage was not abolished in the period surveyed. This fact that marriage was not transformed into a civil institution but remained a Christian union was one of the main hindrances faced by eugenic projects.

Ultimately, the potential of eugenics in Bulgaria to inspire radical, transformative action in the name of the "national organism" remained to a great extent latent; the effects that it produced were more rhetorical and less regulatory. The reasons for this are complex but they should be looked for in the failure of the Bulgarian right-wing project in particular and in the social history of Bulgarian medicine and law in general.

In order to be put into practice, eugenics required a strong institutional basis and high social prestige for biomedicine, which was not the case in Bulgaria. Therefore, the endeavors of eugenic propaganda to realize the implementation of the enacted laws with eugenic relevance was not effective enough. As a result of the legal reception of Western medical models, the laws had a more or less nominal existence. The fact that these aspirations of radical "purification" of the collective "organism" remained unrealized was due to a deficit of conceptual and social legitimacy. In the interwar culture-political situation—particularly in the far-right political regime after 1934—the popularity of eugenic ideas grew. However, there were more urgent measures on the state's agenda. Primarily it was preoccupied with institutional attempts to establish basic, hygienic conditions in the countryside, which is perhaps why eugenic concerns were of secondary importance.

Last but not least, in the collective imagination it was an impossibility that the Bulgarian race become fatally ill for hereditary reasons, i.e. "from within" of its own "nature." Its distressing state was thus most often ascribed to a crisis of modernity and modernization, and not to the Bulgarians as such. The "contamination" of modernization came "from the outside" (from the West), causing a "degeneracy" that became "epidemic" after the Balkan wars and World War One. But it could not "cripple" the authenticity or the substantial "health" of Bulgarian national traits to an extent that would justify more radical eugenic undertakings. Precisely its "epidemic" scale, however, provoked some measures under a broader and more moderate project for public health and hygiene, including legislation. The eugenic ideal of a healthy and "pure" national organism as the most precious cultural capital, infiltrated the public health discourses and influenced social and national politics.

# EUGENIC BIRTH CONTROL AND PRENUPTIAL HEALTH CERTIFICATION IN INTERWAR GREECE

*Sevasti Trubeta*

A world in which maternity is obsolete, with children not born but bred by technical means, is a fiction best described in Aldous Huxley's (1894–1963) utopian novel *Brave New World*. In Huxley's vision, reproduction is a simple act of engineering in an institutionalized, totalitarian system. Children are predestined to be the bearers of certain biological and social qualities. Immediately after "birth," children are "educated" (manipulated by brainwashing) to adhere to society's main aim: that the community be perpetuated.

Despite the fictive and exaggerated character in Huxley's totalitarian dystopia, his basic idea was nevertheless drawn from reality. Neither the concept of breeding humans nor "coercion through persuasion" was fiction; especially in the first decades of the twentieth century, a eugenic movement flourished that strove to put such ideals into practice. According to eugenic criteria at the time, selective breeding was increasingly discussed in international science, politics and the public.[1] Starting either from the Malthusian idea of overpopulation or the preoccupation with the decline in national

* In researching this article I found the personnel at the Parliament Library in Athens to be of wonderful assistance. Furthermore, the generous support of Ms Simone Langner (Bundesarchiv, Berlin) facilitated insights into important data available in the Bundesarchiv. I would like to thank all of them. All translations from Greek are mine unless otherwise specified.

1 Artificial insemination as a measure of positive eugenics was proposed by H. Rohleder at the 1921 First Congress for Sexual Reforms held in Berlin. See Rohleder, "Die künstliche Befruchtung vom Standpunkt der Sexualreform," in Arthur Weil, ed., *Sexualreform und Sexualwissenschaft. Vorträge gehalten auf der I. Internationalen Tagung für Sexualreform auf sexualwissenschaftlicher Grundlage in Berlin* (Stuttgart: Julius Püttmann, Verlagsbuchhandlung, 1922), 213–217.

birth rates, birth control was broadly considered an imperative for population politics. In the first decades of the twentieth century, the best method for "thinking about population" was developed with the desire to improve the biological substance of the population, increasingly on an international scale.[2] Ideas and experiences were exchanged internationally, especially regarding the rationalization of biological reproduction and the ubiquitous population question. Shared concerns and joint activities notwithstanding, each country was mostly concerned with its own betterment.

This chapter is a case study and examines to what extent eugenic birth control was debated and implemented in Greece before the outbreak of World War II. The working hypothesis is that in Greece, demands for control over reproduction were to some extent associated with eugenic thinking, which penetrated scientific and political as well as public spheres. In the Greek case, eugenic ideas and proposals were embedded in social reformist projects. Essentially, eugenicists added their own perspective to contemporary developments concerning the establishment of social and welfare institutions across the country; eugenic conceptions of birth control responded to a generally recognized necessity for coping with the Greek population and social question, especially as these arose in the aftermath of the Balkan Wars and World War I. From the point of view of Greek eugenicists, the population question—one intimately connected to the social question—was closely associated with two major issues: "overpopulation" and "social diseases." This chapter argues that while experts were primarily concerned with the former issue, the preoccupation with social diseases and their intended eradication penetrated several sectors of society and involved intellectuals, politicians, feminists, activists and indeed, the national legislative body. Furthermore, given that "social diseases" were a matter not only for eugenicists but for anyone interested in social or public hygiene, this article traces the interface and differences between eugenic and hygienic reasoning. Within this general setting, the focus here is on

2 Rainer Mackensen, "Ursprünge, Arten und Folgen des Konstrukts 'Bevölkerung' vor dem 'Dritten Reich,'" in Rainer Mackensen, ed., *Bevölkerungslehre und Bevölkerungspolitik vor 1933* (Opladen: Leske und Budrich, 2002), 27–38; and Ursula Ferdinand, "Geburtenrückgangstheorien und 'Geburtenrückgangs-Gespenster' 1900–1930," in Ursula Ferdinand and Josef Ehmer, eds., *Herausforderung Bevölkerung. Zu Entwicklungen des modernen Denken über die Bevölkerung vor, im und nach dem 'Dritten Reich'* (Wiesbaden: VS-Verlag für Sozialwissenschaften, 2007), 77–98.

eugenic arguments for imposing birth control in Greece in addition to attempts to implement eugenic measures, health certification prior to marriage in particular.

## The "Social Question," Overpopulation and the Regulation of Biological Reproduction

While other European countries like Germany and France complained in the interwar period about a declining birth rate, a remarkable population growth in Greece was observed, owing primarily to the relatively high birth rate among the rural population. Following the 1912–13 Balkan Wars, the Greek population increased remarkably as a consequence of the annexation of new territories. Moreover, in the aftermath of the 1922 Greek military defeat suffered against Turkey, approximately 1,200,000 individuals (referred to as "refugees") arrived from formerly Ottoman lands. At the same time, infant mortality, although considerable, was estimated to be manageable. Considering these developments, some of those engaged with the population question in Greece agreed that the country, with a population of approximately 5 million people,[3] was overpopulated. This was mainly asserted on the grounds that the domestic resources were insufficient for nourishing the population on account of major deficiencies in the organization of society and the economy. This already problematic situation further deteriorated in the face of increasing urbanization, which went hand in hand with unemployment in the cities, as well as the decline of agrarian production in the countryside. These transformations had a corresponding impact upon social stratification; a working class was emerging and the middle class was growing.[4]

3 The Greek population was estimated in 1907 to be 2,631,952; in 1920 it increased to 5,016,889 and in 1928 to 6,204,684.

4 On the transformation of social stratification in Greece during the interwar period as well as the accompanying modernization of society and state institutions, see Γιώργος Μαυρογορδάτος, Χρήστος Χατζηιωσήφ, eds., *Βενιζελισμός και Αστικός εκσυγχρονισμός* (Heraklion: Πανεπιστημιακές Εκδόσεις Κρήτης, 1988); Κωστής Μοσκώφ, *Εισαγωγικά στην Ιστορία του κινήματος της εργατικής τάξης. Η διαμόρφωση της εθνικής και κοινωνικής συνείδησης στην Ελλάδα* (Athens: Καστανιώτης, 1988, 3d edition); Πέτρος Πιζάνιας, *Οι φτωχοί των πόλεων. Η τεχνογνωσία της επιβίωσης στην Ελλάδα το μεσοπόλεμο* (Athens: Θεμέλιο, 1993); Athanassios Alexiou, *Zur Frage der Entstehung und Formierung der griechischen Arbeiterbewegung – Eine sozialhistorische Untersuchung* (Frankfurt am Main: Peter Lang, 1994).

The "population question" in Greece was, to a considerable extent, associated with the "social question," itself seen to arise from a contradiction. On the one hand, the precarious situation of the lower social classes (mostly "refugees" and workers) urgently demanded a solution; on the other, existing social and state institutions were in fact unable to meet this challenge. This was the period in which efforts were made to establish an institutionalized public health care system, a system of occupation insurance and related to this, a system of health and social insurance for the working classes.[5] Thus, it is not surprising that the "social question" was also equated with the "labor question."[6] Diverse institutions and political agencies were concerned with managing both the social and labor question. In addition to successive Greek governments, the International Labor Office, political parties of all wings, unionists, labor organizations—such as the Labor Center of Athens [*Εργατικό Κέντρο Αθηνών*, founded in 1910]—were all involved. The solutions proposed were correspondingly varied. Socialist thinkers like Georgios Skliros,[7] and later, leading socialist and communist figures such as Nikolaos Giannios[8] and Pantelis Pouliopoulos,[9] underlined the central role of the working class in solving the "social issue." Furthermore, they queried whether the liberation of the working class could, or indeed should, be realized by means of democratic reforms or class revolution. Nevertheless, socialists and communists, similar to a strong current among liberals, emphasized the necessity for improving living and working conditions; generally speaking, bettering the social situation of the working strata. At the same time, economic experts stressed the urgency for economic and infrastructural reforms. For example, Petros Garoufalias from the Chamber of Commerce in Epirus saw a possible solution to the social question in promoting domestic industry.[10] Others, however, shifted the focus to the "problem of overpopulation," underlining the need to man-

5 See Leda Papastefanaki's contribution to this volume.

6 Αντώνης Λιάκος, *Εργασία και Πολιτική στην Ελλάδα του Μεσοπολέμου. Το Διεθνές Γραφείο Εργασίας και η ανάδυση των κοινωνικών θεσμών* (Athens: Ίδρυμα Έρευνας και Παιδείας της Εμπορικής Τράπεζας της Ελλάδος, 1993), 462.

7 Γεώργιος Σκληρός, *Το κοινωνικό μας ζήτημα* (Εν Αθήναις: Κωνσταντινίδης Α., 1907).

8 Νικόλαος Γιαννιός, *Ο σοσιαλισμός του κράτους* (Athens: Εκδόσεις Σοσιαλιστικού Κέντρου, 1914).

9 Παντελής Πουλιόπουλος, *Δημοκρατική ή Σοσιαλιστική Επανάσταση στην Ελλάδα* (Athens: Γκοβόστη, 1934).

10 Πέτρος Α. Γαρουφαλιάς, "Το δημογραφικόν μας πρόβλημα. Ο υπερπληθυσμός των πόλεων," *Εργασία* 1, 8 (1930), 17. See, also, Πάνος Γραικός, "Το πρόβλημα της αστυφιλίας" (Part 1) *Εργασία* 1, 14 (1930): 10–11; and (Part 2) *Εργασία* 1, 15 (1930): 15–16.

age Greek demography through birth control and the rationalization of human reproduction, specifically by regulating the reproduction of the social classes. It was argued that the institutional framework for both family and marriage be reconsidered and revised in order to meet changing social conditions.

Such conditions certainly helped to shape the international discourse on the decline in the birth rate—or the "ghosts of decline in the birth rate" in the words of Ursula Ferdinand.[11] Even though the Greek case was not congruent with this paradigm, the issue of fertility was deemed to pose a problem extending to Greece, from a dual perspective. The annual growth rate in population notwithstanding, a population decrease was signaled in the long-term. Furthermore, given the shortage in resources and the higher birth rate amongst the poorest classes, the biological quality of the population was inclined to degenerate. To counteract this impending development, measures applied in other countries were examined as possible models for implementation in Greece. A comprehensive account of the international state of affairs on population politics (and "thinking about population") was given by the director of the *General Statistical Survey of Greece* [*Εθνική Στατιστική Υπηρεσία της Ελλάδος*], Ioannis G. Michalopoulos in the Greek daily press on the occasion of the 1931 International Congress for Statistics held in Rome.[12]

Diverse measures for coping with the problem of overpopulation in Greece were proposed by the hygienist and head physician for the municipality of Piraeus, Grigorios P. Chatzivasileiou.[13] He suggested solving the problem of overpopulation by first developing agrarian and industrial economic sectors. Starting from the assertion that different fertility rates between social classes were responsible for an increasingly degenerate populace, Chatzivasileiou pointed to the lurking threat of qualitative decline in the Greek *fili* (race or nation).[14] In his opinion, this had two causes: low

11 Ursula Ferdinand, "Geburtenrückgangstheorien und 'Geburtenrückgangs-Gespenster' 1900–1930."

12 See Michalopoulos' article on the decline of birthrates published over four issues in the Athenian newspaper *Ελεύθερον Βήμα*. Ιωάννης Μιχαλόπουλος, "Η ελάττωσις των γεννήσεων," *Ελεύθερον Βήμα* (22, 23, 24, 26 October 1931).

13 Γρηγόριος Π. Χατζηβασιλείου, "Το πρόβλημα του πληθυσμού εν Ελλάδι," *Αρχείον Οικονομικών και Κοινωνικών Επιστημών* 5, 3 (1925): 257–266.

14 The Greek word *fili* [*φυλή*] means, alternatively, nation or race. In the eugenic and hygienic discourse it typically refers to "nation" with stress upon biological connotations.

natality rates from the upper social classes and diseases plaguing the poorest parts of the population. The former was the result of the intentional abstention from childbearing by the upper classes. This heralded a new model for the modern family, with few children "that caused the degradation of the family in moral terms; both the will as well as the sense of responsibility for the future of the nation and the *fili* do not exist anymore."[15] Drawing upon arguments from German experts such as Fritz Lenz and Sebald Rudolf Steinmetz, Chatzivasileiou advocated the regulation of births rather than their overall reduction. In concrete terms this meant decreasing the birth rate in certain parts of the population, while simultaneously ensuring lower child mortality rates through better health care. As he argued, precarious living conditions, poor nourishment and diseases like malaria and tuberculosis were frequent among refugees from the formerly Ottoman territories and the lower social classes—threatening Greece with degeneration.[16]

Those associating the perceived problem of "overpopulation" with the "social question" were diverse, including many endorsing better care for the poorest classes and the improvement of their living conditions. Emphasis on the quality of the population, however, shifted from social care to managing reproduction through birth control and eugenic selection amongst the population. Eugenic concerns thus emerged at the crossroads of demands for immediate social care extending to all the needy population and a preoccupation with the future of the "national community." Although eugenic ideas were often evident in demands for managing the population question, coherent eugenic concepts were much more rare. One eugenicist with a solid concept, for instance, was the physician and publisher of the journal *Health* [*Υγεία*], Moisis M. Moiseidis, who advanced his ideas through numerous articles and books. Although Moiseidis sympathized with Malthusianism,[17] he nevertheless argued that population politics should comply with the principle that "*quality* may not be sacrificed for *quantity*, all the

[15] Χατζηβασιλείου, "Το πρόβλημα του πληθυσμού εν Ελλάδι," 258.

[16] See another article in which the same author, Chatzivasileiou, approaches hygiene by anthropometry: Γρηγόριος Π. Χατζηβασιλείου, "Ο συντελεστής θνησιμότητος ως δείκτης υγιεινής. Ιδιοστασία και ανθρωπομετρία," *Ιατρικά χρονικά*, 2, 9 (1929): 531–546. On social legislation and social hygiene see Γρηγόριος Π. Χατζηβασιλείου, "Η κοινωνική νομοθεσία παρ' ημίν ως παράγων της κοινωνικής υγιεινής. Η ενδεικνυομένη κατεύθυνσις αυτής," *Ιατρικά Χρονικά* 2, 12 (1929): 784–791.

[17] Μ. Μ. Μωϋσείδης, *Ο Μαλθουσιανισμός άλλοτε και νύν. Έλεγχος γεννήσεων και αποστείρωσις. Μελέτη ιατρική, ευγονική και κοινωνική* (Athens: Τυπογραφικά καταστήματα αδελφών Γεράρδων, 1932), 26.

more when recent statistics undoubtedly bear testimony to evident dysgenic indications in our *fili*."[18] Moiseidis did not stand for an overall decline in births but for a selective reduction of births in the lower social classes, particularly there "where the financial means are insufficient for sustaining and nourishing the descendant and a faulty *quality* is foreshadowed in the bad health of the offspring."[19] He thought the regulation of reproduction among social classes to be a means to eliminate poverty and its side-effects. Birth control would not only prevent degeneration and improve the Greek *fili*, it would further contribute to solving the problem of social inequality by reducing the poor population through restrictions on their biological reproduction.[20] The fight against poverty would thus be effectively realized by the elimination of the poor.

One of the most important Greek eugenicists to propose an elaborated eugenic program of population politics was the pediatrician and liberal politician Apostolos Doxiadis. His case deserves particular attention.

## Controlling Reproduction of the Social Classes through "Biological Politics"

The pediatrician and director of the Pediatric Polyclinic of Athens, Apostolos Doxiadis (1874–1942) also had an outstanding political career. He was Minister of Health (1922–1924), Deputy Minister of Hygiene (1928–1929) and a senator (1932–1933) in various liberal governments. In addition to his involvement in the state apparatus, Dioxiadis was committed to the protection of children, combining professional expertise with political activism. One of his major achievements was to found the "Patriotic Foundation for the Protection of Children" [*Πατριωτικό Ίδρυμα για την Προστασία του Παιδιού*].[21] His field of interest was not restricted to caring for children however; it extended to include more general issues regarding the family as well as the population question. He elaborated concepts of both family

18 Μωϋσείδης, *Ο Μαλθουσιανισμός άλλοτε και νύν*, 26.
19 M. Μωϋσείδης, "Υγιεινή του γάμου," *Βιβλιοθήκη Κοινωνικής Υγιεινής* [special edition on the "Hygiene of Marriage"], vol. 2 (1933), 35.
20 Μωϋσείδης, *Ο Μαλθουσιανισμός άλλοτε και νύν*, 26.
21 See also the contribution by Vassiliki Theodorou and Despina Karakatsani in this volume.

and population politics based on the rationalization of human reproduction through eugenic birth control starting from the assumption that, like all humanity, Greece too was threatened by degeneration. From his point of view, control over reproduction was an integral part of a larger eugenic program that was urgently necessary to impede imminent degeneration. But such radical intervention could also achieve the ultimate goal of the creation of a Greek population of superior biological quality. Indeed, the possible implementation of eugenic measures concerning family, childbearing and marriage gained top priority in Doxiadis' thinking:

> The goal of practical eugenics is to preserve the hereditary capital of families who own it, to improve their living conditions and to exterminate hereditary stigmas that burden so many families today. How may this goal be realized? [...] Providing guarantees for the health of the couple about to get married and evaluating their ethics, by excluding from reproduction all organisms that may endanger the health of the race due to mental or somatic defects, by facilitating marriage by the state through the founding of agencies that will provide all necessary guarantees for the precise estimation of people and things.[22]

An article entitled "Social Biology-Biological Politics" published over two issues of the humanities journal *Greek Letters* [*Ελληνικά Γράμματα*] may be regarded as a programmatic text for his eugenic worldview.[23] This is essentially a biopolitical concept: Doxiadis reasoned that the objective of state politics was the objective of life as a whole. From his perspective, "biology" emerged as a key notion for understanding history and life, for it "teaches us about the history of humanity." State politics should comply with the principles of "social biology," that is, "the science of the conditions of life" forming the basis of society. The conclusions of "social biology" were deemed to provide the basis upon which "state biological politics" should enforce the appropriate size of the population as well as to guarantee its quality, in other words "the value of each individual, that is, the eugenics of a people."[24]

---

22 Απόστολος Δοξιάδης, "Ευγονική," *Βιβλιοθήκη Κοινωνικής Υγιεινής* [special edition on the "Hygiene of Marriage"], vol. 2 (March 1933), 47.

23 Απόστολος Δοξιάδης, "Κοινωνική βιολογία – Βιολογική Πολιτική," *Ελληνικά Γράμματα* 3, 2 (1928): 49–51.

24 Απόστολος Δοξιάδης, "Βιολογική πολιτική με βάση την αύξηση του πληθυσμού της χώρας," *Ελληνικά Γράμματα* 3, 3 (1928): 95–99.

Doxiadis' proposals on managing the population by eugenic means reached the public via numerous articles in several Athenian newspapers and journals. The liberal daily *Free Forum* [*Ελεύθερον Βήμα*], for example, published two serialized articles, the first entitled "Problems of Life" in the autumn of 1930[25] and some months later, a second called "Biological Problems."[26] Being well-informed on developments in the global eugenics movement, Doxiadis buttressed his assertions with arguments and practices from international sources, like Margaret Sanger's efforts to link birth control and eugenics in the United States.[27] He also visited the Second Hygiene Exhibition in Dresden in 1930 and sent a report to the Greek press.[28]

Doxiadis thus proposed a project of human engineering aimed at providing answers to the perceived problem of overpopulation, which he considered closely linked to social stratification. Parts of the overpopulation problem were, he argued, not only insufficient resources and deficiencies in infrastructure; degeneration threatened the whole Greek population, whose biological qualities were in decline. In order to counteract this lurking danger, Doxiadis proposed a rational *modus operandi* for the selective biological reproduction of social classes. According to his model, in Greece there existed four social classes, each with different fertility rates: the ("leading") upper class, consisting of bankers, scientists and officials who tended to bear two children; the middle class, which inclined to three or four children; the *petit bourgeoisie* class, merchants and tradesmen with four to five children; and, finally, a class of workers and farmers, whose families consisted of five to as many as seven children.[29] Although Doxiadis estimated that every year there were 220,000 births and 100,000-120,000 deaths in Greece, (hence an annual net surplus of 100,000 people), he claimed that the overall incongruence in the reproduction by the different social classes adversely affected the general biological quality of the population. Long-term persistence of this development would ultimately endan-

25 *Ελεύθερον Βήμα* (7–14 September 1930) (see details below).

26 *Ελεύθερον Βήμα*, (11, 13–19 January 1931).

27 Απόστολος Δοξιάδης, "Η Αμερικανίς απόστολος κατά της πολυτεκνίας," *Ελεύθερον Βήμα* 9, 9 (1930), 1. On Margaret Sanger's campaign for eugenic birth control, see Christine Rosen, *Preaching Eugenics: Religious Leaders and the American Eugenics Movement* (Oxford: Oxford University Press 2004), 154–157.

28 Απ. Δοξιάδης, "Βιολογία και Εκκλησία," *Ελεύθερον Βήμα* (7 September 1930), 1.

29 Απόστολος Δοξιάδης, "Ο αναπληθυσμός της Ελλάδος," *Ελεύθερον Βήμα* (13 September 1930), 1.

ger the Greek *fili* itself.[30] While increasing natality in the higher classes (the bearers of "hereditary capital"[31]) would be an important task for state politics, the reduction of births in the lower social classes seemed to him, at that point in time at least, premature. Furthermore, he attached top priority to educating the lower classes in a new ethic.[32]

Yet even Doxiadis recognized a contradiction in his proposal: by reducing births in the lower classes, the population might be improved in terms of quality, but at the same time, it would decline in size.[33] In effect, the looming degeneration and, finally, the collapse of the Greek *fili* could not be avoided. In order to counter this contradiction for the benefit of the *fili*, Doxiadis proposed an efficient, rationally organized system of control over childbearing. In doing so, he had in mind a long-term vision as well as a program for immediate implementation, both based on state interventionism: "I do not know if it will ever happen that, like the tax inspector, there will also be a family inspector who, after intensive study and medical certification, will determine the number of children permitted to every married couple."[34] In the short term, however, Doxiadis considered a series of partially institutional measures aimed at reducing child mortality (by means of protection of maternity and childhood) to be more realizable—a project that became the focus of his *Patriotic Foundation for Protecting Children.*[35] Nonetheless, childbearing needed to be controlled and selectively reduced on the grounds of hygienic, social and familial reasons. Anyone who could threaten the health of the Greek *fili* should be excluded from reproduction.[36] Further proposals included the establishment of agencies engaged in providing all necessary guarantees for the health of bridegrooms,[37] ending any propaganda for advocating large numbers of children among 70–80% of the

30 Doxiadis argues much the same point in his column "Biological Problems" [*Βιολογικά Προβλήματα*] in the newspaper *Ελεύθερον Βήμα*.

31 According to Δοξιάδης "Ευγονική," *Βιβλιοθήκη Κοινωνικής Υγιεινής*.

32 Απόστολος Δοξιάδης, "Ποιάν τακτικήν οφείλει να ακολουθήση η κοινωνία και η πολιτεία;" *Ελεύθερον Βήμα* (15 March 1931), 3.

33 Απ. Δοξιάδης, "Περιορισμός ή διαρρύθμισις των γεννήσεων;" *Ελεύθερον Βήμα* (10 September 1930), 1.

34 Απ. Δοξιάδης, "Περιορισμός ή διαρρύθμισις των γεννήσεων;" *Ελεύθερον Βήμα* (11 September 1930), 1.

35 See also Απ. Δοξιάδης, "Η προστασία της μητέρας και του παιδιού," *Εργασία* 1, 37 (1930): 9; and "Η μέριμνα για το παιδί (Τα μεγάλα κοινωνικά προβλήματα)," *Εργασία* 2, 102 (1931): 1316–1318.

36 Απ. Δοξιάδης, "Ο αναπληθυσμός της Ελλάδος, (Βιολογικά Προβλήματα)," (part 7) *Ελεύθερον Βήμα* (13 September 1930), 1.

37 Δοξιάδης, "Ευγονική," *Βιβλιοθήκη Κοινωνικής Υγιεινής*, 47.

population (coinciding with the two lower social classes) and supporting classes where the low birth rate was a result of financial shortages.[38] From his point of view, state support should primarily address the middle classes, for example by establishing institutions responsible for nurturing children from middle class families.

As a prominent politician, Apostolos Doxiadis always had a forum to express his ideas. Nevertheless, he did not escape criticism. Both his warning about the supposed urgent threat of Greek degeneration, as well as his suggestion reducing natality in the poorest parts of the population, met with public criticism. Counter-arguments claimed[39] that the lower social classes needed social care rather than restrictions on birth; not before their living conditions were improved could poor people "acquire the necessary mental clarity for realizing that large numbers of children are a social calamity in the face of the challenge of contemporary life. Today their heads are obfuscated and they alleviate their pain in blind passion that has neither logic nor halt."[40]

Yet it must be stressed that Apostolos Doxiadis' preoccupation with the problem of overpopulation in Greece found little resonance among other eugenicists. Eventually, the majority focused their attention on—and was more sympathetic to—the issue of social diseases than overpopulation. The former, through public dissemination, could reach broader sectors of society than merely specialists in medicine.

## Social Diseases and Vague Boundaries between Hygiene and Eugenics

Eugenic ideas about birth control in Greece were mostly associated with the fight against "social diseases" within the broader context of "social hygiene." The term social hygiene was often used synonymously with "public hygiene", meaning "hygiene of the people" extending to pre-

38 Απ. Δοξιάδης, "Ποιάν τακτικήν οφείλει να ακολουθήση η κοινωνία και η πολιτεία; (Βιολογικά Προβλήματα)," (part 8) *Ελεύθερον Βήμα* (15 March 1931), 3.

39 Δημ. Σαράτσης, "Η Ανθρωπότης εκφυλίζεται," *Ελεύθερον Βήμα* (3 February 1931), 3.

40 [μ.ρ.], "Πολυτεκνία," *Ελεύθερον Βήμα* (10 September 1930), 2.

ventative health politics.[41] The field of hygiene was, however, disputed. Hygienists complained about the confusion caused by the fact that "social hygiene is often misunderstood as restricted exclusively to fighting infectious diseases."[42] Instead, social hygiene included a comprehensive statistical picture of national demography in terms of birth and mortality rates. Moreover, it was concerned with registering the living conditions of the population.[43] Social hygiene was thus afforded to be in a position to shape the fields of social and economic policy in the interests of preventing disease and, more generally, any pathology among the population thought to be immanent. The fields of hygiene and eugenics overlap up to this point; indeed, hygienists and eugenicists essentially used the same vocabulary.

The distinction between "social hygiene" and "eugenics" seems to have preoccupied only declared eugenicists like, for example, Stavros Zurukzoglou (1896–1966), a well-known participant in German and Swiss debates on eugenics.[44] In his Greek publications, Zurukzoglou stressed the differences between hygiene and eugenics: "hygiene protects and promotes any individual irrespective of its value in terms of biology and culture," while "eugenics aims to safeguard and proliferate, for the most part (if not exclusively), those who hold true biological and cultural value."[45]

In practice, hygienists and eugenicists shared a desire to combat social diseases and control population developments through scientific and legislative means. These common goals rendered distinctions between hygiene and eugenics fluid. Furthermore, the majority of those using either hygienic or eugenic rhetoric generally associated population policy with the prevention of social diseases as well. The notion of social disease posed a category for describing "social pathologies" in medical terminology. In this way

---

41 Νικόλαος Β. Οικονομόπουλος, *Κοινωνική υγιεινή: Κοινωνική πρόνοια - κρατική μέριμνα* (Εν Αθήναις: Πετράκης 1922); and Κωστής Χαριτάκης, *Τα νεώτατα δεδομένα επί της κοινωνικής υγιεινής: αρχαί και κριτήρια οργανώσεως της δημοσίας υγιεινής* (Εν Αθήναις: Εκ του Εθνικού Τυπογραφείου, 1929).

42 Χατζηβασιλείου, "Το πρόβλημα του πληθυσμού εν Ελλάδι," 265.

43 See for example, Παναγ. Γ. Παναγιωτάκου, "Κοινωνική Ιατρική ή Κοινωνική Υγιεινή," *Ιατρικά Χρονικά* 2 (1928), 95.

44 On Stavros Zurukzoglu and his scientific and political career in Switzerland, see Sevasti Trubeta, "Anthropological Discourse and Eugenics in Interwar Greece," in Marius Turda and Paul J. Weindling, eds., *Blood and Homeland: Eugenics and Racial Nationalism in Central and Southeast Europe, 1900-1940* (Budapest: Central European University Press, 2007), 123–142.

45 Σταύρος Τσουρουκτσόγλου, "Η Ευγονία" (1st part), *Υγεία*, 1, 11 (1925), 222. The chapter is based on the author's German book; see Stavros Zurukzoglu, *Biologische Probleme der Rassenhygiene und der Kulturvölker* (Munich: J. F. Bergmann, 1925).

the terminological couplet "health vs. disease" became the cornerstone of a holistic worldview and, at the same time, of a reductionist perception of society. Health was defined in terms of moral and social norms themselves derived from the prevalent value system. Consequently, "social diseases" were defined so broadly that they included any kind of supposed social abnormality deemed to threaten the "collective body" with biological, indeed ethical, degeneration. Either in terms of poor health or moral deficits, social diseases were regarded as transmittable to the next generation—which justified the prevention of affected individuals reproducing. Bearers of social diseases were largely thought to reside in the lower social classes on account of their disadvantageous living conditions that so encouraged excess and promiscuity. "How many workers and burly villagers," Moisis Moiseidis asked, "have cachectic children because they combine alcohol and tobacco with sex? Children who are conceived on Saturday (the *enfants du samedi*, according to Legrain), the day when the workers draw their wages, are mentally degenerated."[46]

In interwar Greece, "social diseases" mooted as possible reasons for precluding childbearing and marriage were venereal diseases (first of all syphilis), hereditary propensity to criminality, insanity, alcoholism, drug addiction, leprosy, tuberculosis[47] and, sometimes, cancer.[48] The first three of these held prominent positions. Either interlinked or separately, venereal diseases, criminality and insanity posed the symbols *par excellence* for hereditary social pathology. For instance, "hereditary syphilitics" were a true danger to society, for "most invalids, disabled as well as the most degenerate and criminal individuals stem from them."[49] This claim was advanced by the dermatologist, veneorologist and psychoanalyst Nikolaos Drakoulidis, best-known as a novelist under his *nom de plume* "Angelos Doxas."[50] Moreover, he viewed syphilis, alcoholism and drug addiction among parents to

46 Μωϋσείδης, "Υγιεινή του γάμου," *Βιβλιοθήκη Κοινωνικής Υγιεινής*, 30.

47 [Anonymous], "Η προ του γάμου ιατρική εξέτασις," *Υγεία* 2, 22 (1925), 435–437; Μαρίνος Σπ. Βάθης, "Δια την πράξιν. Ο γάμος των συφιλιδικών," *Ιατρική - Ελληνική Ιατρική Επιθεώρησις* 3, 7 (1925), 178–179.

48 Νικόλαος Δρακουλίδης, "Γάμος," *Μεγάλη Ελληνική Εγκυκλοπαίδεια*, vol. 8 (2nd revised edition - no year), 126–127.

49 Νικόλαος Δρακουλίδης, "Η προ του γάμου ιατρική εξέτασις," *Βιβλιοθήκη Κοινωνικής Υγιεινής* [special issue on the "Hygiene of Marriage"], vol. 2 (1933), 56.

50 Nikolaos N. Drakoulidis (1900-1985) studied medicine in Athens, Vienna and Paris, specializing in venereal diseases and psychoanalysis. He introduced the field of sexology into Greek medicine.

be the main reason for juvenile delinquency, as children of addicted parents, "when they survived, suffer from epilepsy, idiocy, insanity; [they] are criminals, cachectic and degenerate."[51]

Importantly, once health ceased to be solely an individual matter but became a social resource, childbearing emerged as a key issue for physicians as well as for those concerned with the health of Greek society at large. Both hygienists and eugenicists attempted to gain influence on public consciousness through the popular diffusion of their ideas. The popularization of hygienic and eugenic discourses for the public posed a challenge to be met via propaganda—or "enlightenment campaigns," to use hygienists' and eugenicists' rhetoric.[52]

One agency dealing with the fight against social disease through the dissemination and popularization of hygienic and eugenic ideas was the National Council of Greek Women [*Εθνικό Συμβούλιο Ελληνίδων*]. This feminist umbrella association encompassed numerous organizations and enjoyed a great deal of influence in the middle class women's movement. Indicative of the importance the Council attached to the role of social hygiene was the establishment of an internal commission (a "Department for Social Hygiene") engaged with questions of hygiene. The stated aim of the Council's activities was "to combat and reduce the diseases which plague Greece, to improve the health and eugenics of our race [in original: *ράτσα*], to teach mothers how to maintain the physical and mental vitality of their children [ ... ] to combat ignorance and superstition and to provide general hygienic educational propaganda."[53] Additionally the National Council of Greek Women suggested a program of preventive hygiene that included propaganda films, lectures held in factories and other workplaces and the establishment of information centers for the public. Clearly, this feminist organization intended to reach society at large by means of lectures addressed not only to scholars but to the wider Greek populace. The first series of lectures covered social diseases "such as tuberculosis, syphilis, malaria, trachoma and

51 Δρακουλίδης "Η προ του γάμου ιατρική εξέτασις," 59.

52 On the demand to popularize hygiene see, for example, the anonymous "Το υγειονομικόν πρόγραμμα (Κοινωνική υγεινή)," *Εργασία* 1, 27 (1930), 19.

53 See the programmatic opening text of the Library of Social Hygiene by N. N. Δρακουλίδης, "Δύο λόγια για την 'Βιβλιοθήκη Κοινωνικής Υγιεινής," *Βιβλιοθήκη Κοινωνικής Υγιεινής*, special issue on "Social Diseases" ("Τα κοινωνικά νοσήματα"), vol. 1 (1932): 5–13.

drug addiction."[54] These talks were held in spring 1932 at the Athenian philological club "Parnassus" [*Παρνασσός*], the Archaeological Association and in two residential districts of Athens, Petralona and Vyronas.[55] Further lectures on hygiene in marriage and childbearing followed. In 1932, the "Department for Social Hygiene" established the "Library of Social Hygiene," a book series under the editorship of Nikolaos Drakoulidis; the second issue of the series was dedicated to hygiene in marriage. Here, prominent scholars, most of them advocates of eugenics, disseminated their views on eugenic birth control and restrictions on reproduction.

These efforts to popularize and spread eugenic ideas on reproduction were also accompanied by demands on state institutions to take action. By doing so, eugenicists joined other critics of state policy in Greece who complained that the welfare and health care system had serious infrastructural deficiencies. At the very least, policies for preventing diseases by means of improving living and working conditions, ensuring proper nourishment and so forth would benefit the state itself, since the expenses of health care would be reduced while the working classes would be able to work better and longer.[56] Eugenicists, in particular, shifted their focus to evaluating the population, demanding a selection of those who most deserved to benefit from welfare institutions. In a series of efforts to establish a health care system and a system of occupational and social insurance, measures suggested by eugenicists were considered and discussed as possible solutions. Eugenic interventions to save the demographically sinking ship through restrictions on reproduction were negotiated, along with a number of proposed institutional measures for controlling births. Naturally, eugenicists proclaimed concern for the common good, and their humanitarian rhetoric (and often their professional status as experts) made them a legitimate partner in the common fight against social disease. One of most discussed measures in this context was the establishment of premarital medical examinations as a precondition for sanctioning marriage.

54 Δρακουλίδης, "Δύο λόγια για την 'Βιβλιοθήκη Κοινωνικής Υγιεινής," 11.

55 [S.], "Τμήμα Υγιεινής. Διαλέξεις κοινωνικών νοσημάτων," *Ελληνίς-Μηνιαίον Περιοδικόν του Εθνικού Συμβουλίου των Ελληνίδων* 12, 6–7 (1932): 131–132; and 14, 11 (1934), 231. See also N. N. Δρακουλίδης, "Η προπαγάνδα υπέρ της κοινωνικής υγιεινής," *Ελληνίς-Μηνιαίον Περιοδικόν του Εθνικού Συμβουλίου των Ελληνίδων* 12, 10 (1932): 201–202.

56 N. B. Οικονομόπουλος, "Η υγιειονολογική πολιτική εις την χώραν μας," *Ιατρική Εφημερίς* (14 January 1934), 2.

## Health Certification Prior to Marriage

Attempts to control marriage and reproduction were based on assumptions that the purpose of the family was to constitute cells in a healthy society, nation and humanity, providing the collective body with superior human capital. One practical way to meet this goal was by establishing appropriate legislative measures. In interwar Greece this meant health certification for bridegrooms prior to marriage. Physicians occupied a prominent position in debates on this issue as they were deemed to possess expertise and "objective knowledge" about health, prevention and the combating of diseases. Yet this did not stop public discussions. Aside from medical publications,[57] possible restrictions on marriage and reproduction were also discussed in the Athenian daily press and in popular journals. The liberal journal *Work* [*Εργασία*], for example, provided a permanent forum for debates on public hygiene and the population question. The journal also published papers presented at the First Pan-Hellenic Congress for Hygiene held in April 1930 in Athens, one of which specifically addressed premarital medical examinations.[58] The Greek public also became familiar with this issue via numerous articles by Nikolaos Drakoulidis, one of the most fervent proponents of eugenic birth control. Drakoulidis published substantially on marriage, birth control and prenuptial health certification in the Greek popular press and newspapers.[59] Furthermore, he also wrote the entry on "Marriage" in the *Great Greek Encyclopedia* [*Μεγάλη Ελληνική Εγκυκλοπαίδεια*] where he addressed, among other matters, the necessity for state control over reproduction by means of prenuptial health certification.[60]

One of the earliest references to premarital health examinations is found in a 1917 essay entitled "Eugenics," written by the physician Michael Kairis. The text was based on a lecture of the same title delivered by the author on

---

57 See the anonymous "Αντί δελτίου. Οι πρωτόγονοι," *Ιατρική Εφημερίς* (22 November 1936), 1 (reprinted from the newspaper *Τύπος* (22 October 1936). See below for discussion of the legislative institutions.

58 M. Μωϋσείδης, "Η προ του γάμου ιατρική εξέτασις (Εισήγησις εις το Α' Πανελλήνιον Συνέδριον της Υγιεινής, 29 Απριλίου 1930)," *Εργασία* 1, 19 (1930): 17–19.

59 See Drakoulidis' serialized article in the Greek press, Νικόλαος Δρακουλίδης, "Η προ του γάμου ιατρική εξέτασις," *Ελεύθερος Άνθρωπος* (17 February 1932); see also the earlier, similar article in the newspaper *Εσπέρα*, published on 9 and 10 December 1925.

60 "Γάμος," *Μεγάλη Ελληνική Εγκυκλοπαίδεια*, 8th Vol., (2nd revised edition, no year), 126–127.

24 March 1917 at the Sinaia Academy in Athens.[61] Therein Kairis expressed doubts about both the necessity and possibility of establishing this very eugenic measure in Greece. Nevertheless, in the ensuing years references to eugenic premarital examinations proliferated. Several lectures and series were organized, such as those by the National Council of Greek Women, as mentioned above. Another, still later instance was a lecture titled "Health of the Bridegroom and the Medical Certification for Marriage" and delivered by the air force physician K. Ntoulas to the Chamber of Industry and Commerce in Thessaloniki at the end of the 1930s.[62]

Despite sharing sympathies with the idea of establishing medical certification as a means to ensure healthy offspring, opinions varied significantly with respect to its practical implementation. The first point of contention was whether such a form of certification should have a voluntary or obligatory character. A further controversial point was over the severity of measures to be applied in cases of diagnosed diseases or disability. The additional options discussed were manifold, including legal prohibition of marriage, voluntary abstention from marriage by couples, voluntary abstention from childbearing and contraception, voluntary abortion and even sterilization. Sometimes proposed measures offered a combination of different options. For instance, the psychiatrist Simonidis Vlavianos endorsed both a ban on marriage by law as well as voluntary sterilization for the insane, feeble-minded, alcoholics and those suffering from tuberculosis and syphilis.[63]

Irrespective of these divergent views concerning the measures most appropriate to implementation in Greece, all participants in this eugenic debate agreed on the necessity to popularize and diffuse the tenets of social hygiene as widely as possible. This would be managed by enlightening the population, chiefly the lower social classes, about the consequences of alcohol consumption and venereal disease. In short, the public needed to become aware of the importance of social hygiene and eugenics. To do so, a form of health propaganda would aim to persuade couples to let them-

61 Μιχ. Νικ. Καΐρης, "Η Ευγονία" (Athens: Εκ του Τυπογραφείου Αδελφών Β Φραντζεσκάκη, 1917).

62 Κ. Ντούλας, "Η Υγεία των Μελλονύμφων και το Ιατρικό Πιστοποιητικό του Γάμου" (Θεσσαλονίκη: [n.p.] 1936).

63 Σιμ. Βλαβιανός, "Ο νέος νόμος του Χίτλερ δια την στείρωσιν των μεταδοτικών νόσων κληρονομικώς," *Ιατρική Εφημερίς* (31 December 1933), 4.

selves be examined voluntarily and, if deemed medically pertinent, to abstain from marriage and procreation. Generally, top priority was given to winning over the Greek public through argumentation rather than trying to enforce a ban on marriage by legal means. Physicians like Michael Kairis, Moisis Moiseidis, K. Ntoulas and many other participants in eugenics debates underlined the social imperative of choosing spouses according to eugenic criteria and creating ideal conditions for conceiving children. Moiseidis developed a "code of marriage" which included eugenic criteria for choosing a spouse.[64] Yet, propaganda alone was deemed insufficient and needed to be accompanied by institutional measures. Indeed policy proposals came from several eugenicists. In a serialized article entitled "Race and Health," the Professor for Physical Anthropology at the University of Athens, Ioannis Koumaris, advocated voluntary certification alongside the establishment of consultation agencies for couples about to get married.[65] For his part, Nikolaos Drakoulidis claimed that consultation alone was an insufficient measure. Like Apostolos Doxiadis, Drakoulidis considered a broad-based network of agencies charged with the organization and management of educational propaganda for the Greek population to be necessary. Drakoulidis thus proposed founding a Eugenic Association to coordinate these agencies.

Another point of contention was the extent to which conditions in Greece were appropriate for effectively implementing such a certification procedure. This skepticism was in keeping with general reservations about implementing eugenic measures (including sterilization) on account of the insufficient state of infrastructure and scientific progress, not to mention the essential problems in the organization of the Greek state and society. On the other hand, however, restrictions on reproduction in Greece were often considered by eugenicists to give impetus to the establishment of welfare institutions—such as the newly-formed occupational and health insurance system—the social care system and so on. But this could occur only if needy individuals posing a burden to the development of the welfare system could be reduced. As the psychiatrist Simonidis Vlavianos argued, "the

64 Μωϋσείδης, "Υγιεινή του γάμου," *Βιβλιοθήκη Κοινωνικής Υγιεινής*, 20–28.

65 Ιωάννης Κούμαρης, "Φυλή και Υγεία," *Βραδυνή*, 1 part (13 October 1938), 1 and 5; 2 part (14 October 1938), 1 and 5.

ill and cachectic products of such criminal marriages [between alcoholics, madmen, the feeble-minded and idiots] are a terrible burden for the social body in terms of both their appearance as well as material expenditure."[66]

Radical proposals to restrict marriage and reproduction generally came from eugenicists attaching a top priority to state intervention rather than to educational campaigns. Such a case was the psychiatrist and lecturer at the University of Athens, Georgios Vlavianos. In a 1938 lecture given at the club "Parnassus," he advocated establishing by law a health certificate for the bridegroom prior to marriage in order to impede the reproduction of mentally defect offspring.[67] Further measures could extend to committing those affected in an asylum and compulsory sterilization. In his opinion, such procedures should be applied not exclusively but concurrently. Thus, if a couple, after medical examinations, was deemed to be unfit for bearing healthy children, then both measures should take place: the marriage should be forbidden and the couple should be either committed to an asylum or sterilized. Supporting this harsh verdict was Vlavianos' conviction that a ban on marriage would not preclude these individuals from bearing children. Hence, a ban on marriage without intervening in procreation would encourage childbearing outside the institution of marriage, meaning that reproduction would escape state control. Georgios Vlavianos also considered the sterilization of mentally disabled individuals, and especially criminals, to be necessary. He claimed that 75% of criminals were afflicted with "congenital perversion" and a "pathologically heightened sex drive." He regarded such measures to be economically beneficial for the Greek state, which could profit from both the reduction of asylum inmates as well as criminality and the expenses of combating it.[68]

Dissenting voices were, however, not lacking. A prominent opponent of eugenic control over reproduction was the bacteriologist, hygienist and Professor for Social Biology at the Panteion Institute in Athens, Dimosthenis Eleftheriadis.[69] Beside his academic career, Eleftheriadis was also publicly engaged as a sanitary inspector in the Ministry of the Inte-

66 Σιμ. Βλαβιανός, "Ο νέος νόμος του Χίτλερ δια την στείρωσιν των μεταδοτικών νόσων κληρονομικώς."

67 The text of the lecture was later published in the daily press; see Γεώργιος Ι. Βλαβιανός, "Ψυχική Υγιεινή και Ψυχική Ευγονική" (Athens: [n.p.] 1939).].

68 See Βλαβιανός, "Ψυχική Υγιεινή και Ψυχική Ευγονική."

69 Dimosthenis Eleftheriadis (1885–1964) studied medicine in Greece and Germany, specializing in hygiene and bacteriology.

rior (Department of Hygiene) and Secretary General in the Ministry of Hygiene (March–August 1936) in the conservative governments of Konstantinos Demertzis and Ioannis Metaxas.[70] A representative example of his views was his contribution to the aforementioned volume on the *Hygiene of Marriage*, published by the National Council of Greek Women. In his article, entitled "Heredity in the Family," Eleftheriadis argued: "In recent years there has been a great deal of discussion about the improvement of humankind by means of gymnastics or hygiene. Unfortunately, they all have no relation to reality."[71] Although not questioning the importance of heredity upon the individual, family and society, he expressed moderate skepticism about the effectiveness of control over childbearing by means of premarital medical examinations. Drawing arguments from biology, he claimed that medical examinations were unable to expose hereditary, pathological features when these only existed in a latent form. Moreover, "by means of athleticism and hygiene we may improve to some extent the phenotype but not the genotype of the people."[72] Later, Eleftheriadis further radicalized this critical position. Immediately after the Second World War, he spoke out emphatically against birth control by eugenic means. A 1947 article in the Athenian newspaper *Nation* [*Έθνος*] reported his view about the medical certificates for bridegrooms as well as the question of degeneration. In Eleftheriadis' opinion, hereditary diseases were not so pervasive that they could result in degeneration of a whole society. Consequently, he stood against all eugenic measures, especially prenuptial health certification. He characterized the latter as "graver and more unrealistic than sterilization, not so much due to the difficulty of its implementation (about which I may make only the worst and most disgraceful predictions), but rather because neither are such measures necessary nor is science able to exterminate or reduce hereditary diseases."[73] Eleftheriadis continued: "Most of the proponents of the prenuptial health certification are Neo-Malthusians masquerading as eugenicists. I do not need to

70 Konstantinos Demertzis died in April 1936, one month after his appointment as Prime Minister. Ioannis Metaxas succeeded him. Dimosthenis Eleftheriadis was deposed from office at the Ministry of Hygiene after Ioannis Metaxas' putsch on 4 August 1936.

71 Δημοσθένης Ελευθεριάδης, "Η κληρονομικότης εις την οικογένεια," *Βιβλιοθήκη Κοινωνικής Υγιεινής* [special issue on the "Hygiene of Marriage"], vol. 2 (1933), 84.

72 Ελευθεριάδης, "Η κληρονομικότης εις την οικογένεια," 84–85.

73 Νίκος Σταθάτος, "Κοινωνικά ζητήματα. Το προ του γάμου πιστοποιητικόν υγείας," *Έθνος* (13 October 1948), 1.

discuss the economic, demographic and above all ethical consequences of such measures as I consider these self-evident."[74] In his textbook, *Lessons of Social Biology* [*Μαθήματα Κοινωνικής Βιολογίας*], which he wrote for his classes at the university, Eleftheriadis also opposed eugenic means of birth control, advising people to "distrust those who allege to wish their happiness."[75]

Obviously, in interwar debates on birth control and marriage restrictions in Greece, opinions differed remarkably ranging from suggestions to implement radical measures to moderate skepticism and even open rejection of any measure to control marriage and reproduction by eugenic means. To be sure, participants in this debate were manifold, right up to Greece's legislative body.

## Eugenic Birth Control in State Institutions

In 1870 a deputy from the island of Tinos, Dr I. A. Zallonis, requested that the Greek parliament impose a legal ban on marriage for dwarves as well as individuals suffering from tuberculosis, leprosy, elephantiasis and cachexia caused by syphilis. Couples about to marry, he continued, should have their health attested by two physicians and to submit a health certificate issued by a judge. In this 16-page petition, I. A. Zallonis also addressed a problem that could emanate from such legislation, namely, whether marriage restrictions violated individual rights. For Zallonis, "individual freedom is inviolable as far as it does not harm the state."[76]

It is unclear whether the Greek parliament discussed this request as there is no reference in the minutes of the parliamentary debates from that time. However, some decades later similar proposals again cropped up concerning eugenic restrictions on marriage and childbearing. These appeared on the agenda of the legislative body when it debated health legislation. For example, in 1919 during a parliamentary debate over the draft law "On the Establishment of Medical-Practices, Hospitals, Convalescent Homes

74 Σταθάτος, "Κοινωνικά ζητήματα."

75 Δημοσθένης Ελευθεριάδης, *Μαθήματα Κοινωνικής Βιολογίας* 2ον Μέρος (Athens: [n.p.] 1948), 171.

76 I. A. Ζαλλώνης, *Υπόμνημα προς την Βουλήν Περί Κωλυμάτων Γάμου – Υπό Ιατρικήν Έποψιν* (Εν Ερμουπόλει: [n.p.] 1870).

and Sanatoriums against Tuberculosis,"[77] the deputy from Cyclades, S. Papavasileiou, recommended imposing a ban on marriage for those suffering from tuberculosis. He argued that tuberculosis was a hereditary disease, typically a usual side-effect of syphilis. The wording of the draft law was, in his opinion, insufficient for eradicating both the causes and spread of tuberculosis. For Papavasileiou the necessary and most effective measure would entail legal marriage restrictions for consumptives—by means such as the health certification of couples prior to marriage or their sterilization—as was the case in many other countries. His model was the sterilization law passed in the American state of Indiana in 1907, the text of which he read during the 1919 parliamentary session. He nevertheless expressed skepticism about passing a similar law on grounds that the Greek public was likely to disapprove. Ensuing parliamentary discussions are illustrative of the diverse opinions about eugenics among Greek politicians. Michael Katopodis (whose professional background was medicine), for instance, disputed the hereditary character of tuberculosis. Moreover, he declared himself opposed both to marriage restrictions for those who suffered from tuberculosis and against sterilization for any reason. Another deputy and physician, Konstantinos Alavanos, agreed with both the establishment of a medical certification system for bridegrooms and compulsory sterilization. In responding to Papavasileiou's request, Alavanos also underlined the state's need to strengthen intervention in health institutions and limit the actions of private health agencies.

Despite heated discussion the proposed eugenic restrictions on marriage and childbearing did not pass in that parliamentary session. Yet the same issue appeared anew on the agenda of the ministry of Hygiene and Welfare [*Υπουργείο Υγιεινής Προνοίας και Αντιλήψεως*] some years later. As the medical press reported,[78] then Minister K. Filandros, addressed a letter to the Church [Holy Metropolis of Athens/*Ιερά Μητρόπολις Αθηνών*], which

[77] See the debate of the 20th parliamentary session of 17 December 1919, reported in documents of the parliamentary debates: Α' συζήτησις επί των άρθρων του σχεδίου νόμου "Περί ιδρύσεως αντιφυματικών ιατρείων, νοσοκομείων, αναρρωτηρίων και σανατορίων." Πρακτικά Ενάρξεως των Εργασιών της Δ' Συνόδου της Κ' Βουλευτικής Περιόδου. Εφημερίς των Συζητήσεων της Βουλής Συνεδρίασις 20ή της 17ης Δεκεμβρίου 1919, 287-292.

[78] [Anonymous], "Η προ του γάμου ιατρική εξέτασις," 58.
The correspondence between the Ministry and the Church was published by Nikolaos Drakoulidis, see Δρακουλίδης, "Η προ του γάμου ιατρική εξέτασις," 65–67.

queried the possibility of enforcing medical certification as a condition for allowing an Orthodox marriage. If the Church would agree, he asked that it lead the way in defining the diseases that should be considered as impediments to marriage.[79] The eugenic idea in the foreground of this initiative was stated explicitly in Filandros' letter: "One of the main concerns of the State Health Office is the study and possibly the implementation of eugenics." The Minister signing the letter indicated that the ongoing dysgenic tendency among Greeks, the infant mortality rate and the dissemination of venereal diseases (mainly syphilis) all demanded that "a series of hygienic measures for protecting Greek ancestry [in original: *γένος*] from dysgenics and promoting eugenics that are the foundation of every hygienic goal" be immediately established in Greece, "like in the rest of the civilized world." The Church immediately responded to the letter, approving the Minister's proposals.[80] Nevertheless, a law demanding prenuptial health certificates was not passed then either. Still, the initiative attracted the medical press' attention, which emphasized the enormous importance of this matter for society, even going so far as to suggest convening a commission to engage in examining the possible implementation of marriage certificates.[81]

Thereafter, state institutions continued searching for a mode to draft a law of international standards. Immediately following the Church's positive response, Greece's government contacted the German Ministry of the Interior [*Reichsminister des Inneren*], requesting advice based on the German experience of legislative measures to safeguard the health of offspring.[82] The German Ministry of the Interior informed the Greek government about the lack of any recent domestic legislation concerning this matter.[83] This

79 The Church was authorized to issue marriage licenses in Greece prior to the establishment of a Civil Marriage Law in 1980.

80 See the Church's (*Ιερά Σύνοδος της Εκκλησίας της Ελλάδος*) answer on 4 November 1925, reprinted in Δρακουλίδης, "Η προ του γάμου ιατρική εξέτασις," 65–67.

81 [Anonymous], "Η προ του γάμου ιατρική εξέτασις."

82 Das Auswärtige Amt, Berlin. Dem Herrn Reichsminister des Inneren. Abschrift VIZ 9993, Griechische Gesandtschaft, Berlin W8, den 15. Januar 1926 (Request from 18 December 1925). BA, R 1501/109380, Blatt 146.

83 See Reichsminister des Inneren, An das Auswärtige Amt, Berlin 15 Januar 1926. Betreff: Gesundheitszeugnisse für Ehebewerber. BA, R86 N. 2372. On the German case see, Sylvia Kesper-Biermann, "'Ehegesundheit' als bevölkerungspolitisches Problem. Internationale Dimensionen von Diskussion und Gesetzgebung in der Weimarer Republik," in Ursula Ferdinand and Josef Ehmer, eds., *Herausforderung Bevölkerung. Zu Entwicklungen des modernen Denken über die Bevölkerung vor, im und nach dem 'Dritten Reich'* (Wiesbaden: VS-Verlag für Sozialwissenschaften, 2007), 123–132.

negative answer might have effected the temporary interruption in government attempts to realize Greek eugenic ideas. This final failure to gain information notwithstanding, state institutions in Greece appear to have been involved in an international network of countries exchanging experiences and ideas concerning eugenic legislation in the interwar period.[84]

Some years later, marriage restrictions for eugenic reasons again appeared on the agenda of the national legislative body. This happened during the 1930 debate on the revision of the Civil Law, in the third liberal government of Eleftherios Venizelos.[85] One member of the revision commission, the Professor for Private International Law, Georgios Maridakis,[86] pled for a legal ban on marriage for lepers, epileptics and those suffering from either tuberculosis or syphilis.[87] Although his proposition was finally rejected by 11 of 15 members of the revision commission, this debate brought to light controversial views over the possible implementation of eugenic measures in Greece and the practice of eugenics at large. Lepers had already been prohibited from marrying by law since 1920.[88]

The members of the revision commission agreed on the importance of eugenics to ensure the health and welfare of society and the nation. But beyond consensus on this point, their opinions differed remarkably with

84 The German Federal Archive (*Bundesarchiv*) testifies to the correspondence between such state institutions worldwide. The German government itself adapted texts of laws on marriage restrictions passed in other countries, including Brazil, Mexico, Turkey and the Scandinavian countries. This form of interaction and networking gives good reason to approach eugenics as a global movement. See Deborah Barrett and Charles Kurzman, "Globalizing Social Movement Theory: The Case of Eugenics," *Theory and Society* 33, 5 (2004): 487–527.

85 The first Venizelos government lasted from 1910 to 1916, the second from 1918 to 1920 and the third from 1928 to 1933.

86 Georgios Maridakis (1890–1979) had an international career as a member of the International Arbitral Tribunal (The Hague, 1940) and the European Tribunal for Human Rights (Strasbourg, 1959).

87 The proceedings of the commission were published in *Εφημερίς των Συζητήσεων του Νέου Αστικού Κώδικος*. See Maridakis' proposal in Αναθεωρητική Επιτροπή του Αστικού Κώδικος, 6th session, 29 December 1930, *Εφημερίς των Συζητήσεων του Νέου Αστικού Κώδικος*, no. 5, (Athens, 5 May 1931), 91.

88 See the Law No 2450 from 1920 which prohibited the marriage among lepers: Νόμος 2450 "Περί μέτρων προς περιστολήν της λέπρας," ΦΕΚ 182 Α'-18/2/1920. Marriage restrictions for lepers were abrogated in 1955 by virtue of the law ΝΔ 3369/1955 ("Περί μέτρων καταπολέμησης της Λέπρας," ΦΕΚ 258 Α' 23/9/1955), apparently after the discovery that leprosy was not a contagious disease. For references to this legislation see Panagiotis Zeppos's contribution to the discussion on the prenuptial health certificate at the *Greek Society for Eugenics and Human Genetics* proceedings; see Βασίλειος Βαλαώρας, Φαίδων Βεγλερής, Ιωάννης Δενέζης, Παναγιώτης Ζέπος, Νικόλαος Λούρος, Βιργινία Τσουδερού, "Το Προγαμιαίον Πιστοποιητικόν" [meeting of 17 February 1964], published in *Κοινωνική Επιθεώρησις. Μηνιαίον Όργανον Ενημερώσεως του Υπουργείου Κοινωνικών Υπηρεσιών. Πρακτικά της Ελληνικής Εταιρείας Ευγονικής και Γενετικής του Ανθρώπου* 4, 9–12 (1976): 5–29. See Zeppos' contribution, 17–20.

respect to the practical means for achieving this common goal. Controversial matters included whether or not the diseases in question were indeed of a hereditary character and whether radical measures, such as a ban on unhealthy marriages, were appropriate and necessary. Furthermore, members of the commission were skeptical about the feasibility of suggested eugenic measures especially given the lack of technological and scientific infrastructure in Greece as well as severe deficiencies in health care system. In addition to these reservations, a Professor for civil rights, Georgios Balis,[89] underlined the immanent risk of proposed restrictions becoming "repressive and socially dangerous."[90] Another member of the commission, the economist Kyriakos Varvaresos added to this criticism, arguing that prohibiting marriage on the grounds of poor health was nothing but a Malthusian principle that restricted individual freedom.[91] Varvaresos thus recognized an ethical dilemma in the proposal, one deriving from the incompatibility of eugenic care for the community and the protection of individual rights: On one hand, safeguarding the health of the nation and the people was a necessity, yet on the other, restrictions on marriage for health reasons violated individual rights. Varvaresos statement was countered by another member of the commission, the constitutionalist Alexandros Svolos—a man from the liberal wing of the socialist movement who was later actively involved in the resistance during World War II.[92] In Svolos' opinion, the dilemma could be resolved if everyone accepted that the top priority was Greece's common good. He assumed that the highest ideal, the good of the collective body and future generations, prevailed over

89 Georgios Balis (1879–1959) was one of the most important experts on civil rights in Greece. In 1938 he was tasked with the revision of the Civil Law. His concepts formed the basis for the Greek civil code still in force today.

90 Αναθεωρητική Επιτροπή του Αστικού Κώδικος, 6th session, 29 December 1930, *Εφημερίς των Συζητήσεων του Νέου Αστικού Κώδικος* 5 (12 March 1931): 92.

91 Kyriakos Varvaresos (1884–1957) studied law in Athens, as well as economics, political science and statistics in Germany. After completing his studies he was appointed head of the Department for Statistics in the Greek Ministry of the National Economy in 1911. In 1932 he became Minister of Economics in the government of Eleftherios Venizelos and Alexandros Papanastasiou. Varvaresos wrote numerous monographs, among others, "The Theory of Population;" see Κ. Χ. Βαρβαρέσος, *Η θεωρία του πληθυσμού* (Εν Αθήναις: Τυπογραφείον της Β Αυλής Α Ραφτάνη, 1912).

92 Alexandros Svolos (1892–1956) studied jurisprudence in Athens, later becoming professor at the University of Athens (1923). He also was an outstanding politician from the moderate left wing. Under the dictatorship of Ioannis Metaxas, Svolos was persecuted and exiled. During World War II he actively participated in the resistance and the exile government. Among other activities, he was President of the Socialist Party in the 1940s, the Union of the People's Republic and later was a founding member of the "Greek Union for Human and Civil Freedoms" in 1953 (*Ελληνική Ένωση υπέρ των ελευθεριών του Ανθρώπου και του Πολίτου*).

individual welfare, which in turn legitimized any restriction on individual freedom. Correspondingly, Svolos was the only member of the commission who, in addition to premarital health examinations, proposed sterilization in keeping to the American paradigm. He did so despite being familiar with international criticism of sterilization, characterizing it as a "tyrannical and barbarous measure."[93] Still, for Svolos, after careful examination and consultation with experts the Civil Law was the most appropriate framework for imposing such a measure in Greece.

A final rejection of this eugenic proposition resulted in protests from those physicians excluded from the process despite their claim to be the most competent in making such decisions.[94] Nikolaos Drakoulidis was one of these. As he reported, some physicians tried to reverse the decision, addressing a series of protests to the Highest State Hygienic Office [*Ανώτατο Υγειονομικό Συμβούλιο του Κράτους*]. Although claiming some success, no measure seems to have been implemented by state institutions to corroborate this. However, this failed attempt was not the last effort to legally impose medical certification on Greek bridegrooms.[95] In 1958 the Greek parliament debated a very similar proposal that still failed to become law. Four years later, in 1962, the Ministry for Social Welfare submitted a draft law for implementing such a certification procedure to the Faculty of Law at the University of Athens. The Faculty has, however, never considered the proposal.

## Researching Eugenic Birth Control in Greece: A History of Failure?

The Greek state repeatedly failed to establish eugenic birth control despite demands from several physicians, scholars, politicians and social agencies. One can only speculate about the reasons for this failure. It seems largely the result of a centripetal set of factors, such as the balance of power within

93 Αναθεωρητική Επιτροπή του Αστικού Κώδικος, 6th session (29 December 1930); and *Εφημερίς των Συζητήσεων του Νέου Αστικού Κώδικος* 5 (12 March 1931), 94.

94 Δρακουλίδης, "Η προ του γάμου ιατρική εξέτασις," *Βιβλιοθήκη Κοινωνικής Υγιεινής*.

95 As reported by the jurist Panagiotis Zeppos, the state made further attempts to pass a law to enforce premarital health certificates for [engaged/betrothed] couples. *Greek Society for Eugenics and Human Genetics*, proceedings of the meetings: Βασίλειος Βαλαώρας, Φαίδων Βεγλερής, Ιωάννης Δενέζης, Παναγιώτης Ζέππος, Νικόλαος Λούρος, Βιργινία Τσουδερού, "Το Προγαμιαίον Πιστοποιητικόν" (meeting of 17 February 1965), 20.

the state institutions and the controversial views the participants in this discourse. Moreover, skeptics and opponents of eugenics were not lacking. One of the factors often mentioned as an impediment to eugenic measures in Greece was the country's infrastructural backwardness. Yet the impact of this factor seems rather ambivalent. The argument for backwardness alone is unable to explain the failure of establishing Greek eugenics; on the contrary, a national eugenic project could also be regarded as a possible answer to this very state of backwardness. The idea of backwardness could help to advance eugenic claims in Greece in dramatizing the threat of degeneration ascribed to social diseases, dysgenics and the consequences of the wars in the first decades of the twentieth century. Such a view was suggested by Nikolaos Drakoulidis, who justified the proposed establishment of eugenic measures on the grounds that, in Greece, "we have, more than any other country, the greatest proliferation of diseases and in particular of so-called social diseases such as tuberculosis, malaria, syphilis, trachoma and so on."[96] Pessimism for the future and predicting threats to civilization were characteristic of eugenic thinking in interwar Greece—creating further pressure to support eugenic projects.

The reconstruction of this very discourse—with all its controversies, contradictions and peculiarities—brings to light mechanisms by which health is described as both a socialized and socializing resource in Greece. Moreover, it reveals the way in which health and health care became instrumental in attempts to gain control over the population in terms of both social behavior and physical reproduction. As was typical for eugenics more generally, in the Greek case healthiness was associated with middle class values. Eugenicists supported the transformation of external obligations (to be enforced by state institutions) into values to be internalized by the population. As a result the reconstruction of the discourse on eugenic birth control also provides an insight into Greek social history and history of ideas. With respect to ideological background, the majority of eugenic proponents on birth control and other eugenic measures in Greece were liberals.[97]

96 Δρακουλίδης, "Η προπαγάνδα υπέρ της κοινωνικής υγιεινής."

97 Exceptions include the anthropologist Ioannis Koumaris, who published in German, Italian and Greek fascist journals. See for example, Ioannis Kumaris, "Rasse und Gesundheit," *Ziel und Weg* 9, 12 (1939), 386–388. Giovani Koumaris, "La Politica della Razza in Italia enel Mondo," *La Difesa della Razza* [Qusto fascicolo di 64 pagine è dedicato ai Razzisti Stranieri con scriti sul Razzismo Italiano] 3, 1 (1939/1940), 38–43; and Ιωάννης Κούμαρης, "'Ελληνικόν αίμα,'" *Ελληνικόν αίμα* 21, 3 (1945), 1.

This political tradition, in turn, facilitated the rehabilitation of eugenics after World War II. Indeed, a Eugenics Society was founded in Greece in 1953 and continued to discuss delicate topics like eugenic premarital certification until its dissolution in 1973.[98]

A central question remains: the relationship between individual and community. Should common welfare take priority over the welfare of single individuals or not? Greek proponents of eugenic measures, especially eugenic premarital examinations and certifications, pled for priority to be given to the health of the community, identified with the nation and the overall society. Yet in reality, the fictive national homogeneity and unity breaks down when eugenics is implemented in practice. Selection among the population invariably reveals social contradictions, discrimination and the class character of an allegedly unified community, all the more so as the "social question" was viewed as an issue needing urgent resolution. In the Greek case, similar to a wider tendency in eugenic thinking, the demands of eugenic selection among the population revealed the limits of modern ideologies committed to the liberation of the people, such as socialism, egalitarianism and humanitarianism.

98 The *Greek Society for Eugenics* (later renamed *Greek Society for Eugenics and Human Genetics*) was founded in Athens in 1953 and held its last session twenty years later, as indicated by the minutes of meetings as published in three volumes (1965, 1976, 1978). The *Society* was founded and headed by the gynecologist Nikolaos Louros.

# EUGENICS AND PUERICULTURE

## MEDICAL ATTEMPTS TO IMPROVE THE BIOLOGICAL CAPITAL IN INTERWAR GREECE

*Vassiliki Theodorou and Despina Karakatsani*

In Greece the first public debates on the necessity of adopting eugenic measures for 'racial improvement' can be traced back to the 1910s.[1] Nevertheless, ideas about how to educate the public on matters of eugenics and the possible imposition of eugenic measures intensified in the mid-1920s, when the issue of improving the health of the population was raised more broadly. The Ministry of Hygiene and Social Welfare[2] was established in 1923 while attempts were being undertaken to improve conditions for childbearing. Eminent physicians with backgrounds in public health, jurists, academics, governmental functionaries, public health officers and intellectuals all participated in discussions on heredity, eugenics and the biological improvement of Greece that featured in parliament, contemporary journals and scientific publications. Related articles were often printed in such periodicals as *Health* [*Η Υγεία*], *The Child* [*Το Παιδί*], *Pedol-*

---

[1] The earliest text on Greek eugenics dates to 1917, when *Eugenics* [*Η Ευγονία*] by Michalis Kairis was published. See Μιχ. Νικ. Καΐρης, "Η Ευγονία" (Athens: n.p., 1917). However, the physician Ioannis A. Zallonis, had maintained in his 1870 petition to the Greek parliament on the occasion of the constitution of the Civil Code that with regard to diseases that should lead to interdiction of marriage, legislative action should be taken on behalf of the state to protect the robustness of the race. See I. A. Ζαλλώνης, *Υπόμνημα προς την Βουλήν περί κωλυμάτων γάμου υπό ιατρικήν έποψιν* (Εν Ερμουπόλει: n.p., 1870). By the twentieth century, references to state intervention in citizens' private life became more frequent, especially regarding prenuptial health certification, as well as anti-tuberculosis and anti-venereal campaigns. In literature from the period the concept of race is referred to as *phili/philetic* or *nation* [γένος]. The deterioration of the race denotes persisting Greek public health problems, whereas the amelioration of the race usually implies improving the general health of the nation rather than the various ethnic groups living in Greece.

[2] The name of this ministry changed several times in 1920s and 1930s.

ogy [*Παιδολογία*]; as well as in the *Proceedings of the Greek Anthropological Society* [*Πρακτικά Συνεδρίων της Ελληνικής Ανθρωπολογικής Εταιρείας*], established in 1924.[3] Although various views covering a wide range of influences were expressed, the debate became more pointed in the 1930s with regard to the institution of prenuptial health certificates and sterilization—whether imposed or voluntary. No legislation on negative eugenic measures in Greece was ever passed. However, these debates remain interesting for two reasons: first, because they allow us to causally approach the development of eugenic thought in Greece during the interwar period; and second, because they played an especially important role in reinforcing social welfare policies for mothers and children.

Given that debates over eugenics were linked to efforts to modernize the public health system during the interwar years, it is important to outline the political and social conditions in which these debates took place. From September 1922, in the wake of the disastrous Asia Minor campaign, some 1,300,000 refugees arrived on mainland Greece, laying bare enormous deficiencies in the public health system, making its reform an urgent imperative. Wretched living conditions in refugee settlements contributed to the crisis in health care and increased the death rate from infectious diseases, including typhus, smallpox and tuberculosis. High rates of infant mortality and the spread of tuberculosis, trachoma and malaria—recorded in statistics compiled since 1923 by the newly-established Ministry of Hygiene—attracted the interest of pediatricians intent on rearing a robust young generation of Greeks.[4]

Thus influenced by various factors, the discourse on eugenics in Greece, primarily advanced by physicians, focused on improving the quality of the "biological capital," adopting mild eugenic measures to do so. At one point, a number of prominent pediatricians who exerted considerable influence in political circles in interwar Greece, often linking eugenics to *puericulture*

[3] On the Greek Anthropological Society, see Sevasti Trubeta, "Anthropological Discourse and Eugenics in Interwar Greece," in Marius Turda and Paul J. Weindling, eds., *Blood and Homeland: Eugenics and Racial Nationalism in Central and Southeast Europe 1900-1940* (Budapest: Central European University Press, 2007), 123–145.

[4] For the organization of statistics services at the Ministry of Health, see Φωκίων Κοπανάρης, *Η Δημόσια Υγεία εν Ελλάδι* (Athens: Τύποις Χ. Χρονοπούλου, 1933). Νικ[όλαος] Γ. Μακρίδης, *Αι Υπηρεσίαι υγιεινής εν Ελλάδι. Από της ιδρύσεως του Ελληνικού Βασιλείου μέχρι των ημερών μας* (Athens: Τυπογραφικά καταστήματα αδελφών Γεράρδων, 1933); for the medical supervision of childhood, see Βασιλική Θεοδώρου and Δέσποινα Καρακατσάνη, *'Υγιεινής Παραγγέλματα': υγεία και κοινωνική πρόνοια για παιδιά τις πρώτες δεκαετίες του 20ού αιώνα* (Athens: Διόνικος , 2010).

[παιδοκομία], made it their task to tackle infant mortality, to train mothers in post-natal care and to establish social hygiene institutions whose goal was to strengthen the overall health of Greece's youth.

This chapter considers the contribution made by pediatricians to the wider eugenic discourse in Greece during the interwar years. We want to explore the intellectual framework within which the combination of eugenics and "puericulture" became possible, as well as analyzing the social hygiene measures adopted by successive Greek governments in order to improve children's health more generally. We further aim to make sense of the various ramifications in this discussion through the concept of "biological capital." Relying on primary sources, such as material from state archives and articles published in contemporaneous popular hygiene journals, we concentrate on the role certain pediatricians played in linking puericulture to eugenics, on views regarding the value of children as biological capital, on the possible influence of other European paradigms and, finally, on those groups and individuals playing key roles in the enactment of social hygiene measures for the amelioration of this putative biological capital.

As the history of Greek eugenics has only recently become the object of scholarly attention, a detailed study of the movement is not yet possible. In recounting efforts made to protect motherhood and childhood, we tackle a number of questions raised within the framework of eugenics. By elaborating upon the suggestions formulated in previous studies with regard to the international framework of influences and dissemination of eugenic models, we attempt to further enrich the comparative study of eugenic movements by exploring a relatively overlooked case in Southeastern Europe.[5]

## Measures to Protect Children's Health, 1911–1930

The end of World War I signaled an explosion of eugenic discourses throughout Europe. In Greece, debates over adopting eugenic measures reflected contemporaneous political changes, leading to the necessity of restructur-

[5] See "Introduction" by Turda and Weindling to *Blood and Homeland*, 1–16, and concluding chapter by Mark B. Adams, "Toward a Comparative History of Eugenics," in Mark B. Adams, ed., *The Wellborn Science: Eugenics in Germany, France, Brazil and Russia* (New York: Oxford University Press, 1990), 217–231.

ing the health services. An especially prolonged period of war —including the Balkan Wars (1912–1913), World War I and the Asia Minor campaign (1920–1922)—had serious repercussions for Greece. These included continuous waves of refugees exacerbating public health problems[6] while the country was assuming new territorial responsibilities provided for in the Sevres Treaty (August 1920) and a surge in the population. The population growth, in conjunction with calls to build a healthy nation, formed a new basis for public health considerations; Greek successes in foreign affairs made this promotion of national efficiency imperative.[7] As may be deduced from parliamentary debates and the contemporary press, the issue of race during this period—even if it continued to be identified with the Greek nation or populace—also started to take on a biological dimension, influenced by ongoing medical discourses.[8]

The earliest references to children's health in the context of improving the overall quality of the race date back to 1921, when parliamentary discussions took place on the parameters for establishing a Ministry of Public Hygiene and Social Welfare. The bill put before parliament in November 1920 and debate on its passing began in early 1921. On account of major political upheavals, as well as objections by the Medical Association,[9] the law was passed with a two year delay. It was never implemented due to the consequences of the Greek defeat in the Asia Minor campaign.[10]

Two health officers, Constantinos Savvas (1861–1929), professor of hygiene at the University of Athens, and Emmanuel Lambadarios (1885–1942), head of the School Medical Service of the Ministry of Education, drew up the preamble of the bill providing for the Ministry's establish-

6 Moreover, repeated mobilizations of Greece's male population between 1912 and 1922 revealed a large number of physically disabled individuals, further making citizens' physical and mental health a priority issue.

7 On the need of the newly-established states to rely upon a strong population, see Iris Borowy and Wolf Gruner, eds., *Facing Illness in Troubled Times. Health in Europe in the Interwar Years, 1918–1939* (Berlin: Peter Lang Verlag, 2005).

8 According to Greek pediatrician, Kostis Charitakis, "the vision of a new Greek civilization could hardly come true if the population is plagued by malaria, tuberculosis and infectious diseases." See Κωστής Χαριτάκης, "Η ίδρυσις υπουργείου Δημόσιας Υγείας και Κοινωνικής Πρόνοιας και η προστασία του παιδιού," *Παιδολογία*, 1, 5 (August 1920): 145–147.

9 Physicians serving as members of parliament probably objected out of fear of becoming state-salaried clerks. For the animosity of the medical world towards the proposals see Μακρίδης, *Αι Υπηρεσίαι υγιεινής εν Ελλάδι*, 30–32 and 67–72. See also Vassiliki Theodorou and Despina Karakatsani, "Health policy in interwar Greece: the intervention by the League of Nations Health Organisation," *Dynamis* 28 (2008): 53–76.

10 Namely Law 2882, "Περί μεταρρυθμίσεως και συμπληρώσεως του Υπουργείου της Περιθάλψεως, μετονομαζόμενου εις Υπουργείον Υγιεινής και Κοινωνικής Προνοίας," *Εφημερίς της Κυβερνήσεως* 1, 122 (22 July 1922): 577.

ment. Both were alarmed at the terrible state of public health in Greece; tuberculosis, malaria, typhus, trachoma and children's diseases accounted for high mortality rates.[11] Once the bill was passed, social hygiene institutions for the protection of childhood and the reduction of infant mortality were planned to be implemented for the first time in Greece. The new law also provided for the establishment of health services to combat tuberculosis, malaria and venereal disease as well as public health statistics and publication services. Regulations for urban hygiene and food inspection would be introduced, measures were to be taken to improve working class housing and modern social welfare institutions were to be adopted. Planned reforms of the country's hygiene services prioritized those services intended to protect motherhood and children, including a system of childhood care from conception to school-age. The first measures to ensure hygienic conditions for natality were greeted enthusiastically by contemporary pediatricians, as may be gleaned by from contributions to the journal *Pedology* [*Παιδολογία*], first published in 1920 by the *Hellenic Pedological Society* [*Ελληνική Παιδολογική Εταιρεία*]. This periodical was published by Kostis Charitakis (1888–1956) and hosted articles on the relationship between eugenics and puericulture.

## The First Steps toward Health Care for Children

The establishment of the first Ministry of Hygiene was expected to lay the foundation for dealing with children's health problems effectively, coordinating fragmentary policies that had been adopted to that point. The most substantial step taken before a decade of intervening warfare was the establishment the School Hygiene Services in the Ministry of Education in 1911, which attempted to arrest the spread of transmittable and childhood diseases. In certain parts of Greece, these continued to claim many victims

11 Constantinos Savvas asserted: "The state of public health in Greece is pitiful in all respects [ … ] general mortality is high compared to other European countries [ … ] People are decimated on the one hand by diseases ruining their health, such as tuberculosis, malaria, etc. and on the other hand by various epidemics such as smallpox, scarlet fever, diphtheria, typhoid fever, dysentery, cerebra-spinal meningitis and many others [ … ]. Due to various and chronic diseases, the race is in the process of gradual degeneration;" Κωνσταντίνος Σάββας, "Περί ιδρύσεως Υπουργείου Υγείας και Κοινωνικής Προνοίας," *Αρχεία Υγιεινής* 16, 3 (1922): 65–72.

among schoolchildren. In 1914 the first school physicians were appointed to this area, while the health of primary school students began to be monitored more systematically. Measures included the vaccination of schoolchildren, the introduction of a personal health card where developmental indices and potential health problems were recorded and finally, the compilation of statistics and morbidity rates among schoolchildren.[12]

In the most comprehensive step taken towards the development of pedology, a new branch of science promoted by Emmanuel Lambadarios, was the measurement of children's bodies, conducted by school physicians in the early 1920s. They used special tools such as cephalometers, thoracometers, stadiometers and ergographs in order to record the children's physical development. Lambadarios attempted to promote studies related to the development of a child's body and great emphasis was laid upon the combination of data on weight, height and circumference of the thorax since it was thought that school physicians would be able to calculate "a coefficient of robustness" based on such data.[13] This latter term, derived from military medicine, allowed for the classification of bodies into various health categories. Converting biological phenomena into figures and mathematic formulae and recording the results on growth charts and body indices made objective calculations of means and variations possible, that is, it allowed for the introduction of criteria to distinguish healthy children (the "eugenics") from those affected by disease or physical deformities (the "dysgenics").[14]

Based on anthropometric measurements conducted by the *School Hygiene Service* and the *Pedological Institute* [*Παιδολογικό Ινστιτούτο*] founded in the 1920s, Lambadarios intended to record "the physical development of the Greek child," classifying measurements depending on sex, race, nationality and age.[15] Moreover, Lambadarios seized every opportu-

12 For the organization of the School Medical Service, see Θεοδώρου and Καρακατσάνη, *'Υγιεινής Παραγγέλματα.'*

13 In order to draw up his somatometric tables, Lambadarios utilized Paul Godin's growth bulletins, *Manuel d'anthropologie* (Neuchatel and Paris: Delachaux and Niestle, 1915).

14 Εμμανουήλ Λαμπαδάριος "Η σωματική ανάπτυξις του παιδίου και ιδία του έλληνος μαθητού," *Παιδολογία* 1, 8 (November 1920), 241–247; and with the same subject and title, *Παιδολογία*, 1, 9 (December 1920): 274–279; and *Παιδολογία*, 2, 1 (January 1921), 2–7. Also by the same author, see "Η σωματική εξέλιξις του Έλληνος μαθητού. Ανθρωπολογική αυξησιολογία," *Ιατρικά Χρονικά*, 6 (December 1928), 354–356.

15 These findings were comparable with measurements conducted on children from other European as well as North American countries. Νικόλαος Εξαρχόπουλος, "Οι Ελληνόπαιδες εν συγκρίσει προς τους παίδας ξένων λαών ως προς την σωματικήν των κατάστασιν και την εξέλιξιν της νοημοσύνης των," Εθνικόν και Καποδιστριακόν Πανεπιστήμιον, *Εκατονταετηρίς 1837–1937. Επιστημονικαί Συμβολαί* (Αθήνα: Διεύθυνσις Δημοσιευμάτων του Πανεπιστημίου, 1938): 155–177.

nity to stress the importance of providing children with information on hygiene, which would also be communicated to their parents. Already by the 1910s, on Lambadarios' initiative, school poly-clinics, children's summer camps and an open-air school for children purportedly prone to consumption had all been established. Such early social relief projects funded by voluntary societies sought to strengthen the constitution of sickly children, especially from the lower social classes.[16]

The foundation for an infrastructure to protect infants and young mothers were also laid by voluntary organizations at the beginning of the twentieth century. Through largely privately funded, the childcare centers of the *Patriotic Union of Greek Women* [*Πατριωτικός Σύνδεσμος Ελληνίδων*][17] established in 1914 attempted to answer the social needs of the working classes. Supported by a group of volunteer, urban middle class ladies and select physicians, they provided soup kitchens and health care, particularly to infants. The Patriotic Union was interested in spreading hygienic habits among families through lectures and pamphlets in addition to running social hygiene institutions that were beyond the limited means of the Ministry of Education.

## The Intensification of the Efforts after the Refugee Influx, 1922–1935

A period of upheaval following Greece's unsuccessful Asia Minor campaign meant that health plans laid out in the 1921 bill had to be put on hold indefinitely. Although the Ministry of Public Health and Social Welfare was finally set up in 1923, there were many shortages. Notwithstanding, significant steps were taken towards establishing institutions for the improvement of children's health over the 1920s. Yet the resettlement of

[16] Vassiliki Theodorou and Despina Karakatsani, "Ecoles de plein air et éducation nouvelle au début du XXe siècle en Grèce: influences et limites d'une tentative," *Carrefours de l'éducation* 23 (2007): 187–204.

[17] The name of the foundation underwent the following changes during its operation: Patriotic Union of Greek Women [*Πατριωτικός Σύνδεσμος των Ελληνίδων*: 1914–1917], Patriotic Relief Foundation [*Πατριωτικόν Ίδρυμα Περιθάλψεως*: 1917–1921], Patriotic Union of Greek Women [*Πατριωτικός Σύνδεσμος των Ελληνίδων*: 1921–1922], Patriotic Relief Foundation [*Πατριωτικόν Ίδρυμα Περιθάλψεως*: 1923–1929], Patriotic Foundation for Child Protection [*Πατριωτικό Ίδρυμα Προστασίας του Παιδιού*: 1929–1936] and Patriotic Foundation for Social Protection and Welfare [*Πατριωτικόν Ίδρυμα Κοινωνικής Προνοίας και Αντιλήψεως ΠΙΚΠΑ*: 1936–2003].

the Asia Minor refugees revealed the inadequacies of the public health system; death tolls due to malaria, typhus, dysentery and tuberculosis in refugee camps highlighted the pressing need for public health reform. After 1923 infant mortality was on the increase, particularly in areas where refugees had settled; in certain instances infant mortality rates doubled. The same period also witnessed an increase in infant abandonment. Such findings were alarming for the perceived future of the race; combined with falling fertility rates, this became an issue of prime national importance. Since the robustness of the nation began to be considered indispensable in a hostile geopolitical environment, the protection of mothers came to be viewed as a prerequisite for ensuring racially robust descendants.

Following the example set by Mussolini, the authoritarian regime of Theodoros Pangalos (1878–1952) lasting from 1925 to 1926 adopted the first measures for the protection of motherhood. To be sure, this was done with a view to restricting the phenomenon of abortion and infanticide, which became all the more frequent after the influx of refugees. In 1926, legislation "for the protection of nursing infants" was introduced in which the state attempted to both increase birth rates and control birth, placing infants up to the age of two along with their indigent mothers under its protection.[18] Moreover, to complete the protection of motherhood, the state undertook to popularize knowledge of infant care through a law passed a few months later, creating a Model Child Care Centre and a Museum of Eugenics and Child Care [*Πρότυπος Παιδοκομικός Σταθμός* και *Μουσείο Ευγονίας και Παιδοκομίας*].[19] Yet these plans did not materialize.

Kostis Charitakis, the Director of the Social Hygiene Service of the Ministry of Hygiene, attempted repeatedly since 1925 to organize state institutions for the protection of motherhood. His attempts failed to bear the anticipated results since institutional changes were a regular feature of this period characterized by political unrest. Be that as it may, a series of social hygiene measures were subsequently adopted by the Liberal Venizelos government between 1928 and 1932. This again aimed to strengthen the constitution of weak children from the lower classes, thus placing the protection

18 "Περί προστασίας του θηλάζοντος βρέφους," *Εφημερίς της Κυβερνήσεως* 1, 137 (14 April 1926): 1025–1026.

19 "Περί οργανώσεως Εθνικού Παιδοκομικού Ινστιτούτου," *Εφημερίς της Κυβερνήσεως* 1, 392 (6 November 1926): 3139–3140.

of childhood largely under the state's jurisdiction. These measures, indicative of an ongoing modernization process—including provision for school meals, open-air schools and children's summer camps—met the need for confronting tuberculosis within the framework of a sociologically oriented medicine; indeed, the latter placed great emphasis on the living conditions of families with children attending school. Such care indicates that influential, high-ranking members of the Liberal Party favored the social orientation of childhood protection.[20]

This period also saw the introduction of two laws for the protection of motherhood and children.[21] First, an agency was established in the Ministry of Hygiene to supervise the institutions involved in the care of infants and expectant mothers. Second, in 1929 the Patriotic Union of Greek Women was promoted from a charity foundation to a semi-state organization, the Patriotic Foundation for Child Protection [*Πατριωτικό Ίδρυμα για την Προστασία του Παιδιού*],[22] and assigned the task of children's health care. Childcare centers run by the Foundation in major cities were to provide expectant mothers with advice and health care, distribute food and milk to destitute mothers, monitor infant health and organize soup kitchens and camps for younger children. The main aims of the Foundation were to decrease infant mortality and disseminate new hygienic practices among poverty-stricken women. To attain this goal, home visits by volunteer visiting nurses were planned, with expectant mothers' classes to be run by the Foundation as well. Educating mothers on their maternal duties was considered to be crucial in spreading eugenic ideas. Moreover, in order to dispel superstitions about hygiene during pregnancy and infant rearing, Apostolos Doxiadis (1874–1942), a distinguished pediatrician educated in Vienna, Berlin and Paris at the turn of the twentieth century, in his capacity as the Chairman of the Foundation, published special books and popular pamphlets with advice for young mothers.[23]

20 Most struggled for the adoption of steps to protect motherhood and childhood and their articles frequently appeared in journals such as *Pedology* [*Παιδολογία*], *The Child* [*Το Παιδί*], *Health* [*Υγεία*] and *Labor* [*Εργασία*].

21 These were Laws 4061 "Περί υγιεινής και προστασίας της μητρότητας και παιδικών ηλικιών" and 4062 "Περί Πατριωτικού Ιδρύματος Προστασίας του παιδιού" both in *Εφημερίς της Κυβερνήσεως* 1, 94, (9 March 1929), 914–918.

22 For the foundation's various names, see note 17.

23 See Apostolos Doxiadis Archive, File/Box 1, 22/256, Benakis Archive. See also Κώστας Δαφνής, *Απόστολος Δοξιάδης. Ο αγωνιστής και ο άνθρωπος* (Athens: Αθηναϊκό Κέντρο Εκδόσεων, 1974).

In his book *Letters to Young Mothers*,[24] Doxiadis attempted to eradicate superstitions concerning child rearing and to imbue a sense of biological duty to society in general.[25] Written in epistolary form, the book attempted to convince mothers that their contribution was decisive for both a robust nation and in creating a heritage able to play a major role in the competition between nations: "in this fight, that group shall win that excels both in terms of numbers and robustness," underlined Doxiadis.[26] This Social Darwinian struggle was exceptional in the case of Greece since "after the recent terrible disaster [the defeat in Asia Minor], nature must be reconstituted, both to heal the wounds of our nation and to imbue us with new strength."[27] By screening films and distributing informational leaflets, the Foundation attempted to complete the education of mothers in their duties. Following practices initiated in 1925, weekly health festivals were dedicated to children alongside prizes for the best-reared babies, serving as a means for rewarding mothers who had met their physician's expectations with regard to the spread of the new hygienic models in Greece.[28]

Interestingly, the joint preamble to the two laws drawn up by Apostolos Doxiadis highlighted the relationship between social hygiene measures, fertility and the "nation's viability." Given that children's health was held to be vital for the biological quality of the nation, the state was obliged to ensure optimum conditions for expectant mothers. Doxiadis, who introduced the term "biological capital" to 1920s Greece, highlighted the importance of health and child rearing knowledge for mothers, for health depends on the "living capital inherited and acquired, both of which are nurtured in childhood."[29]

---

24 Απόστολος Δοξιάδης, *Γράμματα προς νεαράς μητέρας* (Athens: Εκδόσεις Γκρέκα, 1926).

25 According to Doxiadis, pre-natal child rearing advice enlightened future mothers on their role in the well-being of the nation. The role of motherhood thus gained an historical mission: "As mothers you must contribute to increasing the prestige of our country, which means having good citizens who will be able to go forward and proceed with its intentions." See Δοξιάδης, *Γράμματα προς νεαράς μητέρας*, 19.

26 Δοξιάδης, *Γράμματα προς νεαράς μητέρας*, 21.

27 Δοξιάδης, *Γράμματα προς νεαράς μητέρας*, 22.

28 Weekly health festivals were an American institution initiated in 1912, which quickly spread throughout Europe.

29 Απόστολος Δοξιάδης, Εισηγητική Έκθεσις επί του σχεδίου νόμου "Περί Πατριωτικού Ιδρύματος Προστασίας του Παιδιού" και "Περί Εθνικού Συμβουλίου Προστασίας της Μητρότητος και των παιδικών ηλικιών," *Αρχείον της Βουλής της Α΄ Συνόδου της Β΄ περιόδου από 17-10-1929 έως 30-12-1929* [*Archive of the Hellenic Parliament*, vol. A, [Second Period] from 17 October 1928 to 30 December 1929] (Athens: n.p., 1930), 2833–2844. In 1926 the physician Nikolaos Makridis also referred to the relationship between the population's health and the nation's ability to wage war; see, Νικόλαος Μακρίδης, "Η υγεία του ανθρώπου ως κεφάλαιον," *Υγεία*, 3, 7 (July 1926): 149–151.

Aiming to improve environmental conditions, (in other words, to affect acquired characteristics), Doxiadis proposed that eugenic measures be instituted in order to increase viability and decrease mortality in Greece's population. Decreasing infant and child mortality was to be one of the main measures for reinforcing "the Greek race."[30] Doxiadis stressed the fact that high rates of infant and early childhood mortality posed a risk to the Greek race. One fifth of all children died before reaching the age of one year, while childhood diseases, though curable, plagued the population. This figure was more than double that in other European countries, where it ranged from 7% to 10%.[31] The repercussions of child mortality were becoming all the more obvious in the case of army recruitment. By 1925, the percentage of those judged unfit to join the army had reached 47%. As Prime Minister Venizelos' consultant on health issues and Health Minister in successive Liberal governments, Doxiadis was the most established authority for tailoring policies for the protection of motherhood and children.

## Linking Eugenics to Puericulture

Doxiadis, Lambadarios and Charitakis had previously familiarized themselves with eugenic thought in Western Europe and held high-ranking positions in the public health sector during the interwar years. As such, they were licensed to take important decisions with regard to motherhood and children. All three had studied pediatrics abroad and had analyzed the way corresponding institutions were organized in other European countries, played a leading role in establishing scientific societies for children's health and attempted to transplant international eugenic theories to Greece. It therefore comes as no surprise that it was these three who struggled to establish social hygiene institutions for children and to link eugenics to pedology in Greece.

30 In the preamble to the bill, Doxiadis as Undersecretary of Hygiene underlined: "Man's health depends on the hereditary and acquired vital capital gained during his childhood. Between the two, the acquired is mainly influenced by environmental conditions. And these conditions it is our debt to render them favorable to childhood." See Δοξιάδης, "Εισηγητική Έκθεσις επί του σχεδίου νόμου," 2833.

31 Δοξιάδης, "Εισηγητική Έκθεσις επί του σχεδίου νόμου," 2833. The high infant mortality rate and its consequences for the population are mentioned by other physicians [who held public offices] in articles published during this period. See, for example, Ε[μμανουήλ] Λ[αμπαδάριος], "Το υγειονομικόν πρόγραμμα," *Εργασία*, 1, 28 (19 July 1930): 21–23; and Σόλων Βέρας, "Η προστασία του βρέφους," *Εργασία*, 1, 35 (1930): 12–13.

The earliest references to the relationship between eugenics and pedology can be traced in articles by Doxiadis and Lambadarios published in the journal *Pedology*. The journal had been created upon the initiative of hygienists and pediatricians who argued that knowledge of heredity laws was necessary in order both to predict and avoid the reproduction of dysgenics and create favorable conditions for the production of healthy offspring. Lambadarios was the first to raise an issue previously taboo in Greek society, namely, how "it would be possible to hold in check the reproduction of individuals who gave birth to children that were delicate, degenerate and harmful to society."[32] He also stressed the eugenic origins of pedology, suggesting the diagnosis of defective children with the use of special tools, some already in use in pedology centers. In fact, similar measurements had already been conducted by the Experimental Laboratory of the University of Athens, set up by Nikolaos Exarchopoulos (1874–1960), professor of experimental pedagogy. These measurements concerned the children's intellectual and mental development, their classification into various levels of intelligence and the diagnosis of intellectual diversity among children from different social classes.[33] Indeed, it was Exarchopoulos who had introduced experimental pedagogy in Greece, a scientific field where, under the influence of experimental psychology, various discourses and questions on normality were interlinked.

For his part, Lambadarios held the view that where eugenics was concerned, importance should be placed on establishing puericulture centers to improve preparation for pregnancy, alongside making provisions for training scientific staff. His model was the Institute of Infant Care at the University of Paris [*L'Institut de Puericulture de l'Université de Paris*], a Franco-American institution for the theoretical and practical education of physicians, midwives and nurses.[34] An ardent admirer of Adolph Pinard

---

32 The issue of physically and mentally impaired children was taboo for Greek society of the 1920s. See the notes on a lecture delivered by Lambadarios on 15 January 1922 for the Lyceum of Greek Women: Εμμανουήλ Λαμπαδάριος, "Αι ευγονικαί απόψεις της παιδολογίας," in *Αρχείο Εμμανουήλ Λαμπαδάριου*, Ελληνικό Λογοτεχνικό και Ιστορικό Αρχείο [ΕΛΙΑ], unclassified archive.

33 Students at the Experimental School of the University of Athens [*Πειραματικόν Σχολείον του Πανεπιστημίου Αθηνών*] had been used as research subjects by the researchers of the Experimental Laboratory [*Εργαστήριον Πειραματικής Παιδαγωγικής του Πανεπιστημίου Αθηνών*] from its inception in 1923. Research carried out during its fifteen-year existence (1923–1938) was solely devoted to the physical, intellectual and mental development of children. Research relevant to their physical development was based on measuring fourteen physical attributes of each child.

34 Λαμπαδάριος, "Αι ευγονικαί απόψεις της παιδολογίας."

(1844–1934), a respected obstetrician, professor of pediatrics and Chairman of the French Eugenics Society, Lambadarios adopted his views on the role puericulture could play in Greece's racial improvement.[35]

Moreover, in his 1921 article entitled "Pedology and Eugenics," Doxiadis postulated that hereditary predisposition was a factor that could account for the incidence of pathological tendencies in children.[36] Also following Sicard de Plauzoles (1872–1968), a French public health physician well-known for his eugenic ideas, Doxiadis claimed that progenitor diseases—such as syphilis, alcoholism, malaria and typhus—were responsible for the deterioration of Greek youth. He further argued that Greeks needed to pay far closer attention to moral depravity as well as to miscegenation with non-Greeks, a phenomenon all the more frequent after a decade of dislocating wars. Similar to the views expressed by Kairis, Moiseidis and others, Doxiadis argued that raising awareness of an individual's biological duty to society remained of primary importance. In order to wipe out those pathological tendencies resulting in the production of defective individuals he suggested a health card containing information on every Greek citizen's heredity background. This card could be used to ensure the production of healthy offspring. Since this health card bore many similarities with the data cards already being filled out by school physicians, Doxiadis suggested that the latter be assigned this task.

In advocating this view, Kostis Charitakis, in a work entitled *Social Hygiene* asserted that eugenics constituted a branch of puericulture in the wider tree of social hygiene.[37] And in light of his view that puericulture was the most important attainable measure for raising a robust new generation, Charitakis favored an aggregate puericultural policy, one encompassing pre-natal puericulture, pregnancy hygiene and puericulture proper. Like the aforementioned pediatricians, Charitakis considered that edu-

35 Lambadarios did not make major distinctions between puericulture and eugenics as the former was defined as "a science whose purpose was to seek and apply the relevant knowledge for the multiplication, preservation and improvement of the human species." Included therein was the study of inherited traits, implementation of new eugenic theories, hygiene during gestation and hygiene during both nursing and early childhood. See File "Α΄ Γυναικείον Συνέδριον. Γυναίκες και παιδολογία," *Αρχείο Εμμανουήλ Λαμπαδάριος*, Ελληνικό Λογοτεχνικό και Ιστορικό Αρχείο [ΕΛΙΑ], unclassified archive.

36 Απόστολος Δοξιάδης, "Παιδολογία και Ευγονία," *Παιδολογία* 2, 12 (April 1921): 14–22.

37 Κωστής Χαριτάκης, *Κοινωνική Υγιεινή* [Special Issue in *Έρευνα*, Year 2, (April-May 1928)], (Athens and Alexandria: Εκδοτικός Οίκος Α Κασιγόνη, 1928): 15–16.

cation in infant hygiene should begin even before conception. Ensuring healthy infants raised a series of considerations bearing upon politics, economics and the feminist movement, including the investigation of parents' hygienic condition before marriage, couples' age, provisions for both family allowances and pregnancy leave for working mothers, medical care during labor, education for mothers and, finally, the establishment of puericulture centers.

Contributing to the public debate over the prenuptial certificate, Charitakis underlined the importance of informing future parents about the dangers posed by hereditary diseases—for he did not advocate the establishment of this kind of certificate.[38] For him, eugenics was considered a form of "acculturation for the reproductive drive," one without the necessity of imposing mandatory medical checks on would-be parents: "Just as a florist or an animal breeder has managed through successive cultivation to produce a superior breed of flowers or animals, in the same way do eugenicists intend not only to avert morbid heredity but also further support the refinement of human beings."[39]

To bring about this "refinement," Charitakis accentuated the role mothers could play in the campaign for racial refinement: mothers had to be informed of their eugenic duties to society, especially in realizing the value of breast feeding and hygiene. In his view, lack of hygienic knowledge and proper nutrition in large part accounted for the high rates of infant mortality from dysentery. He perceived most mothers to be illiterate and that their superstitious beliefs led them to treat medical advice with suspicion.

In his capacity as Head of the Department of Social Hygiene in the Ministry of Hygiene, Charitakis had visited infant care centers in many European countries. As a result he put forward a number of proposals for the refinement of "the Greek race," including the establishment of consultation services, dissemination of public health propaganda to mothers and the training of visiting nurses.[40] To a great extent Charitakis, like Doxia-

---

38 Κωστής Χαριτάκης, *Τα νεώτατα δεδομένα επί της κοινωνικής υγιεινής : αρχαί και κριτήρια οργανώσεως της δημοσίας υγιεινής* (Εν Αθήναις: Εκ του Εθνικού Τυπογραφείου, 1929), 82, 167.

39 Χαριτάκης, *Κοινωνική Υγιεινή*, 18.

40 See the serialized article by Κωστής Χαριτάκης, "Κοινωνική υγιεινή," *Ελληνίς*, 6, 11 (November 1927), 233–235; 6, 12 (December 1927), 259–262; 7, 1 (January 1928), 10–12; 7, 2 (February 1928), 40–42; and 7, 3 (March 1928), 63–65.

dis, adopted positions previously put forward by Sicard de Plauzoles, which had been reprinted in 1931 in the popular hygiene periodical *Health* [*Υγεία*]. Here, de Plauzoles argued that "a child has the right to a hygienic life and thus we owe it to that child to ensure the best possible conditions for childbearing, which depend on its parents' health."[41] State intervention deemed necessary for the "production of a healthy new generation" extended to mandatory medical examinations for all couples proposing to marry, medical supervision of pregnant women, with social protection provided by the state from the fifth month of pregnancy, enforced rest for pregnant women—at least during the final months of pregnancy—remuneration for motherhood and "biological education for parents."

The obstetrician Moisis Moisidis also referred to the relationship between eugenics and puericulture in his 1924 "*Eugenics and Puericulture among the Ancient Greeks*."[42] Moiseidis was then a member of the Gynecological Society of Paris, the French Eugenics Society and the *Institut International d'Anthropologie de Paris*, editor of the journal *Health* [*Υγεία*] and author of several books popularizing medical knowledge for mothers.[43] Moiseidis, citing Pinard, equated the aims of puericulture with those of eugenics, claiming that "eugenics is puericulture before conception."[44]

By the end of the 1920s the absence of any working health policy for children—from the mother's womb until they enter the army"—was linked by pediatricians to the risk of racial degeneration.[45] The importance of protecting motherhood and childhood was also highlighted in the debate concerning prenuptial certificates at the time. Although most participants in this debate—mostly physicians and jurists—were in favor of passing such

41 These are proposals put forth by Sicard de Plauzoles to the "Committee for the Protection of the Right to a Healthy Life" [Επιτροπή δια την προστασίαν του δικαιώματος εις την υγιεινήν ζωήν] and the Central Committee of the "League for the Rights of Man" [*Κεντρικόν Κομιτάτον του Γαλλικού Συνδέσμου υπέρ των Δικαιωμάτων του Ανθρώπου*], reprinted in Sicard de Plauzoles, "Ευγονισμός και Παιδοκομία" in *Υγεία* 7, 7 (July 1931): 153–154.

42 In this work, which was honored by the Medical Society of Athens in 1924, Moisis Moiseidis argued that ancient Greeks had discovered the principles of both eugenics and puericulture; see M. Μωυσείδης, *Ευγονική και Παιδοκομία παρά τοις Αρχαίοις Έλλησιν. Συμβολή εις την ιστορίαν της παιδοκομίας* (Athens: Εκ του Τυπογραφείου Κ. Γ. Μακρίδου και Ι. Α. Αλευροπούλου, 1925).

43 Isidor Fischer, ed., *Biographisches Lexikon der hervorragenden Ärzte der letzten fünfzig Jahre*, vol. 2 (Munich & Berlin: Verlag von Urban & Schwarzenberg, 1962), 1057–1058.

44 In 1911, Moiseidis reiterated views first expressed by Pinard before the Medical Academy; see Μωυσείδης, *Ευγονική και Παιδοκομία παρά τοις Αρχαίοις Έλλησιν*, 5.

45 Χαριτάκης, "Η ίδρυσις υπουργείου Δημόσιας Υγείας και Κοινωνικής Πρόνοιας και η προστασία του παιδιού," 18.

a measure, they were met with doubts anchored in emotional, moral and social principles. Many claimed that a period of time would be required to prepare public opinion in order to facilitate the practical implementation of such measures. Educating the public on their eugenic duties was considered to be the main measure that would gradually lead to a voluntary medical examination by couples seeking to marry. Instead of mandatory examinations potentially causing family tragedies and putting the measure in ill repute, these physicians proposed to cultivate a sense of responsibility in couples intending to marry, that is, a sense of duty to the collective biological capital.

The Jurist and director of the state prisons, Ilias Lagakos (1899–1958), in his 1930 article "The Prenuptial Certificate. Legal and Sociological Perspectives" published in the magazine *Child* [*Το Παιδί*] considered that three measures could be implemented in the Greek instance: the distribution of propaganda material to couples seeking marriage, the creation of centers staffed by experts providing advice for couples proposing to marry and lectures at universities, army barracks, workers' clubs and schools.[46] In keeping with these ideas, Charitakis further argued that eugenics need not be supported by laws or coercion.[47] Similarly, Doxiadis claimed that citizens should be properly inculcated at school so as not to choose their number of offspring "on the grounds of their own satisfaction based on their bestial impulse and sexual drive while being totally indifferent to the destinies of their offspring."[48] It was expected that through instruction rather than enforcement, a new ideal would be cultivated, one ultimately leading the individual away from personal to future family happiness. Most participants in the eugenic debate were aware of the serious deficiencies existing in the social hygiene sector and considered that significant results could only be achieved in Greek eugenics with intensive propaganda.

46 Ηλίας Λαγάκος, "Το προγαμιαίον πιστοποιητικόν. Νομικαί και κοινωνιολογικαί του απόψεις," *Το Παιδί* 1, 2 (1930): 4-23.

47 He was nevertheless in favor of forbidding lepers to marry. See Χαριτάκης, *Κοινωνική Υγιεινή*, 17.

48 As the public had not been properly prepared in Greece, Doxiadis regarded the cultivation of eugenic ideas among young people, whose education he proposed should begin at school, as a first, feasible step to be taken. For this reason he believed in the necessity of teachers being informed about eugenic issues; see Απόστολος Δοξιάδης, "Ευγονία," *Το Παιδί*, 4, 28 (1934): 5–15, and 9.

## The Influence of French Eugenics and the Quality of Biological Capital

French eugenic traditions exerted considerable influence on the proposals put forward by Doxiadis, Charitakis and Moiseidis.[49] In fact, during the early twentieth century, the French Eugenics Society, shaped by the ideas of Adolph Pinard, turned towards "la puericulture avant la procreation" [puericulture before procreation]. Pinard was responsible for reviving the concept of puericulture in France, which he defined as "knowledge relative to the reproduction, the conservation and the amelioration of the human species."[50] Having spent many years working on issues relating to pregnancy, pre-natal care, infant mortality and especially the effects of alcoholism, tuberculosis and venereal diseases, Pinard placed great emphasis on heredity as well as upon environmental influences. It should be noted here, however, that Pinard did not regard the well-being of infants as simply deriving from the health of the pregnant mothers. Rather, it was intertwined with the health of previous generations. Thus, his formulation "la puericulture avant la procreation" denotes the dominant influence procreators could exert on the health of their offspring.[51] Without doubt, the neo-Lamarchian heredity of puericulture was attractive to a wide array of specialists concerned with infant health and in particular to Greek physicians who had studied obstetrics and pediatrics in France.

49 The diversity of views formulated by French eugenicists as well as the variety of neologisms adopted (viriculture, elevage de l'home, selection artificielle, sociobiologie, hominiculture) led certain scholars to accept that French eugenics never constituted a school; see Jacques Leonard, "Les origines et les conséquences de l'eugénique en France," in *Annales de Démographie Historique* (Paris: École des hautes études en sciences sociales, 1985), 203–214.

50 Pinard brought back the term puericulture, introducing it in the Académie de Medicine in 1895. This was an ambitious plan, one that included medical checks conducted at three stages: in the period before procreation, during pregnancy and in the period immediately after birth; see A. Pinard, "De la dépopulation de la France," *Revue Scientifique* (30 July 1910): 30, cited in Anne Carol, "Médecine et eugénisme en France, ou le rêve d'une prophylaxie parfaite. XIXe - Première moitié du XXe siècle," *Revue d'histoire moderne et contemporaine* 43, 4 (1996): 618–631, 621; see also M. Μωυσείδου, *Ευγονική και παιδοκομία παρά τοις Αρχαίοις Έλλησι*, 5.

51 William H. Schneider, "The Eugenics Movement in France 1890–1940," in Adams, ed., *The Wellborn Science*, 69-109. For the French Eugenics Society, see also William. H. Schneider, *Quality and Quantity: The Question for the Biological Degeneration in Twentieth–Century France* (Cambridge: Cambridge University Press, 1990) and Andrés Horracio Reggiani, *God's Eugenicist. Alexis Carrel and the Socio-Biology of Decline* (Oxford: Berghahn Books 2007).

Neo-Lamarckism gradually became the key component of French eugenics, not least because it was consistent with the social and political philosophy of the French Third Republic. This entailed acknowledging that both environmental and social influences played an important role in heredity through acquired characteristics and were subject to improvement. Reinforced by France's social and political community, neo-Lamarckism tightened its hold on the scientific community. Yet French eugenicists remained skeptical about the imposition of sterilization, justifying this stance on grounds of individual freedom, humanism and medical ethics.[52] Their cautious attitude towards the adoption of negative measures in the first two decades of the twentieth century was dictated by three factors, namely, the dispute over hereditary laws as the sole determinant of physical impairment, the insistence on safeguarding doctor/patient confidentiality and the belief in individual liberty and, lastly, the dominant trend among puericulture advocates that social leadership should be in the hands of pediatricians.[53]

The demographic decline in France following World War I (in which 1.3 million French soldiers were killed) caused nationally-minded eugenicists to increasingly favor natalism and consequently endorse social hygiene measures in order to increase the population. But World War I only made the links between social hygiene and eugenics, which was already provided with a theoretical basis by neo-Lamarckism, appear all the more evident. For this reason the majority in the French Eugenics Society, being either natalists or pediatricians, invested their efforts in the development of social hygiene policies, at least until the 1930s. Often functioning as members of the Chamber of Deputies or the National Office of Social Hygiene, which was established in 1924, they forcefully steered government policy in this direction. In the early 1920s, for example, the French Eugenics Society spearheaded campaigns against alcoholism, tuberculosis and syphilis. Through lectures, radio broadcasts and conferences the Society attempted

52 Gwen Terrenoire, "Eugenics in France before 1945," in *Cloning, Gene therapy, Human Behavior, Eugenics;* booklet published on the occasion of the international Conference of the NGOs UNESCO, Paris, 14–16 December 2005, available online at: www.unescp.org (last accessed 16 September 2008).

53 Appalled by steps taken in other countries, they proposed that the disabled either be removed or confined in asylums or be altogether removed from the country; see Leonard, "Les origines et les consequenses de l'eugenique en France," 209.

to convince not only the educated public but also intellectual, scientific and political decision-makers of the need to adopt social policy health measures. Critically, because of their interest in increasing the population, members of the French Eugenics Society were far from adopting negative measures for racial improvement, which they felt would lead to a population decline. The Society also declared itself against birth control, abortion and all accompanying contraception propaganda. "In order to have the right quality, we should first secure the numbers," was the slogan in the French Eugenics Society.[54]

Notwithstanding different trends, one may assert that until 1930, French eugenics when compared with its German and English counterparts remained a far less Darwinian affair.[55] In 1926, Pinard's introduction of a bill concerning prenuptial certificates for French citizens in the Chamber of Deputies was received with skepticism by many eugenicists, and was later delayed by counterproposals and disagreements among physicians. As proponents of positive eugenics, French eugenicists were in favor of providing instruction, stressing the duty of the individual to their society and race, which they now defined as a form of political duty. They further argued that ensuring a generation of healthy and strong children able to guarantee the prolongation of life and progress of the race were a citizens' most important duty.[56] In consequence, educating families on racial hygiene was a crucial factor in increasing the national birth rate. Along the same lines of thought, Greek eugenicists participating in their country's discussion on sterilization and prenuptial health certificates in the 1930s agreed with their French colleagues: a people's biological awareness could be raised only by means of the spread of information.

Yet in the early 1930s a shift is recognizable in the positions of French eugenicists. Influenced by the catastrophic effects of the 1929 economic depression, the growth in the number of immigrants and higher rates of population growth, certain physicians began to view existing French

54 William H. Schneider, "Towards the Improvement of the Human Race. The History of Eugenics in France," *Journal of Modern History* 54, 2 (1982): 268–291.

55 Certain eugenicists, influenced more by the United States than Germany, did not, however, hesitate to propose negative eugenics, like Charles Richet in *La sélection Humaine* (Paris: Félix Alcan, 1919).

56 On the popularization of eugenic ideas in France, see Anne Carol, *Histoire de l'eugénisme en France* (Paris: Le Seuil, 1995), 285–293 and Leonard, "Les origines et les conséquences de l'eugénisme en France," 211.

eugenic policy in a different light, formulating views more favorable to birth control. The main proponent of these negative eugenic measures, Sicard de Plauzoles, helped to shift the interest of eugenicists from a focus on quantity one of quality of births. De Plauzoles, disagreeing with those seeing the value of France's population based on birth numbers, argued that the nation's power depended on the quality of its biological capital. Indeed, the concept of biological capital was central to de Plauzoles' reading of eugenics. This was defined as the accumulation of biological characteristics passed on by means of heredity from generation to generation. Consequently, the value of the nation, in de Plauzoles' view, resided in the number of healthy adults able to work, to produce and to reproduce healthy offspring. In attempting to calculate the population's value, de Plauzoles resorted to terminology borrowed from veterinary sciences, which understood as the art of creating and perfecting the individual as a working machine. The role for eugenics, then, was to assure the production of human capital of superior quality, with social hygiene helping to promote this invested human capital in order to achieve the best racial health possible. The National Health Service was thus called to assist in both increasing the country's population and bringing about its perfection.

In a series of lectures held with the support of the Eugenics Society in the early 1930s, Sicard de Plauzoles attributed France's decline in the quality of births to higher rates of proliferation in the lower classes.[57] Sordid living conditions, combined with alcoholism, syphilis and tuberculosis were the main factors responsible for racial degeneration. Characterized by its high proliferation, the French working class was represented at a higher ratio in the overall population as high as 5 to 1. De Plauzoles predicted that this trend would lead the entire population toward degeneration, for the offspring of the poor were inferior to those of the well-to-do in terms of their physical, mental and moral condition. While the population increased in terms of numbers, it thus declined in terms of quality. De Plauzoles, as opposed to natalists and the Catholic Church, dismissed the demographic policy of the French government as unrealistic. At that time it favored families with many children, and in doing so it encouraged the reproduction of the lower classes and, consequently, degeneration. Instead,

---

57 Schneider, "The Eugenics Movement in France, 1890–1940," 88.

he regarded any measure as beneficial that was able to reduce the proliferation of the lower classes. Thus, birth control was acceptable as a means of artificial selection as it preempted the dangers posed by defective or excessive fertility. While at a theoretical level the policy on the population question was revised, at a practical level no legislative steps to change population policy were taken in France up to the eve of World War II.

The influence of French eugenics on the adoption of positive eugenic measures in interwar Greece may be discerned by the insistence of pediatricians on educating citizens on their biological duties, which was considered to be the most important and attainable measure for creating robust offspring without recourse to coercion. Most Greek publications on the role of the mother, written by pediatricians and obstetricians first educated in France, highlighted the difficulties posed by sterilization and the mandatory prenuptial certificate in Greece since up to that time, the public had not been properly educated and social policy regarding health issues was still in its early stages.[58] Greek physicians therefore put forward arguments similar to those of French eugenicists: the difficulties in diagnosing hereditary characteristics, medical confidentiality, material and social interests of family physicians and the Hippocratic respect for all human beings.

Doxiadis' views on the value of the people as biological capital also bear the distinct mark of interwar French eugenicists. Besides cultivating a consciousness of citizens' eugenic duties, according to Doxiadis, the success of eugenic policy depended on the state making use of the biological capital of the Greek race. The view held by Sicard de Plauzoles—that the future of the race was not a matter of numbers but rather a matter of the quality of the biological capital—can also be found in Doxiadis' concerns over racial degeneration in interwar Greece.[59] According to him, the nation's future rested upon the biological ability of the families comprising the nation. He argued that the demographic issue was a matter of quality rather than quantity. Greece had not faced the problem of low birth

58 See, for example, the conclusions drawn by M. M. Μωϋσείδης, *Ευγονική αποστείρωσις: αρχαί, μέθοδοι, εφαρμογή* (Athens: Εκδοτικός οίκος Α. Κασιγόνη, 1934), 67–68.

59 In another article, Doxiadis noted that people living in urban areas were biologically inferior to those living in rural areas; furthermore, the fertility and biological value of individuals depended on the social class to which they belonged; see Απόστολος Δοξιάδης, "Βιολογική Πολιτική με βάση την αύξηση του πληθυσμού της χώρας," *Ελληνικά Γράμματα* 3, 27 (16 July 1928), 96–97.

rate, as was the case in other European countries in the aftermath of World War I, since the influx of refugees following the Asia Minor campaign had infused new blood into the population.[60] Yet this was dangerous since the population consisted of individuals who were not only mentally and physically wretched but biologically vulnerable and living in squalid conditions as well[61]. Consequently, the biological value of these citizens' descendants could be expected to be low. Although many eugenicists attributed the danger of degeneration to the high fertility of the lower classes, Doxiadis instead held the refugees responsible.

Child mortality rates also remained high given that the lower classes were severely afflicted with tuberculosis, malaria and malnutrition. Although Greece had a 25% surplus in births, many children never reached reproductive age. According to Doxiadis' estimates, if the nation was to stay racially viable, each family needed to have an average of 4 children and moreover needed to raise them in hygienic conditions until at least the age of 5. His research findings pointed to a level of infant mortality among working and agrarian classes of approximately 30% while it ranged from 10% to 12% in the middle classes. Therefore, in his view the quality issue had been negatively impacted after 1922. After that point the lower classes with lesser biological value became ever more dominant, while the upper classes began to dwindle.[62] Thus, he concluded that "although Greece does not face a problem of childlessness, we lack the human capital that would render our race powerful."[63]

By the end of the interwar years in Greece, state intervention was deemed necessary in order to improve the biological value of reproduction. Just as a gardener selects his seed and a farmer the best animals for reproduction, so too should the state select parents able to guarantee their children's bio-

60 For his views on the Greece's fertility rates, see Απόστολος Δοξιάδης, "Η προστασία της μητέρας και του παιδιού," *Εργασία* 1, 38 (20 September 1930): 9.

61 As Doxiadis noted: "Refugees are biologically strong and give birth to many children. But what does this population consist of? Of individuals mentally and physically wretched, biologically vulnerable, living in unfavorable conditions and therefore the biological value of future generations is not worthy"; see Personal Notes in the Apostolos Doxiadis Archive, Benaki Archive F.256: 18. He employed the same argument in the following article: [Απόστολος Δοξιάδης,] "Το δημογραφικό ζήτημα," *Νέα Πολιτική* 4 (April 1939): 418–419.

62 Similar views regarding high fertility rates for defective families can be found in an article entitled "Fertility Problems," which featured in the journal *Υγεία* in 1926, translated from an article published in the Italian journal *Social Defence* in January 1926; see [Anonymous], "Υγιειονομική κίνησις. Προβλήματα ευγονίας," *Υγεία* 3, 5 (1926): 96–99.

63 See Δοξιάδης, "Το δημογραφικό ζήτημα."

logical value.[64] According to Doxiadis, public health policy needed to be established on a biological basis. He underlined the role of citizens' biological capital as this related to both class differences and productivity.[65] To achieve such a radical eugenic restructuring, he proposed that as a first step, the biological value of families should be assessed. In other words, statistics should be compiled on each family. He envisaged these statistics as personal data files including information on the health of family members and any hereditary and infectious diseases as well as addictions to alcohol or drugs. This catalogue would also contain information on individuals' profession and living conditions, the age of parents, number of childbirths, employment status of mothers and incestuous relationships, if any. The combined results of these factors on the health of individuals would then be evaluated and families classified according to their biological value. Based on this Taylorist approach, the state would either subsidize or prevent reproduction.[66]

In order to avoid the risk of degeneration, the state needed to offer financial support to poor families of high biological value as a way of encouraging them to have children. Doxiadis argued that when states and governments need children, they must create favorable conditions for births. Being well aware that the finances behind the Foundation for Child Protection were meager, Doxiadis proposed that a tax be levied on the unwed who therefore did not contribute to the eugenic maintenance of society and that taxes in general should be proportionate to family obligations.[67] Moreover, it should be obligatory to devolve the property of those dying without offspring to this fund. According to Doxiadis, the unmarried "had to pay money so that the nation could reap the benefits from the blood

64 By the end of the 1930s, Doxiadis argued that hereditary factors determine the biological capital quality far more than environmental factors. Yet he was a little hesitant to suggest any form of selection, as such a proposal would run counter to common morality. For this reason he only proposed the education of parents in order to advise their children on the right choice of partners; see, Απόστολος Δοξιάδης, "Το παιδί ως κληρονομικόν κεφάλαιον," *Το Παιδί* 2, 55 (1939), 3–5. These views diverge from those formulated in 1928 in the preamble to the laws for the protection of motherhood and children.

65 In 1934 Doxiadis argued that "class difference would not be based on the resources a family has but on the latter's biological value;" see Δοξιάδης, "Ευγονία," 15.

66 Δοξιάδης, "Αρχαί Ευγονικής," *Το Παιδί* 1, 2 (1930): 32–33; see also A. Δοξιάδης, "Το παιδί ως κληρονομικόν κεφάλαιον," 3–4. Doxiadis was also familiar with the positions on birth control held by Margaret Sanger, an activist and leader of the American birth control movement; see Απόστολος Δοξιάδης, "Η Αμερικανίς απόστολος κατά της πολυτεκνίας," *Ελεύθερον Βήμα* (9 September 1930): 1.

67 See Δοξιάδης, "Ευγονία," 14.

of children of those who raised a family."[68] Social policy would have to be entirely revised as well as tax obligations and wage scales in order to take into account not only numbers and materials but also the biological value of each family.

Following the example set by Fascist Italy—which had introduced a taxation policy "based on the biological value of each citizen"[69]—Doxiadis, in his capacity as Undersecretary of Hygiene, tried to convince Prime Minister Venizelos to impose additional taxes on the unwed in order to effectively fund social policy on motherhood.[70] In fact, the relevant bill had been drawn up and approved by Venizelos. But this law never passed in Greece for reasons that remain unclear.[71] It should be noted that after 1932, political unrest and related events never allowed the Liberals' social policy on public health reform to be instituted.

## Conclusion

Some ten years of conflict between 1912 and 1922, alongside 1,300,000 refugees to interwar Greece, set public health and the issue of eugenically improving the nation on a new basis. In the 1930s the public debate on Greek eugenics, as reflected in the works of pediatricians and social thinkers, nonetheless failed to result in any negative eugenic measures being adopted. This was largely the result of questions over how improving the biological quality of the race could be reconciled with rights to privacy. While, various views were put forward across a broad ideological spectrum, emphasis was placed upon educating citizens as to the value of eugenics in addition to establish-

68 [Anonymous], "Φορολογία των αγάμων," *Η Ελληνική* 4, 311 (20 September 1928): 1.

69 On social policy in Italy, see Aristotle A. Kallis, "Racial Politics and Biomedical Totalitarianism in Interwar Europe," in Turda and Weindling, eds., *Blood and Homeland*, 389–415. Indeed, there were many similarities between the Greek case and different models of Latin eugenics, especially those suggested by French and Italian eugenicists. Another important issue worth noting is the impact that natalism of the Italian Fascist period had on the liberal political circles in interwar Greece. For the Greek approach to Fascist Italy, see Χρ. Χατζηιωσήφ, "Κοινοβουλευτισμός και Δικτατορία," in Χρήστος Χατζηιωσήφ ed., *Ιστορία της Ελλάδας του 20ού αιώνα. Ο Μεσοπόλεμος, 1922–1940*, vol. B, Part 1 (Αθήνα: Βιβλιόραμα 2002), 108-109.

70 For Doxiadis' opinion on this subject; see [Anonymous], "Φορολογία των αγάμων," *Η Ελληνική* 4, 311 (20 September 1928): 1. As is clear from this article, Doxiadis' taxation policy was strongly criticized and mocked in the press.

71 On preparations for drawing up a list of unwed individuals, see the 1929 correspondence between Doxiadis and the Deputy Minister of Finance; Doxiadis Archive, *Benaki Archive*, F 256, 12.

ing social hygiene institutions for mothers and children for a healthy, young generation. Arguably, such a "mild" approach to eugenics was linked to influences already informing such pediatricians, to the political conditions of the time and the developments pedology had already experienced in Greece. High rates of infant mortality and child morbidity—attributed to the lack of knowledge and to the poverty of the lower classes—in particular after the end of World War I, and the admission of refugees from Asia Minor, helped to place the issue of healthy births on a racial basis.

Under the influence of French eugenics and puericulture, especially the earlier positive eugenics movement, influential Greek pediatricians now occupying governmental positions attempted to link eugenics with puericulture and social hygiene. Following the neo-Lamarckian trend of French eugenics, they attached considerable importance to environmental factors, which they believed influenced both the quality of parental health and the quality of births, which as a consequence led them to advocate social hygiene measures as a means to improve the eugenic conditions for the new generation. In the fight against threatening national degeneration, Greek pediatricians were more likely to convince interwar Liberal governments to allocate funds for the adoption of measures to better care for infants and mothers. Yet at the same time they attempted to cultivate a eugenic consciousness in prospective mothers through the popularization of hygiene principles.

Nevertheless, in attempting to locate resources to fund a birth control policy able to guarantee improvement in the quality of Greece's biological capital, some Greek pediatricians declared their support for measures privileging those considered to be biologically powerful, as already adopted by Fascist Italy. Proponents of such measures never hesitated to suggest, especially at the end of the 1930s, that state intervention in the private life of citizens and the selection of progenitors who would be worthy of the state's financial aid due to their high biological value were initiatives worth backing. Therefore, while cultivating a eugenic consciousness in citizens was overwhelmingly approved by pediatricians participating in the debate on eugenics, they were not unanimous in terms of using the state to engineer biological capital. Despite any divergence from the dominant current of skepticism, the above findings highlight the various versions of eugenic thought while leaving open questions as to whether the Greek case corresponded with the wider Latin eugenics movement of the interwar years.

# CONTROLLING THE NATIONAL BODY

## IDEAS OF RACIAL PURIFICATION IN ROMANIA, 1918–1944[*]

*Marius Turda*

In an oft-quoted lecture delivered at the Collège de France in 1976, Michel Foucault situated biopolitics at the intersection of knowledge and power, one which emerged in the second half of the eighteenth century. Biopolitics, Foucault argued, was a modern discipline trying "to rule a multiplicity of men to the extent that their multiplicity can and must be dissolved into individual bodies that can be kept under surveillance, trained, used, and, if need be, punished."[1] Since then, scholars in numerous disciplines have employed Foucault's innovative theory on "biopolitics of the human race"[2] to explain socio-political phenomenon as diverse as Nazism, China's one child policies and contemporary terrorism.[3] As a term, biopolitics has become ubiquitous in both scholarly and general parlance. As Corrina Treitel recently remarked: "Aimed at producing a more powerful and productive society by regulating, optimizing, and even exterminating specific human populations, biopolitics has encompassed everything from housing reform, anti-smoking campaigns, and child vaccination programs

* I am grateful to Oxford Brookes University and the Wellcome Trust, London, for providing the financial support for my research in Romania. Thanks are also due to Mioara Georgescu and Răzvan Pârâianu, who ensured that my constant search for primary sources in Romanian archives and libraries was always successful.

1 Michel Foucault, *"Society Must be Defended:" Lectures at the Collège de France, 1975–1976* (New York: Picador, 1997), 242. See also Michel Foucault, *The History of Sexuality: An Introduction* (New York: Vintage Books, 1980), 136–140.

2 Foucault, *"Society Must be Defended,"* 243.

3 Among the most recent works worth mentioning are Roberto Esposito, *Bios: Biopolitics and Philosophy* (Minneapolis: University of Minnesota Press, 2008) and Stephen Morton and Stephen Bygrave, eds., *Foucault in an Age of Terror: Essays on Biopolitics and the Defence of Society* (Basingstoke: Palgrave Macmillan, 2008).

to pro- and anti-natalist tax policies, national census talking, and the science of industrial hygiene."[4] Though such a broad description may be difficult to apply to specific historical contexts across Europe, by the end of the 1930s biopolitics—both as concept and ideology—had come to represent in most European countries the synthesis of eugenic reflections about the social, cultural and biological improvement of the nation.

Not surprisingly, then, much has been written about the elaboration of a biopolitical and eugenic discourse during the interwar period, especially in Germany.[5] The present chapter shall contribute to this burgeoning scholarship by directing attention to Romania, which has, until very recently, been virtually ignored.[6] In an attempt to counter this historiographic oversight, I shall concentrate on debates on eugenic sterilization in interwar Romania as well as how these debates fed into larger biopolitical visions regarding the ethnic future of the Romanian nation. For supporters of "negative" eugenic measures like sterilization, what mattered, ultimately, was the creation of a healthy Romanian nation, dominant within the borders of the Romanian state.[7] But this was a discriminatory process. Those perceived as ethnically different or socially and biologically inadequate were gradually categorized as potential enemies and were eventually excluded from the Romanian national community.

The relationship between eugenics and nationalism was symbiotic in that they shared the same interest in the body of the nation and its future. During the 1930s and 1940s especially, the eugenicists found their nationalist agenda massively promoted in the popular and scientific press. The publicity

4 Corinna Treitel, "Max Rubner and the Biopolitics of Rational Nutrition," *Central European History* 41, 1 (2008): 1.

5 For a good summary of this literature, see Edward Ross Dickinson, "Biopolitics, Fascism, Democracy: Some Reflections on Our Discourse about 'Modernity'," *Central European History* 37, 1 (2004): 2–21

6 The only study to date is Vladimir Solonari, *Purifying the Nation: Population Exchange and Ethnic Cleansing in Nazi-allied Romania* (Baltimore: The Johns Hopkins University Press, 2009).

7 The scholarship on biopolitics and eugenics in interwar Romania includes Maria Bucur, *Eugenics and Modernization in Interwar Romania* (Pittsburgh: Pittsburg University Press 2002) and Marius Turda, *Eugenism şi antropologie rasială în România, 1974–1944* (Bucharest: Ed. Cuvântul, 2008). See also Viorel Achim, "Romanian–German Collaboration in Ethnopolitics: The Case of Sabin Manuilă," in Ingo Haar and Michael Fahlbusch, eds., *German Scholars and Ethnic Cleansing, 1919–1945* (New York: Berghahn Books, 2005), 139–154; Vladimir Solonari, "An Important New Document on the Romanian Policy of Ethnic Cleansing during World War II," *Holocaust and Genocide Studies* 21, 2 (2007): 268–297; Marius Turda, "The Nation as Object: Race, Blood and Biopolitics in Interwar Romania," *Slavic Review* 66, 3 (2007): 413–441; and Marius Turda, "'To End the Degeneration of a Nation': Debates on Eugenic Sterilization in Interwar Romania," *Medical History* 53, 1 (2009): 77–104.

served successfully to promote ideas of biological rejuvenation, ideas which in turn intersected with the program of national purification advocated by the Antonescu regime after 1941. The promise of biological perfectibility offered by eugenicists during the interwar period was therefore one which some Romanian politicians found attractive in their quest for a nation purified of both dysgenic individuals and ethnic minorities deemed either inferior or degenerate. Yet political support for eugenic sterilization failed to materialize. Ultimately, the eugenic metamorphosis of the Romanian nation was only one of the several routes to the perfect ethnic state to which Romanian nationalists aspired. The other route involved population exchanges and the extermination of the Jews and the Roma in concentration camps in Transnistria.

## Pre-World War I Antecedents

Methods to improve the biological qualities of the nation through negative selection had preoccupied Romanian physicians since the mid-nineteenth century. In 1874, the founder of Romanian psychiatry, Alexandru Sutzu (1837–1919), established a direct relationship between heredity and the degeneration of nations.[8] In his 1876 *Filosofia medicală: Despre ameliorațiunea rasei umane* [Medical philosophy: on the improvement of the human race] the dermatologist Mihail Petrini-Galatzi (1847–1926) endorsed this novel argument. Yet he went even further in advocating that members of communities suffering from hereditary diseases ('morbe') should be discouraged from reproduction.[9] That such ideas were increasingly gaining adherence among physicians in Romania is documented in a 1908 note published in *Archives d'anthropologie criminelle, de medicine légale et de psychologie normale et pathologique*. Here Romania is listed for its intention to legislate marriage restrictions for those "suffering from epilepsy, tuberculosis and syphilis."[10]

8 S. [Alexandru Sutzu], "Evoluțiunea și hereditatea," *Gazeta medico-chirurgicală a spitalelor* 5, 12 (1874): 182–187; all translations from the original are my own unless otherwise cited.

9 Petrini-Galatzi, *Filosofia medicală: Despre ameliorațiunea rasei umane* (Bucharest: Tipografia D. A. Laurian, 1876).

10 "Restriction du marriage," *Archives d'anthropologie criminelle, de medicine légale et de psychologie normale et pathologique* 23, 169 (1908): 96.

A more elaborate program for introducing pre-nuptial certificates and selective sterilization was offered by the gynecologist Constantin Andronescu (1859–1943). In 1911 he visited the Dresden Hygiene Exhibition and was thereafter convinced of the importance of racial hygiene for the future of the Romanian nation. This commitment to pursuing the negative consequences of racial hygiene also helps explain Andronescu's inclination towards punitive eugenic policies. Upon his return to Romania in 1912, Andronescu suggested the introduction of pre-nuptial certificates and eugenic sterilization, ideas derived from his trip to Germany. The First International Eugenics Congress held in London in 1912 occasioned another intervention in the Romanian medical press on the theme of degeneration and the eugenic means of preventing it. I. S. Mendonidi and C. Andronescu, both contributors to the journal *Higiena* [Hygiene], noted sympathetically that there were participants at the Congress who advocated the introduction of negative eugenic measures in order to counteract the effects of social and biological degeneration.[11]

In 1914, at the National Congress of Romanian Physicians, Andronescu further argued in favor of pre-nuptial certificates in addition to the sterilization of the feeble-minded and mentally ill, as well as those suffering from goitre and terminal tuberculosis. These proposals were considered "inhuman" by other participants, as Andronescu himself acknowledged. However, such a hindrance did not curb his enthusiasm for negative eugenics; on the contrary, he contentedly remarked that the eugenic sterilization legislation being gradually introduced in the United States since 1907 had confirmed the timing of his ideas.[12] But the attempt to circumscribe the biological improvement of the Romanian society to an adequate eugenic environment needed more than just erratic support from a few physicians attuned to Western developments in heredity and evolution. What

11 I. S. Mendonidi, "Eugenia sau ştiinţa de a naşte copii voinici," *Higiena* 1, 19 (1912): 2–3; Constantin Andronescu, "Eugenia," *Higiena* 1, 21 (1912): 2; and D.A. Olaru, "Eugenia: Congresul eugenic de la Londra," *Universul* 30, 209 (1912): 5. Andronescu referred, especially, to Bleecker van Wagenen's report on sterilization presented at the First International Eugenics Congress in 1912. See Bleecker van Wagenen, "Preliminary Report of the Committee of the Eugenic Section of the American Breeders' Association to Study and to Report on the Best Practical Means for Cutting off the Defective Germ-Plasm in the Human Population," in *Problems of Eugenic. Papers Communicated to the First International Eugenics Congress held at the University of London, July 24th to 30th, 1912* (London: The Eugenics Education Society, 1912), 460–479.

12 Constantin I. Andronescu, *Pentru ce ne îmbolnăvim? Noţiuni de patologie socială* (Bucharest: Tip. 'Cartea Românească, 1943), 15 and 44.

was needed was a strong political commitment to the emerging Romanian eugenic discourse and a way to disseminate the eugenic ethos into mass culture. Both were achieved after World War I.

## Popularizing Eugenics in the 1920s

It was the creation of Greater Romania in 1918 which prompted Romanian health officials and physicians to engage in an unprecedented program of institutionalization in the field of eugenics, social hygiene and public health. The dermatologist Aurel Voina (1896–1967), for instance, regarded hygiene and eugenics as the two sources of regenerative nationalism and scientific progress.[13] Popularizing eugenics was an integral part of this process alongside establishing institutes for research and university departments in the leading Romanian cities. Shortly after the foundation of a Romanian university in Cluj, the physician Valeriu Bologa (1892–1971) explained in the newly founded journal *Revista sănătății* [The health review] that eugenics was, theoretically, "the discipline which studied the laws of evolution and heredity, and thus the causes for the degeneration of the human species." In practice, however, Bologa continued, eugenics should concentrate on "obstructing the marriage, or at least the procreation, of inferior individuals."[14] Essential for this reference to negative eugenics was a growing awareness among physicians that new social policies were needed in Romania in order to address the health predicaments of the newly established state.

Hygienists and eugenicists played a major role in educating their contemporaries in the intricacies of the alleged social and biological degeneration of their society. At the same time some, however, many advanced depressing predictions, portraying degeneration as a national calamity all but certain to endanger the future of the Romanian race. In his 1921 *Crâmpeie de eugenie şi igienă socială* [Fragments of eugenics and social hygiene], the gynecologist Ioan Manliu (1886–?), for instance, suggested that:

> 1) Every degenerate individual should be sterilized and, if possible, returned to society. 2) Every degenerate and sterilized individual should be kept in iso-

[13] Aurel Voina, "Doi factori de progres: igiena şi eugenia," *Societatea de mâine* 1, 8 (1924): 183–184.

[14] Valeriu Bologa, "Eugenezia," *Revista sănătății* 1, 3 (1921): 83.

> lation in asylums and colonies until he/she can be returned to society as a useful member. 3) Only those individuals who still pose a danger to society after their sterilization should be isolated for life, while they should sustain themselves and society through work in gardens, workshops, and so on.[15]

While this was a radical eugenic program, Manliu firmly believed, "[i]t is in this direction that we must orient our efforts to protect superior elements and prohibit, without mercy, inferior elements from paternity." The most effective way to regain control over the body of the nation, he concluded, was the "mass sterilization of degenerates."[16] Although Manliu's crude prognostications were not widely shared, his contemporaries nevertheless also stressed the importance of controlling reproduction and correcting the social fabric of the nation through interventionist eugenic policies.

In addition to monographs and specialized journals, the media was also involved in disseminating eugenic ideas for human improvement. In 1923, for example, perhaps the most important Romanian newspaper in Bucharest, *Adevărul* [The truth], hosted a debate on abortion and birth control, aiming to clarify some of the uncertainties surrounding sexual reproduction.[17] Two contrasting points of view emerged: one put forward by a surgeon, Constantin Poenaru-Căplescu (1874–1948), was against abortion; the other, articulated by a physician, Iosif Glicsman (1871–1938), writing under the pseudonym Dr Ygrec, endorsed it.[18] If some Romanian eugenicists, like Manliu, looked first and foremost to the North American model of eugenics, the debate quickly extended to the experiences of other European countries like Germany and France. Glicsman, for instance, informed his readers about the sterilization debate in Saxony, basing his arguments on the report presented by the criminologist Erich Wulffen to the *Landrat*

---

15 Ioan Manliu, *Crâmpeie de eugenie şi igienă socială* (Bucharest, Tip. 'Jockey-Club', 1921), 21.

16 Manliu, *Crâmpeie de eugenie*, 21.

17 For the French case, see William Schneider, *Quality and Quantity: The Quest for Biological Regeneration in Twentieth-Century France* (Cambridge: Cambridge University Press, 1990); for the British case, see Richard A. Soloway, *Demography and Degeneration: Eugenics and the Declining Birthrate in Twentieth-Century Britain* (Chapel Hill: University of North Carolina Press, 1990); for the German case, see Atina Grossmann, *Reforming Sex: The German Movement for Birth Control and Abortion Reform, 1920–1950* (Oxford: Oxford University Press, 1995).

18 See, for example, Ygrec, "O problemă importantă: Păreri din public în chestiunea provocatorilor şi a provocărilor de avorturi," *Adevărul* 36, 12141 (1923): 1–2; C. Poenaru-Căplescu, "Medicii avortori şi făcătoarele de îngeri!," *Adevărul* 36, 12144 (1923): 1–2; and Ygrec, "Iarăşi despre avorturi provocate şi medicii avortori," *Adevărul* 36, 12152 (1923): 1–2.

of Saxony. Wulffen's report on sterilization was explicitly in favor of radical eugenic measures; criminals, so-called idiots, the feebleminded, epileptics, the blind, deaf and dumb should all be sterilized, alongside women and girls who gave birth to more than one illegitimate child.[19] That Glicsman subtly endorsed German notions of eugenic sterilization did not escape the attention of many domestic readers, as indicated by the prompt reaction by the renowned neurosurgeon, Dimitrie Bagdasar (1893–1946). In taking a diametrically opposed view, Bagdasar stressed that "repeated sterilization could only lead to the numerical reduction of the race but does not guarantee its future improvement, not even a relative one."[20] Yet if Bagdasar expressed doubts about negative eugenic practices like compulsory sterilization, this was not on account of humanistic objections. Instead, he believed that the current state of medical knowledge provided inconclusive evidence on the hereditary nature of most degenerative diseases. Without a clearly established connection between heredity and biological degeneration, Bagdasar warned, eugenicists should abstain from endorsing the sterilization of those suffering from mental illnesses.

This paradigm-shift toward discussing degeneration in terms of heredity and genetics occurred in the wake of an amplified focus on national projects of ethnic protectionism across Europe. Here Romania was not exceptional. The notion that there was a strong relationship between dysgenic individuals and the broader national community also emerged during this period. In his 1925 *Igiena naţiunii: Eugenia* [The hygiene of the nation: Eugenics], the prominent eugenicist and founder of the School of Hygiene and Public Health in Cluj, Iuliu Moldovan (1882–1966), defined the nation as "a biological reality, a human structure with its specific biology and pathology."[21] Correspondingly, eugenics was closely connected to biopolitical interventionism and radical measures to regulate health, including "preventing defective individuals from procreating."[22] Moldovan placed

19 Doctorul Ygrec, "Cum vor nemţii să-şi îmbunătăţească rasa? Se cere sterilizarea celor anormali, a epilepticilor, surdo-muţilor, demenţilor etc.," *Adevărul* 36, 12143 (1923): 1. See also Stefan Kühl, *The Nazi Connection: Eugenics, American Racism, and German National Socialism* (Oxford: Oxford University Press, 1994), 23; and Sheila Faith Weiss, "The Race Hygiene Movement in Germany, 1904–1945," in Mark B. Adams, ed., *The Wellborn Science: Eugenics in Germany, France, Brazil and Russia* (Oxford: Oxford University Press, 1990), 39.

20 Dumitru Bagdasar, "Sterilizarea," *Adevarul* 36, 12170 (1923): 1.

21 Iuliu Moldovan, *Igiena naţiunii: Eugenia* (Cluj: Institutul de Igienă şi Igienă Socială, 1925), 12.

22 Moldovan, *Igiena*, 37.

the family at the centre of his theory of national eugenics and biopolitics, envisioning measures to protect "acceptable" families from both social and biological threats. No medical prophylactic methods, however, Moldovan claimed, would be sufficient unless the Romanian population—especially those affected by hereditary diseases—acquired "a racial consciousness, a sentiment of biological responsibility."[23] What should unite the members of the community was not merely a political ideology acting in the name of the nation, but a fusion of new nationalist and eugenic ideals into a comprehensive biopolitical state.

Various eugenicists and politicians seized upon Moldovan's writings to demonstrate the necessity for a state-controlled eugenic program of national rejuvenation. The 1927 launch of the *Buletin eugenic şi biopolitic* [Eugenics and biopolitics bulletin] in Cluj is a further indicator of the absorption of eugenic ideas into a larger national and biopolitical agenda.[24] As editor-in-chief, Moldovan became immersed in the difficult task of constructing a eugenic philosophy adapted to the Romanian milieu. The biopolitical control of the national body, too, needed systematic rationalization, and Moldovan invited contributions from the leading supporter of eugenic sterilization, Harry Laughlin (1880–1943), Deputy Director of the Eugenics Record Office in the United States. Given Laughlin's public successes in advising various sterilization committees in the United States (including the famous Carrie Buck case in Virginia[25]), the *Bulletin* introduced Laughlin as "the most appropriate person to express opinions on eugenics, and especially on its practical application."[26]

Although there were Romanian eugenicists sharing Laughlin's belief in the necessity of eugenic sterilization, many thought that negative eugenics

---

23 Moldovan, *Igiena*, 46.

24 That one of the first lectures given to the newly established Biopolitical and Eugenics Section of the 'Astra' Association in Cluj was by prominent politician Alexandru Vaida-Voevod (prime minister of Romania between 1919 and 1920 and again between 1933 and 1934) is illustrative. The lecture was then published as "Politica naţională şi capitalul biologic naţional," *Buletin eugenic şi biopolitic* 1, 7–8 (1927): 199–211.

25 In 1927, while a patient at Virginia State Colony for Epileptics and Feebleminded, Carrie Buck (1906–1983) was compulsory sterilized in accordance with Virginia's sterilization law. Recommending the procedure, the Supreme Court Justice Oliver W. Holmes, Jr. declared infamously: "Three generations of imbeciles are enough." See Paul A. Lombardo, *Three Generations. No Imbeciles. Eugenics, the Supreme Court and Buck vs. Bell* (Baltimore: The Johns Hopkins University Press, 2008).

26 Harry Laughlin, "O privire asupra sterilizării eugenice în America," *Buletin eugenic şi biopolitic* 1, 9–10 (1927): 17. See also Philip K. Wilson, "Harry Laughlin's Eugenic Crusade to Control the 'Socially Inadequate' in Progressive Era America," *Patterns of Prejudice* 36, 1 (2002): 49–67.

needed to be modified in order to reflect a Romanian context. Eugenicists of this persuasion, like the pediatrician Gheorghe Popovici (1895–1946), continued to argue for improvements in education, housing and public health as prerequisites for a healthy nation,[27] but supporters of biopolitical rejuvenation regarded these methods as inefficient and representative of a merely transitional phase in the biological transformation of Romanian society—one which would inexorably be replaced by a new scientific ethos based on eugenics and heredity. By the late 1920s, there were an increased number of Romanian eugenicists who referred dismissively to the generalized socialization of health care attempted by the state after 1918 as upholding an outmoded utilitarian liberal model incapable of biologically resuscitating Romanian society. Following developments elsewhere in Europe, these eugenicists offered to counsel the state to adopt new techniques to bolster health improvement and biological strength. Eugenic sterilization was one such measure.

## A Radical View on Sterilization

Considering this burgeoning eugenic impulse by the late 1920s, it is worth devoting some closer attention to Ioan Manliu's conception of sterilization. Manliu considered the Romanian nation as threatened by widespread degeneration. In 1931, he energetically asserted the need for eugenic sterilization. "How can we fight degeneration?" The answer was as unequivocal as was his question, namely "by stopping the reproduction of degenerates and encouraging the fertility of superior individuals."[28] Manliu assessed several methods for neutralizing the proliferation of "degenerate individuals." The first, "life imprisonment," was deemed unattainable due to the high social costs expected; others, like the prohibition of marriage, abortion or the use of birth-control techniques, he considered difficult to put into practice. In consequence, Manliu advocated eugenic sterilization.

Manliu envisioned a racial crusade against national degeneration. He mentioned a wide range of individuals affected by social and clinical con-

[27] Gheorghe Popovici, "Biopolitica, puericultura şi schimbarea de sistem în conducerea statului," *Societatea de mâine* 5, 22–24 (1928): 443–444.

[28] Ioan Manliu, "Sterilizarea degeneraţilor," *Revista de igienă socială* 1, 5 (1931): 375.

ditions, including psychopaths, epileptics, criminals and alcoholics, those suffering from hemophilia or cancer as well as Romanians he considered "remarkably ugly," all of whom he wanted to have sterilized. Manliu thus translated his personal fears of biological degeneration into a national crisis, broadening the reasons for eugenic sterilization to include anyone he deemed injurious to the body of the nation: "From the sterilization of a few thousand individuals, interned or free, one cannot expect a transformation of our hereditary constitution. To end the degeneration of a nation we need to sterilize *en masse*."[29]

Hyperbolic though the escalation of Manliu's position on eugenics may appear, his arguments were neither new nor was his rhetoric something confined to Romania. Eugenicists in both the United States and Germany had expressed similar positions. Quoting Laughlin, for instance, who claimed that "10% of each generation should be sterilized in order to obtain practical results"—not to mention the German social hygienist Alfred Grotjahn (1869–1931), who considered "inferior a third of the [German] population"—Manliu proposed that "we should sterilize *5½–6 millions*."[30] But how to implement such a radical plan in Romania? "As a practical foundation," Manliu continued unabatedly:

> Sterilization should be declared legal in all prisons, mental asylums, schools for juvenile delinquents, colonies of vagabonds and for homeless people found during police raids, etc. For other abnormal individuals (psychopaths, mentally insane, alcoholics, etc) sterilization must be conducted at their request and initiated immediately without the need of a special law. It is sufficient to have the consent of the person [the physician] performing the operation.[31]

As far as medical authority was concerned, Manliu elevated the role of the physician to an empowering status. Sterilization would be performed "only after [the patients] have been examined by at least two physicians (one for physical, the other for mental illnesses) and found that they exhibited the probability of degenerate offspring."[32] In the name of scientific progress, and as a means for Romania to acquire international recognition, Manliu

29 Manliu, "Sterilizarea degeneraților," 378. (emphasis in the original)
30 Manliu, "Sterilizarea degeneraților," 381. (emphasis in the original)
31 Manliu, "Sterilizarea degeneraților," 381.
32 Manliu, "Sterilizarea degeneraților" 381.

celebrated the prospect of eugenic sterilization: "If Romania has not yet contributed to the solution of problems of human biology, she should at least now sterilize 10. 000 degenerates in one year in order to accumulate a greater experience *which could contribute greatly to the generalization of this principle in all of Europe*."[33]

Like his eugenic contemporaries, Manliu believed that Romanian society was dominated by a liberal egalitarianism detrimental to the future of the nation. This imported ideology needed to be replaced by a new biopolitical ethos, and Manliu invited schools, churches, the judiciary, bureaucracies and the armed forces to share in the "biological rejuvenation" of the Romanian nation. That Manliu enlisted the Romanian Orthodox Church to assist in disseminating eugenic thought is particularly important. Compared to the public announcements of the Catholic Church, which condemned negative eugenic methods such as abortion and sterilization, such as in the 1930 papal encyclical, *Casti Connubii* [On Christian marriages],[34] the Orthodox Church was rather ambiguous on the subject and did not issue any such unequivocal official statements.[35] Although some ecclesiastical journals did voice their concern about eugenics, this anxiety was never endorsed or even explored by the leading prelates of the Orthodox Church.[36] In a country where Orthodox Christians amounted to 72.6% of the population, the Orthodox Church played a central role in all aspects of life.[37] Whether Manliu's eugenic vision of an extensive biological transformation of Romania could be embraced was ultimately dependent on support from the Orthodox Church. Not surprisingly, then, Manliu called on the Orthodox Church "to use its overwhelming moral authority, declare itself in favor of the biological purification [of the nation] and act accordingly."[38]

Manliu had grasped the essential precondition for any successful project of social and biological engineering in Romania: its acceptance by the

33 Manliu, "Sterilizarea degeneraţilor," 382. (emphasis in the original)

34 See Etienne Lepicard, "Eugenics and Roman Catholicism: An Encyclical Letter in Context: *Casti connubii*, December 31, 1930," *Science in Context* 11, 3–4 (1998): 527–544; and Monika Löscher, *"...der gesunden Vernunft nicht zuwider..."? – Katholische Eugenik in Österreich vor 1938* (Innsbruck: Studienverlag, 2009).

35 See Mirel Banică, *Biserica Ortodoxă Română. Stat şi societate în anii '30* (Iaşi: Polirom, 2007), 98–105.

36 See, for example, I. D., "Despre Eugenie," *Glasul Monahilor* 14, 482 (1936): 4.

37 The statistical data is from Joseph Rothschild, *East Central Europe between the Two World Wars* (Seattle: University of Washington Press, 1974), 284.

38 Manliu, "Sterilizarea degeneraţilor," 382.

Orthodox Church or, instead, a state powerful enough to circumvent the influence of religious authorities, as was the case in Nazi Germany (and, after 1947, communist Romania).[39] Predating the more extreme, racially-motivated eugenicists, Manliu's conclusions nonetheless echoed their apocalyptic visions of biological degeneration. To this end, he proposed: "The struggle against the inferior man must be continued acrimoniously. *Sterilization is a decisive factor in the extermination of sub-humans and towards the biological ennoblement of mankind.*"[40]

Eugenic sterilization and the elimination of dysgenic individuals were two principles at the heart of Manliu's biopolitical philosophy. Indeed, he was the first to argue for widespread sterilization of Romania's "inferior" population. Moreover, subsequent contributions to the Romanian debate on eugenic sterilization indicate that, at the time, Manliu was acting as an isolated enthusiast, one whose theoretical concerns and rhetorical usages were not generally shared by other eugenicists. Rather, most of them preferred to direct their attention toward specific medical and social categories—particularly the feebleminded—than speculate on the racial destiny of the Romanian nation.

## Other Opinions on Sterilization

The first Romanian text entirely devoted to eugenic sterilization, co-authored by the jurist and legal adviser Eugen Petit (1882–1959) and the laryngologist Gheorghe Buzoianu (1901–1974), was published in 1935. Petit reviewed European and North American experiences of eugenic sterilization, devoting extensive analysis to the 1933 Nazi "Law for the Prevention of Hereditarily Diseased Offspring" [*Gesetz zur Verhütung erbkranken Nachwuchses*]. While sympathetic to the idea of penal and therapeutic sterilization, Petit expressed concerns about compulsory sterilization, especially its putative introduction in Romania: "No matter how many serious arguments are invoked in favor of compulsory sterilization, we believe that, in our case, such

39 For the case of communist Romania, see Gail Kligman, *The Politics of Duplicity: Controlling Reproducing in Ceausescu's Romania* (Berkeley: University of California Press, 1998).

40 Manliu, "Sterilizarea degeneraților," 384. (emphasis in the original)

a measure cannot be introduced as an ordinary law."[41] The existing Romanian penal code, Petit explained, expressly outlawed the rejection of individual liberties. Accordingly, a modification of the penal code was needed, one whereby a sterilization law could be introduced. But was eugenic sterilization actually needed in Romania? Despite finding no reasons why voluntary sterilization should be rejected, Petit added the caveat that such a measure was to be directed at individuals with hereditary diseases, but only after a commission of specialists had already agreed to the procedure. When it came to compulsory sterilization, however, Petit was decidedly opposed.[42]

This legal discussion was followed by the complete Romanian translation of Germany's new sterilization law, which subsequently served as the primary source of information for both legal and medical analysis in interwar Romania. Buzoianu—the author of the second part of the book—provided a lengthy medical discussion of various sterilization techniques for both men and women. Buzoianu was well-acquainted with surgical procedures surrounding sterilization and offered a balanced synthesis of the latest developments in medical practice. Most importantly, he disseminated this information to the general public in accessible prose, dismissing complex reservations and legitimate anxieties about the impact of sterilization on individual health and especially on sexual performance.

Petit and Buzoianu's book on sterilization represented a major response to the increased currency of the topic in Romania; the statutes of the Royal Romanian Society for Heredity and Eugenics was another. Under the chairmanship of the renowned neurologist Gheorghe Marinescu (1863–1938), this society, established in 1935, undertook to explore all possible venues for the popularization of eugenics.[43] With respect to negative eugenics, one of the society's statutes stipulated that members "will propose a sterilization law for degenerated individuals, idiots, imbeciles, the feebleminded and criminals."[44]

41 Eugen Petit, Gheorghe Buzoianu, *Sterilizarea din punct de vedere juridic şi chirurgical* (Bucharest: 'Adevărul', 1934), 17.

42 Petit, Buzoianu, *Sterilizarea*, 21.

43 See Sabin Manuilă, "Societatea regală română de eugenie şi studiul eredităţii," *Sociologie românească* 1, 5 (1936): 31–32; and "Autorizarea de funcţionare, actul constitutiv şi statutele Societăţii Regale Române pentru Eugenie şi Studiul Eredităţii," *Revista de igienă socială* 6, 4 (1936): 271–278. See also Ion Vasilescu-Bucium, "Eugenia şi studiul eredităţii," *Mişcarea medicală română* 8, 3–4 (1935): 169–174.

44 Gheorghe Marinescu, "Despre hereditatea normală şi patologică şi raporturile ei cu eugenia," *Memoriile Secţiunii Ştiinţifice* 3, 11 (1936): 84. See also Marius Georgescu, *Protecţia medico-socială a maternităţii. Constatări. Studii comparative. Propuneri de organizare* (Bucharest: Institutul Grafic 'Luceafărul', 1937), 143–155.

This was an important commitment to eugenic legislation. Until then, none of the other eugenics societies in Romania had included sterilization in their program, although many members were ardent supporters.[45] Iordache Făcăoaru (1897–?), for instance, was a member of the Eugenics and Biopolitical Section of the "Astra" Association in Transylvania, and taught eugenics at the Institute for Hygiene and Social Hygiene in Cluj. For his part, Făcăoaru considered sterilization "one of the means unanimously indicated by eugenicists" to prevent "the increase of dysgenics of all categories."[46]

In subsequent years leading to the outbreak of the Second World War, Făcăoaru published further articles on degeneration, dysgenics and sterilization. Echoing Manliu's warnings on national degeneracy—while combining it with extensive knowledge of the existing literature on heredity, eugenics and racial hygiene—Făcăoaru declared that "10% of the [Romanian] population should not have offspring."[47] This contention related to his view that such an increased number of dysgenics jeopardized the biological future of the nation, leading Făcăoaru to throw his weight behind negative eugenic measures. Basing his analysis on famous dysgenic cases, like the history of the Kallikak family in the United States, and on the more well-known Nazi racial laws, Făcăoaru contemplated various solutions in order "to make it impossible for dysgenics to reproduce: to isolate them or their eugenic sterilization."[48] Yet Făcăoaru had more to offer than reproducing the ideological principles of National Socialism's biopolitical revolution. In addition to his support for Nazi Germany's sterilization law, Făcăoaru was also simultaneously attracted to the model of national eugenics ear-

45 The Romanian Royal Society of Eugenics and Heredity had two sections in the provinces: one in Craiova (president Ion Vasilescu-Bucium); the other in Chişinău (president Iosif Lepşi). The other eugenic societies in Romania were the Eugenics and Biopolitical Section of the "Astra" Association in Transylvania (established by Iuliu Moldovan in 1927), and the Demographic, Anthropological and Eugenics Section of the Romanian Social Institute (established by Sabin Manuilă in 1935). In 1939 these societies formed the Union of the Eugenics Societies in Romania under the presidency of Constantin Parhon, at the time also the president of the International Latin Federation of Eugenics Societies. The Union was formed with the intention of organizing the second congress of the Federation, planned to be held in Bucharest between 25 and 30 September 1939. See the announcement in *Revista de ştiinţe medicale* 28, 3 (1939): 247–248. For reasons unknown, the congress did not take place.

46 Iordache Făcăoaru, *Curs de eugenie* (Cluj: Institutul de Igienă şi Igienă Socială, 1935), 67.

47 Iordache Făcăoaru, "Înmulţirea disgenicilor şi costul lor pentru societate şi stat," *Buletin eugenic şi biopolitic* 6, 4-5-6 (1935): 179–180.

48 Iordache Făcăoaru, "Familiile degenerate şi costul lor pentru societate şi stat," *Buletin eugenic şi biopolitic* 6, 4-5-6 (1936): 220.

lier advocated by his mentor, Iuliu Moldovan. Like many other eugenicists working with Moldovan at the Institute of Hygiene in Cluj in the interwar period, Făcăoaru was convinced of the supremacy of nation over race, even if he remained open to racial arguments. This was an inspired tactic, as it was this racial imagery of the Romanian nation in decline that opponents of eugenic sterilization deemed especially problematic.

## Eugenics for the Romanian Nation

However, there also existed eugenicists who, invoking a specific form of "Romanian eugenics," rejected sterilization. For example, the founder of animal genetics in Romania, Gheorghe Constantinescu (1888–1951), warned that in the case of sterilization, "we need to wait for the results of many years of experience to be able to formulate a final point of view."[49] The physician Iosif Leonida (1903–1942) also expressed his reservations about the efficiency of eugenic sterilization in a rural country like Romania. While Constantinescu spoke on behalf of experimental biology, Leonida alluded to the general character of Romanian society, opposing Romanians' "Latin mentality" to the "Anglo-Saxon mentality of countries where sterilization was introduced."[50]

Responding to eugenicists claiming that the introduction of sterilization would contribute to the betterment of society, Leonida denied that such a radical eugenic measure would actually improve the Romanian nation's social life. Leonida insisted that only through the development of the country's economic situation would Romania witness a decrease in illiteracy and alcoholism. Once prosperous, the rural population of Romania would then become aware of the importance of racial hygiene, meaning that eugenic sterilization would then be "superfluous." Ultimately, Leonida minimized

49 Gheorghe K Constantinescu, *Ereditate şi eugenie* (Bucharest: Torouţiu, 1936), 91.

50 Iosif Leonida, "Ce poate realiza practic eugenia la noi," *Mişcarea medicală română* 8, 5–6 (1935): 367. For the differences between "Latin" and "Anglo-Saxon" eugenics, see Nancy Leys Stepan, *"The Hour of Eugenics": Race, Gender, and Nation in Latin America* (Ithaca: Cornell University Press, 1991); Michael Richards, "Spanish Psychiatry c. 1900–1945: Constitutional Theory, Eugenics, and the Nation," *Bulletin of Spanish Studies* 81, 6 (2004): 823–824; Yolanda Eraso, "Biotypology, Endocrinology, and Sterilization: The Practice of Eugenics in the Treatment of Argentinian Women during the 1930s," *Bulletin of the History of Medicine* 81, 4 (2007): 793–822; and Andrés H. Reggiani, "Dépopulation, fascisme et eugénisme 'latin' dans l'Argentine des années 1930," *Le Mouvement Social* 230 (2010): 7–26.

the importance of heredity, characterizing mental and physical illnesses as occurrences likely to cause immediate biological damage to the individual and as having no influence on the genetic heritage of generations to come.

Similar skepticism over sterilization pervaded the eugenic theories offered by the psychiatrist Grigore Odobescu (1881–1958). In his 1936 *Eugenie pentru neamul românesc* [Eugenics for the Romanian nation], Odobescu argued that in Romania, "neither the voluntary sterilization practiced in Switzerland, nor the social prophylactic sterilization practiced in the United States will be received favorably."[51] He did, however, accept that eugenic sterilization would be justified in "some cases, such as for those suffering from neuropsychotic disorders (especially incurable variants), those who must be interned for life," and "those suffering from critical hereditary diseases."[52] As a general rule, however, eugenic sterilization was discouraged. "Degenerates" in Romania, Odobescu observed, were largely the result of ruinous economic and hygienic conditions. Among the most important causes of degeneration, Odobescu identified poor nutrition, a total lack of hygiene and rampant contagious diseases. Significantly, hereditary diseases were not included. Contrary to Manliu's dire diagnosis of a nation crippled by hereditary degeneration, Odobescu believed that there were numerous examples proving the "wonderful quality of the biological substance" suffusing the Romanian nation. What was needed, therefore, was "the education of the masses." Such an educational program would not only increase economic and social standards, more importantly, it would "augment the biological condition" of the population. These were, Odobescu concluded, the "best-suited eugenic policies for our country and nation."[53] Ultimately, Odobescu believed that Romania could be improved through educational, social and agricultural reforms rather than by introducing biologically-motivated eugenic policies.

Odobescu pursued the same arguments in his *Politica eugenică* [Eugenic politics], deliberately connecting negative eugenics in Germany with Nazism, arguing that "such radical measures cannot be generalized."[54]

---

51 Grigore Odobescu, *Eugenie pentru neamul românesc* (Bucharest: Monitorul Oficial şi Imprimeriile Statului, 1936), 12.

52 Odobescu, *Eugenie*, 12.

53 Odobescu, *Eugenie*, 15.

54 Grigore Odobescu, *Politica eugenică* (Bucharest: Institutul de Arte Grafice 'Eminescu,' 1936), 4.

Accordingly, any introduction of eugenic policies in Romania should take local specificities into account. It thus emerged that "none of the methods applied elsewhere would bring us the desired results; they would be inefficient due to our specific conditions."[55] In place of sterilization Odobescu proposed another solution to social decline; he insisted that the physical and mental health of the Romanian peasantry could be the source of national rejuvenation. In other words, her rural and agrarian environment protected Romania from forms of urban degeneration experienced by the far more industrialized countries of Western Europe.[56] Accentuating the social and economic differences between Romania and other European countries, Odobescu downplayed the significance of eugenic thinking among contemporaneous health reformers and physicians.

Yet Odobescu was rather premature in his criticism. By the late 1930s, the Legionary Movement (established in 1927 by Corneliu Zelea Codreanu) not only attracted significant segments of the Romanian cultural, scientific and politic elite but also persuasively disseminated its message about a revitalized Romania. Creating biological and spiritual 'new men' became one of the Legionary Movement's dogmas.[57] By thinking in terms of individual purification and collective redemption, many of the Legionary Movement's ideologues created a cultural edifice ideally suited to the biopolitical society promoted by the eugenicists. The social philosopher and playwright, Nicolae Porsenna (1892–1966), for example, deplored "today's Romanian, who is weak, stupid, superficial, indifferent to progress and egotistic," in other words, "a semi-barbarian." But the Romanian of the future, he hoped, would be "re-educated, regenerated;" he will be a "man of iron, a pioneer, a creator."[58] National regeneration, Porsenna insisted, was not "a frivolity, a caprice" but "a necessity."[59] Thus, the idea of a new, reborn nation became a cultural symbol in Romania, linking an assortment of ideological positions and personalities. With few exceptions, most eugenicists publishing during this period adopted this cultural and political discourse.[60]

55 Odobescu, *Politica eugenică*, 10.

56 See the analysis offered by Daniel Pick, *Faces of Degeneration: A European Disorder, c. 1848–1918* (Cambridge, Cambridge University Press, 1989).

57 Corneliu Zelea Codreanu, *Pentru legionari* (Bucharest: Editura Scara, 1999) [first published 1936].

58 Nicolae Porsenna, *Regenerarea neamului românesc* (Bucharest: Cugetarea, 1937), 216.

59 Porsenna, *Regenerarea*, 219.

60 See, example, Ovidiu Comşia, "Neamul regenerat," *Buletin eugenic şi biopolitic* 8, 10–12 (1937): 304–316.

## The Normalization of Race

The themes of ethnic revival and national regeneration figured prominently in the work of Gheorghe Banu (1889–1957). Other eugenicists, most notably two mentioned earlier, Ioan Manliu and Iordache Făcăoaru, were also influenced by biopolitical concepts of racial hygiene, but Banu consistently created a corpus of writing in which eugenics, hygiene and racial nationalism were deeply connected. In 1939, Banu published *L'hygiène de la race,* arguably the most sophisticated account of racial hygiene ever produced in Romania.[61] Here, Banu offered both an informed theoretical discussion of heredity and concrete solutions for the biological improvement of the race. The sixth section of Banu's book focused on "principles and methods for the normalization of the race." Banu's eugenic philosophy centered on the notion that hygienic values and racial improvement were closely linked. "To normalize the race" was, according to him, to employ various methods by which to "maintain and increase the normal elements of the race, and eliminate from the heart of the social organism elements which are deficient, physically and mentally." Special emphasis was therefore placed on certain methods, including "practical and theoretical investigations of heredity; biological and hereditary statistics; the study of family genealogies, the biological and hereditary status of the population and the demographic evolution of communities."[62]

Preventive sterilization, like pre-nuptial certificates and compulsory segregation, was deemed one of these "socio-biological measures" to be adopted to ensure the "normalization of the race." Banu also engaged two dissenting positions: one based on rationality, which condemned sterilization as an "encroachment on the human rights of the individual;" and one based on Christian morality, which "opposed the control of heredity."[63] While some of the objections raised by "moralists and the representatives of the church" were legitimate, Banu nevertheless contended that the scientific arguments justifying preventive sterilization were overwhelming.

61 Gheorghe Banu, *L'hygiène de la race. Étude de biologie héréditaire et de normalisation de la race* (Paris: Masson, 1939).

62 Banu, *L'hygiène de la race,* 256.

63 Banu, *L'hygiène de la race,* 290–293.

In consequence, it was necessary for penal codes to be devised according to "the principle of social protection" instead of reflecting the "dogmas of liberal orthodoxy."[64] Eugenic sterilization, by its very nature, bore significant implications for the state. In particular as it offered a means by which to cut expenditures and re-invest in other public sector areas, diverting funds for the treatment and protection of perceivably dysgenic social groups. Moreover, Banu continued, preventive sterilization was, "first and foremost, of biological importance: it [was] about the purity and the vital value of the race."[65] His overriding aim was thus to work towards a program of biological rejuvenation in which relationships between the individual and his or her racial community were mutually advantageous. Banu therefore stressed that sterilizing "pathological individuals" such as "imbeciles, idiots, epileptics, criminals and those affected by diverse psychoses, syphilis and tuberculosis as well as hemophiliacs and diabetics," should be accepted as the necessary formula for the conservation and betterment of the race.[66]

Banu's discussion of eugenic sterilization is symptomatic of the theoretical and ideological views propounded by eugenicists in Romania and elsewhere in Southeastern Europe, not only in illustrating the weaknesses and inconsistencies of previous proposals but by conceiving of eugenics as a totalizing solution that was just as medical and social as it was political and national. Closely associated with this eugenic quest for comprehensive answers to social questions in interwar Romania were debates on authoritarian projects for national survival—especially after the territorial losses of 1940.[67] In the highly unsettling wartime years, these concerns took a distinctively racial turn: if previous eugenic sterilization targeted individuals suffering from diverse medical conditions, focus now turned to the alleged source of national degeneration posed by ethnic minorities.[68] What Michelle Mouton termed the "inability to function in the *Volkgemeinschaft*"[69]

64 Banu, *L'hygiène de la race*, 294.

65 Banu, *L'hygiène de la race*, 297.

66 Banu, *L'hygiène de la race*, 297–298.

67 In 1940 Romania lost Bessarabia and northern Bukovina to the USSR; northern Transylvania to Hungary and southern Dobrudja to Bulgaria.

68 Marius Turda, "Fantasies of Degeneration: Some Remarks on Racial Antisemitism in Interwar Romania," *Studii iudaice* 3 (2003): 336–348 and *Eugenism şi antropologie rasială în România* (Bucharest: Cuvântul, 2008).

69 Michelle Mouton, *From Nurturing the Nation to Purifying the Volk: Weimar and Nazi Family Policy, 1918–1945* (Cambridge: Cambridge University Press, 2007), 147.

became, with the installation of Marshal Ion Antonescu's regime in 1941, synonymous not so much with conscious racial extermination (like Nazi Germany) as with assumed racial nationalism.

## The Roma and Eugenic Sterilization

Similar to Fascist Italy and Nazi Germany, various forms of radical biopolitics emerged in Romania between 1940 and 1944 that also endorsed the totalitarian state as the quintessential expression of Romanian ethnic supremacy. Like contemporaneous racial hygienists across Europe, Romanian eugenicists adopted and championed principles of ethnic re-engineering and social segregation. Although many eugenicists opposed marriages between Romanians and national ethnic minorities (especially in the Banat and Transylvania) none argued for the sterilization of Jews, Hungarians or Germans.

Yet one ethnic group was singled out for its perceived otherness and the "dysgenic" danger it posed to the Romanian majority: the Roma.[70] Outlining the "racial problem in Romania" to 1940, demographer and director of the Central Institute of Statistics in Bucharest, Sabin Manuilă (1894–1964), accused both Jews and Roma of existing outside, and in opposition to, the Romanian national body. This was identified almost exclusively through a racial representation of each group's social and ethnic functions. The Jews, for instance, were, "the most important social problem, the most sensitive political problem and most serious economic problem in Romania"; however, they did "not constitute a racial problem as racial mixing between Romanians and Jews occurs very rarely."[71]

The Roma, on the other hand, represented "the most important, sensitive and serious racial problem for Romania."[72] This ethnic minority had

70 According to the 1930 census 262,501 Roma lived in Romania. Of these, 221,726 (84.5 %) lived in rural areas and 40,775 (15.5 %) in urban areas. See Sabin Manuilă, *Studiu etnografic asupra populației României* (Bucharest: Editura Institutului Central de Statistică, 1940), 34–37. For a discussion of the "Gypsy problem" in Nazi Germany, see Guenther Lewy, "Himmler and the 'Racially Pure Gypsies,'" *Journal of Contemporary History* 34, 2 (1999): 201–214; for a less eugenic-oriented policy toward the Roma in neighboring Bulgaria, see Christian Promitzer's contribution to this volume.

71 Sabin Manuilă, "Problema rassială a României," *România nouă* 7, 41 (1940): 5. See also Sabin Manuilă, "Das Judenproblem in Rumänien zahlenmäßig gesehen," *Deutsches Archiv für Landes- und Volksforschung* 5 (1941): 603–613.

72 Manuilă, "Problema rassială," 5.

mixed with Romanians in villages and urban slums, thus creating a new racial hybrid which, in turn, infiltrated all spheres of social life in Romania. Unsurprisingly, Manuilă's assessment was meant to indicate a racially textured national drama: "The mixing of Gypsy with Romanian blood is the most dysgenic occurrence affecting our race."[73] A year later, Manuilă framed his ideas of miscegenation through the familiar trope of eugenic sterilization: "Obstructing dysgenics, the unwanted, should be pursued until their complete sterilization."[74]

Other authors agreed, at least with respect to the Roma. The Orthodox theologian Liviu Stan (1910–1973) thus complained that contrary to their racial philosophies, "neither National Socialism nor Fascism" had introduced a "racial policy towards the Gypsies," assuming erroneously that in Germany and Italy the "centre of infection and degeneration represented by the Gypsies" was non-existent.[75] Such a policy was, however, imperative in Romania, where "racial promiscuity between Gypsies and Romanians," especially in the southern regions, was resulting in the moral and biological degeneration of the dominant group. Similar to Manuilă, Stan perceived the Roma to have caused more "biological damage" to the Romanian racial body than Jews, suggesting "prophylactic measures" for their "segregation" that included the "prohibition of marriage between Gypsies and Romanians."[76] This "racial policy towards the Gypsies" was intended to serve both moral and biological purposes, and Stan unhesitatingly presented it as part of the glorious destiny that God had planned for the Romanians.

Apparently in response to such "racial fears" the sterilization of the Roma was put forward as a political aim. Gheorghe Făcăoaru, Iordache's brother, suggested, for instance, that:

> Nomadic and semi-nomadic Gypsies are to be interned in forced labor camps. There their clothes will be changed; they will be shaved, receive a haircut and *sterilized*. To cover the costs of their maintenance, they should be used for forced labor. We will be rid of them from the first generation. Their place will be taken by national elements, capable of disciplined and creative work. Sed-

73 Manuilă, "Problema rassială," 5.

74 Sabin Manuilă, "Acţiunea eugenică ca factor de politică de populaţie," *Buletin eugenic şi biopolitic* 12, 1 (1941): 2.

75 Liviu Stan, "Rasism faţă de ţigani," *Cuvântul* 18, 92 (1941): 1.

76 Stan, "Rasism," 2.

> entary Gypsies will be sterilized at home, so that within a generation [Romania] will be cleansed.[77]

In addition to this explicit program of racial purification, sterilizing the Roma was presented as a cost-saving solution in a period of economic depression: "The state spends almost a third of its budget on the maintenance of hospitals and various institutions of social assistance and vice squads, yet the social dirt increases daily. There is an explanation and easy solution to this: *evil must be cut at its roots and not cultivated.*"[78]

These examples indicate how eugenic sterilization became intertwined with a version of ethnic nationalism at the center of the biopolitical program envisioned by the Iron Guard. The sociologist and Legionary Movement supporter, Traian Herseni (1907–1980), made this connection clear: "Dysgenics must not be allowed to reproduce; inferior races should be completely isolated from the [Romanian] ethnic group. The sterilization of certain categories of individuals must not be conceived stupidly as a violation of human dignity but as a tribute to beauty, morality and perfection."[79]

The support given to radical biopolitics by prominent intellectuals like Herseni and Manuilă was consonant both with the new ideological objectives for Romanian eugenics emerging after 1940 and with the general tendency in Antonescu's regime towards national homogenization and ethnic purification.[80] Yet the new beginning, in both politics and culture, one author warned, must however not repeat the mistakes of the past; superior biological leadership was therefore recommended, one born equally from nationalist political ceremonies and eugenic engineering.[81] Antonescu's national biopolitics represented a direct attempt to fuse these diverse opinions about the biological improvement of the Romanian nation. On the one hand, there was the view that the Romanian nation was a biological entity defined as much by history as by blood; on the other, there was the view that the purpose of biopolitics was to create a future Romania, one purified of unwanted ethnic elements.

---

77 Gheorghe Făcăoaru, *Familia şi statul biopolitic* (Bucharest: Bucovina, 1941), 17.

78 Făcăoaru, *Familia*, 18.

79 Traian Herseni, "Rasă şi destin naţional," *Cuvântul* 18, 91 (1941): 7.

80 See Turda, *Eugenism şi antropologie rasială* and Solonari, *Purifying the Nation*.

81 Matei Gr. Peucescu, *Superiorităţile şi transmisiunea lor ereditară* (Bucharest: Tip. Cartea Românească, 1942).

Although Jews were to become the main target of these policies,[82] the Roma were also subjected to deportation and starvation by the Romanian government.[83] In 1941, Mihai Antonescu (1904–1946), for instance, the Deputy Prime Minister and Minister of Foreign Affairs of Romania spoke of the "ethnic and political purification" of Bessarabia and Bukovina. The Romanian nation, Antonescu explained, found itself at a "the most favorable historical moment to achieve its complete ethnic freedom, [and] for our people to be purified of all those elements foreign to its soul." Insisting upon defining the Jews as a "race," one alien to the Romanian nation, Antonescu was unambiguous in stating the aims of the "policy of ethnic purification," namely "elimination or isolation of all Jews in work-camps."[84]

That said, there is no documentary evidence to suggest that Roma or Jewish communities were subjected to sterilization, either in Romania or in Transnistria, (which Romanian troops occupied in 1941 and where the extermination of the Jews from Bessarabia and Bukovina took place). In 1943, Andronescu sternly remarked that in contrast to other European states and the United States, eugenic sterilization had still not been introduced in Romania—an impediment towards the racial improvement of the Romanians he hoped would soon be completed.[85] This was also the thinking of the ethnographer Ion Chelcea (1902–1991) in 1944, when he suggested that certain Roma groups should be "settled in an isolated region, transferred to Transnistria and, if necessary, sterilized."[86]

Ultimately, Antonescu's policy of national purification was a perfect example of biopolitics with its polysemy of elements drawn from Romanian history, anti-Semitism, theories of degeneration and military politics. It spoke of many things: honesty, return to glory and national expansion. To the eugenicists it spoke of the need for a biologically strong nation. "We

82 A series of anti-Semitic laws were introduced in Romania between 1938 and 1941. See Radu Ioanid, *The Holocaust in Romania: The Destruction of Jews and Gypsies under the Antonescu Regime, 1940–1944* (Chicago: Ivan R. Dee, 2000)

83 See Viorel Achim, *The Roma in Romanian History* (Budapest: Central European University Press, 2004) and Viorel Achim, *Documente privind deportarea ţiganilor în Transnistria* (Bucharest: Enciclopedică, 2004).

84 Mihai Antonescu, "Directive şi îndrumări date inspectorilor administrativi şi pretorilor trimişi în Basarabia şi Bucovina," in *Martiriul evreilor din România, 1940–1944. Documente şi mărturii* (Bucharest: Hasefer, 1991), 139.

85 Andronescu, *Pentru ce ne îmbolnăvim?* 46.

86 Ion Chelcea, *Ţiganii din România. Monografie etnografică* (Bucharest: Editura Institutului Central de Statistică, 1944), 101.

need a pure race," answered one contributor to the biopolitical questionnaire initiated by the journal *Societatea de mâine* [The society of tomorrow] in 1943. To improve the Romanian race, he continued, a number of measures were necessary, including mandatory "pre-nuptial medical check-ups;" the prevention of the "procreation of those suffering from incurable diseases;" and "drastic punishment for those suffering from venereal diseases." Investing biopolitics with the sacred mission of "improving the race" was, consequently, "the most important, existential even, quandary facing our state."[87] The ideal nation under Antonescu's regime was not one of political propaganda but was based on a biopolitical transformation of the nationalist ideology through eugenics. The Romanian of tomorrow will, it was hoped, recreate the glorious days of the past lost under foreign domination and cosmopolitan decadence.

## Conclusions

Between 1940 and 1944, ideas for a healthy nation voiced by Romanian eugenicists since the early twentieth century ultimately became encoded in the racial nationalism expressed by Antonescu and other officials.[88] Debates over biopolitics and eugenics discussed here were predicated upon totalizing socio-cultural responses to a number of perceived biological problems allegedly threatening the ethnic fabric of the Romanian national community. Plans for a new, national order based on scientific knowledge—one firmly in accordance with eugenic principles—were constantly enunciated during the interwar years, reaching a crescendo between 1940 and 1944. Even before Antonescu's rule the Romanian ethnic state had become a cen-

87 Petru Suciu, "Drumul spre omul nou," *Societatea de mâine* 20, 8–10 (1943): 137.

88 To give another example: in 1935, the eugenicist Petru Râmneanțu (1902–1981) declared "blood" to be "the real, perhaps the unique, source which remained untouched by the vicissitudes of time," an assumption confirmed in 1940 by the professor of Law at the University of Bucharest and Minister of Justice, Ioan V. Gruia (1895–1952), when he declared on the occasion of the introduction of anti-Semitic racial laws in Romania: "We consider Romanian blood as a fundamental element in the founding of the Nation." See Petru Râmneanțu (în colaborare cu Petru David), "Cercetări asupra originii etnice a populației din sud-estul Transilvaniei pe baza compoziției serologice a sângelui," *Buletin eugenic și biopolitic* 6, 1 (1935): 40 and Ioan V. Gruia, "Expunere de motive la decretul lege nr. 2650/1940 privitor la reglementarea situației juridice a evreilor din România," *Monitorul Official* 183 (9 August 1940): 1. For a similar rhetoric in the Independent State of Croatia during the war, see Rory Yeomans' contribution in this volume.

tral trope for the eugenic imagination. Indeed, the relationship between eugenics and this form of nationalism provides, ultimately, an invaluable insight into modernist ideas of national renewal in post 1918 Romania. In this way, Zygmunt Bauman's metaphor for the totalitarian "gardening state," seen not only as the symbol of modernity but, most importantly, as the repository of the biological qualities of the nation, poignantly expresses the eugenic utopia cultivated by Romanian physicians and scientists—vigilant, scientific "gardeners"—in attempts to rejuvenate their ethnic community.[89]

Despite intense debates, lecturing and lobbying, eugenicists failed to secure the widespread support necessary for a sympathetic government to enact legislation sanctioning sterilization. But eugenic ideas were to some extent, albeit distortedly, put into practice. The concern with degeneration and the health of the nation was as dominant in Romania as it was in Western countries where eugenic legislation was introduced. If in the 1920s and 1930s eugenics was predominantly defined as medical praxis, one centered upon individuals suffering from physical and mental illnesses, by the 1940s it had became thoroughly identified with Romanian racial nationalism. The focus thus shifted from the ill individual to ethnic categories deemed detrimental to the body of the Romanian nation and its racial future. Yet the support for eugenics exhibited by the Romanian professionals and politicians discussed in this chapter was not simply a symptom of their racism and anti-Semitism; it was moreover the expression of their desire to protect the national body by controlling its biological functions. The ultimate biopolitical goal was to create a new Romanian nation through both individual and collective regeneration.

Like elsewhere in Southeastern Europe, eugenic discourses in Romania—expressed through an obsession with the body of the nation—were much more influential and diffused than has been previously assumed. Returning to Foucault's discussion of biopolitics, eugenics made it possible for Romanian nationalists to speak of harnessing the national body, of cultivating and weeding out extraordinary individuals and of purifying the race. The eugenic vocabulary thus overlapped with an adjacent set of

89 See Zygmunt Bauman, *Modernity and Ambivalence* (Cambridge: Polity Press, 1991), especially, 26–39; Roger Griffin, *Modernism and Fascism: The Sense of a Beginning under Mussolini and Hitler* (Basingstoke: Palgrave Macmillan, 2007); and Marius Turda, *Modernism and Eugenics* (Basingstoke: Palgrave Macmillan, 2010).

fears over racial and national decline, which served to amplify the vision of Romania as a country under siege by internal and external enemies. The biopolitical axiom shared by Romanian eugenicists and racial scientists—"The individual is nothing; the nation is everything"—forcefully illustrates how the national body was planned and subsequently controlled between 1918 and 1944.

# THE EUGENIC FORTRESS

## ALFRED CSALLNER AND THE SAXON EUGENIC DISCOURSE IN INTERWAR ROMANIA

*Tudor Georgescu*

Current historiography has largely neglected the Transylvanian Saxon eugenic discourse as it emerged and evolved in early twentieth century Romania.[1] It is a somewhat surprising omission given the substantial influence eugenic population policies came to exert over this ethnic minority's socio-political life in interwar Romania. While this exploration of Saxon eugenics will chart Alfred Csallner's (1895–1992) particular influence in the 1920s and his agenda's gradual institutionalization in the 1930s, the Saxon experiment with eugenics in the service of national renewal, towards engineering a biologically purged and ethnically exclusive "eugenic fortress," is in dire need of further study. Despite the plethora of studies of the interwar period more generally,[2] hardly any attention has been paid to the various rich lines of enquiry the Saxon eugenic movement offers, including the import and adaptation of racial hygienic thought and literature from Germany and Austria; the nature and extent of German financial commitments; the medical establishment's involvement in these and subsequent debates

[1] For a broader overview of Transylvanian Saxon history see, amongst others: Ernst Wagner, *Geschichte der Siebenbürger Sachsen: Ein Überblick* (Innsbruck: Wort und Welt Verlag, 1981) and Harald Roth, *Kleine Geschichte Siebenbürgens* (Cologne: Böhlau, 2003). Please note that all translations from the original are my own unless otherwise cited.

[2] See, for example: Johann Boehm, *Die Deutschen in Rumänien und die Weimarer Republik, 1919–1933* (Ippesheim: AGK-Verlag, 1993); *Die Deutschen in Rumänien und das Dritte Reich, 1933–1940* (Frankfurt a/M: Peter Lang, 1999); Hildrun Glass, *Zerbrochene Nachbarschaft: Das deutsch-jüdische Verhältnis in Rumänien, 1918–1938* (Munich: Oldenbourg, 1996); and Walter König ed., *Siebenbürgen zwischen den beiden Weltkriegen* (Cologne: Böhlau, 1994); Harald Roth, *Politische Strukturen und Strömungen bei den Siebenbürger Sachsen, 1919–1933* (Cologne: Böhlau, 1994) and *Minderheit und Nationalstaat: Siebenbürgen seit dem Ersten Weltkrieg* (Cologne: Böhlau, 1995).

on the physician's role as the guardian of a biologically re-defined Saxon identity; the ever evolving and adapting role of the Saxon protestant Church [*evangelische Landeskirche A.B.*] and its murky relationship with the eugenic discourses that courted it; or even the question of how viable the prospect of an ethnic minority designing, implementing, and enforcing eugenic population policies was in the absence of their own centralized nation-state and despite increasingly assertive Romanian nation-building projects.

The discussion of the interwar Transylvanian Saxon eugenic discourse offered here scrutinizes the ideological and methodological interfaces between the German minority's fascist "Self-Help"[3] Movement for Renewal's experiment with national regeneration on the one hand and Alfred Csallner's life and work in the service of the eugenic cause on the other. This text explores his role in not only redefining the interwar eugenic discourse along increasingly radical and politicized lines but the population policy proposals he advanced to counter a perceived dysgenic crisis in the 1920s and 1930s. How did Csallner assess and address the nature of this perceived threat to the Saxon "national body" and its underlying "hereditary worth"? What measures did he advocate, both via private initiatives and through the Church, in the 1920s to counteract this looming crisis? How and why did Csallner turn to Fritz Fabritius and the fascist "Self-Help" in 1929 to pursue these visions of national regeneration within a no longer clerical, but party-political framework? And to what extent was Csallner's eugenic agenda empowered?

3 While the nature of Saxon fascism in general (understood here in terms of Roger Griffin's *Modernism and Fascism: The Sense of a Beginning under Mussolini and Hitler* [Basingstoke: Palgrave, 2007]) and its involvement with National Socialism in particular are highly contested fields of study, the "Self-Help"s' embrace of eugenic thought is as astonishing as its 11-year metamorphosis from a "workers" book club cum building society to dominant political force. While spatial limitations bar any substantive analysis of its ideological core, chronologically speaking, "Self-Help" was founded in 1922 at a time when public opinion was increasingly turning into opposition to an unreceptive political status quo. The movement was to explicitly define itself in terms of German National Socialism by May 1932 (at the very latest), when it appended its building society with a political party tellingly called the *Nationalsozialistische Selbsthilfebewegung der Deutschen in Rumänien* [National Socialist self-help movement of the Germans in Romania, or NSDR]. The movement's political empowerment was, essentially, a three stage process: In October 1933 the NSDR managed to legislate a watered down version of its National Program at the Saxon National Day [*Volkstag*]. A month later elections handed the NSDR the Saxon "National Council." In June 1935, the Self-Help arrived at its ultimate pan-German destination when Fabritius was elected head of the "Association of Germans in Romania" ["Verband der Deutschen in Rumänien"]. During this remarkable accent to political empowerment, the Self-Help underwent several structural transformations: The NSDR, banned in late 1933, re-emerged nine days later as the *Nationale Erneuerungsbewegung der Deutschen in Rumänien* [National movement for renewal of the Germans in Romania, or NEDR], only to be banned again in July 1934 and succumb to three years of bitter infighting in the wake of the movement's complete disempowerment in 1935.

This chapter hence examines the core themes and sense of existential crisis that molded and guided Csallner's discourse. It will also illustrate the extent to which eugenics and racial hygiene were seen to offer an embattled ethnic minority a novel methodology with which to redefine their national sense of self, to diagnose the nature of the cancer afflicting its organic national body and to formulate a scientifically verifiable cure. Third, it explores the manner in which these ideological trends are revealed by Alfred Csallner's ravenous embrace of a eugenic discourse that could translate his sense of a statistically-evidenced, impending national extinction into a viable program of national salvation. Finally, this discussion of Saxon eugenics will hopefully yield useful insights into the wider mechanisms of how and why segments of an ethnic minority in general, and a clergyman in particular, turned to eugenics—and ultimately fascism—in an increasingly desperate attempt to avert an increasingly despairing battle for national renewal both internally, and vis-à-vis its host nation.

## Who Was Alfred Csallner?

Born to Dr Ludwig and Frieda Csallner in the northern Transylvanian town of Bistritz in 1895, Alfred Csallner's (see fig. 1) student years were largely typical in that he, like virtually all of his peers, studied abroad. Dedicating himself to the fields of geography, history and theology, Csallner studied in Budapest, Vienna and Berlin before prematurely returning to Transylvania in 1918 to tend to his sick father. After returning to Bistritz, Csallner worked as a teaching assistant [Supplent] for both high and girls schools between 1918 and 1922, then lectured at a teaching vocational college [*Lehrerinnenbildungsanstalt*] in Schässburg in 1923. Crucially—and although he was later to reflect on whether his ideological stance justified it—Csallner was then ordained a priest and successively employed in Roseln, Stolzenburg and Kleinscheuern prior to laying down his priestly robes in favor of the offer made by the *Verein für das Deutschtum im Ausland* (VDA) to pay him a salary in 1936.[4]

[4] Alfred Csallner, *Meine wissenschaftlichen Arbeiten* (Traunau: Selbstverlag, 1975), 7. In fact, Csallner enjoyed substantial financial assistance from Germany as he received funding from the *Deutsche* and *Südostdeutsche Forschungsgemeinschaften* as well as from the *Verein für das Deutschtum im Ausland*.

Fig. 1: Csallner Portrait (ca. 1940)

Csallner avidly believed the Saxons had unwittingly found themselves in the advanced stages of a cataclysmic degenerative crisis, one which had already simmered for decades before its scientific diagnosis by Heinrich Siegmund's (1867–1937) "national-biological" research in the early 1900s. In equal measure fundamental to and forgotten by Saxon historiography, Siegmund not only introduced and guided Csallner's understanding of eugenic themes and means after their first meeting in 1922, he was in many ways the founding father of Saxon eugenics more widely.[5] Commencing with his 1901 pamphlet *On Transylvanian Saxon Racial Hygiene*,[6] Siegmund

[5] That said, the secondary literature is surprisingly sparse as the three existing articles are, regrettably, merely short summaries rather than studies: Fritz Behrwerth, "Dr. Heinrich Siegmund," *Südostdeutsche Vierteljahresblätter* 2 (1970): 104–8; Wagner, Ernst, "Heinrich Sigmund und die 'Volksbiologische' Forschung in der Zwischenkriegszeit," *Zeitschrift für Siebenbürgische Landeskunde* 2 (1983): 177–86; and Heinrich Linger, "Sitte, Moral und Volksreinheit bei Heinrich Siegmund," *Siebenbürgische Semesterblätter* 2 (1992): 169–72. Despite this shortfall, several invaluable biographies were published by Siegmund's close associates after his death, the best of which is: J. Rehner, *Dem Andenken Dr. Heinrich Siegmunds weiland Grosstemplers von Rumäniens Grosslaube 2 (deutsch) des I.O.G.T.* n.p., n.d. [1937]. But also see the biography and partial bibliography offered by Julius Ernst Gyurgyevich, who took over Siegmund's "Welfare Exhibit" in 1937: Julius Ernst Gyurgyevich, "Vorarbeiten zu einem Schriftenverzeichnis des Landeskonsistorialrats Dr. med. Heinrich Siegmund," *Medizinische Zeitschrift* 10 (1937): 315–26; and "Aus gesegneter Ernte: Zum schriftstellerischen Schaffen des Dr. Heinrich Siegmund," *Evangelischer Fürsorger* 8 (1937): 31f.

[6] Heinrich Siegmund, *Zur sächsischen Rassenhygiene* (Hermannstadt: Drotleff, 1901) [also reprinted in: *Medizinische Zeitschrift* 9 (1937): 273–81].

produced a veritable flood of publications spanning over 40 books, several journals and literally hundreds of articles until his death in 1937. Siegmund embarked on a relentless, if largely unsuccessful, quest to convince the Saxon establishment that his "Displacement Theory" [*Verdrängungslehre*] could account for how and why the Saxon population had declined numerically and retreated geographically over past centuries far better than the dominant "Extermination Theory" [*Vernichtungslehre*]. While the latter argued that external factors, such as wars and plagues, had emptied Saxon settlements, which were subsequently re-populated by Romanians and Hungarians, Siegmund insisted that the Saxon nation had declined for internal reasons, namely its failure to actively protect its *Lebensraum*—the ongoing loss of which he thought contributed to the decline in Saxon fertility, which amounted to an impending "national death" [*Volkstod*] or, as the title of his 1931 book threatened, a *Twilight of the Germans in Transylvania.*[7] The measures he proposed to revitalise Saxon settlements diagnosed in his "National Biology" [*Volksbiologie*] discourse as either "endangered," "ill," "seriously ill" or "dying" [*gefährdet, krank, schwerkrank* or *sterbend* respectively] were subsumed under a "Science of National Defence" [*Wissenschaft der Volksverteidung*] and formed the bedrock upon which Saxon eugenicists hoped to build their re-homogenised eugenic fortress. Indicative of the ideological core that was to define Saxon eugenics, Siegmund called for a significant qualitative increase in Saxon fertility, the betterment of the nation's racial stock by discouraging the hereditarily ill and mentally infirm from bearing children, a decrease in emigration, a battle against the corrosive effects of tobacco and alcohol on societal health and the nation's collective purse and the promotion of an internal colonisation schemes to create a living belt of racially vibrant settlements enclosing an enlarged and re-homogenised Saxon *Lebensraum.*

Siegmund was also instrumental to Saxon eugenics' search for institutional means and initially attempted to translate his agenda through private initiatives before turning to the Church for organisational support in the early 1920s. In short, the 1906 creation of the "Land Preservation Society" [*Bodenschutzverein* or BSV] was a crucial benchmark in the evolution of Saxon eugenics, the first of the host of eugenically-minded organ-

[7] Heinrich Siegmund, *Deutschen-Dämmerung in Siebenbürgen* (Hermannstadt: Honterus, 1931).

isations that followed. Subsuming the various key tenets of his "science of national defence," the BSV set out to employ a "defence fund" [*Wehrkasse*] filled by donations and annual membership fees towards acquiring non-Saxon properties. These, in turn, were to be reallocated to hereditarily sound settlers. But the BSV did not survive World War I, and Siegmund turned increasingly to the Saxon national Church. In fact, he became the first 'medical member' of the Church's governing council, the *Landeskonsistorium*, in 1920 in recognition of his role in the introduction of "School Nurses" in Protestant schools in 1908. But Siegmund had an ambiguous relationship with the Church he complained was too dogmatic and rigid an institution on the one hand, while courting it for its societal influence and structural network on the other. Siegmund persuaded the Church to create a "Welfare Committee" [*Fürsorgeausschuss*] under his direction in 1922. Not only did this organisation offer him a novel avenue down which to pursue his assessments of the Saxon nation's racial health through annual censuses, it was also instrumental in disseminating his views amongst the clergy and wider population. Crucially, in the late 1920s the "Welfare Committee" also became a forum in which Siegmund, Fritz Fabritius' Self-Help, and Alfred Csallner's "Society of Child Enthusiasts" collaborated (in fact, the "Welfare Committee" joined the "Self-Help" in 1927).

Thus heavily influenced by Siegmund both personally and ideologically, Csallner quickly concluded that "this danger of degeneration is the greatest of all threats. We, of course, don't see it yet, not even our leaders want to recognize and acknowledge it. And we who have must therefore be all the braver and call all the louder."[8] And call loudly he did. Csallner, brief as his career might have been, unleashed a torrent of publications, set up and ran several eugenic organizations and even resorted to penning poetry and plays[9] in a bid to warn the Saxon body politic of the existential threat they faced. But considering that Csallner was, at heart, always a priest—fervently believing in the clergy's duty to lead and shape their flock—he also dedicated a signifi-

8 Alfred Csallner, "Die Grösste Gefahr: Ergebnisse einer erb- und gesellschaftsbiologischen Untersuchung über die Siebenbürger Sachsen," *Volk im Osten* 8 (1935): 257.

9 Csallner actually wrote a number of rather interesting pieces, including a play entitled: *Um der Zukunft Willen: Volksstück in drei Akten aus der Vergangenheit der Siebenbürger Sachsen* (Bistritz: Gustav Zikeli, 1927). Furthermore, a number of poems, the most interesting of which are Alfred Csallner, "Achthundert Jahre...!," *Selbsthilfe* (8 August 1931): 4–5; "Du und Dein Volk," *Selbsthilfe* (15 June 1931): 5–6; and "Höher Hinauf!" which is found on the final page of his 1940 version of the Saxon *Ahnenpass* (Hermannstadt: Bruckner, 1940).

cant portion of his time to lobbying the Protestant Church, calling upon it to embrace the racial hygienic measures he believed could reverse this impending eugenic meltdown and usher in a new period of national regeneration.

Csallner's characterization of the various ills afflicting the Saxons can generally be divided into the following four categories:

1) Lost Children: Family Planning and the Impending Demographic Collapse
Living up to his nickname "the child enthusiast" [*Kinderfrohe*], Csallner insisted that the dysgenic crisis was born out of the catastrophic proliferation of a one and two child system, which robbed the nation besieged by ethnic rivals of the offspring necessary to secure its *Lebensraum*.[10] But apart from opening a door to the "flood" of, in particular, Romanians supposedly streaming from the hills into Saxon settlements, family planning had also produced generations of inherently inferior offspring who missed out on the beneficial influence of siblings in larger families. What was worse, Csallner also believed these lonely children to be inferior physically and ultimately characterized them as pampered weaklings unsuitable to the great challenges of their time.[11] Worse still, this shortage of vigorous human capital was magnified in light of Csallner's "socio-biological" discourse on the varying fertility rates exhibited by the nation's eugenically "best" and "worst."

2) Socio-Biology: The Extinction of Inherited Talent
Csallner was obsessed with the study of differential fertility rates amongst what he viewed to be the nation's best and least hereditarily-endowed professions and stipulated that the gradual loss of racial fitness was the key motor in a looming socio-economic collapse. He insisted that it was not only the nation as such that would suffer from a demographic crisis but that its most valuable eugenic stock was teeter-

10 While virtually all of Csallner's texts inevitably problematize fertility and demographic trends at some point, the singularly most interesting article, and one of his first, would be his tabulation of literally how many offspring the Saxons had 'lost' in "Deutsche Verluste im Nösnerland," *Bistritzer Kalender* (1921): 48–59.

11 See Alfred Csallner, "Unsere einzigen Kinder," in Alfred Csallner, *Kleine Aufsätze* (Hermannstadt: Selbstverlag, 1928).

ing on the cliff of an abyss. For one, Csallner became obsessed with the increasing number of non-Saxon apprentices in Saxon workshops and businesses. He was adamant that this trend, if continued unabated, would eventually lead to the complete dissolution of the local economy—with more and more non-Saxons taking over Saxon stores, factories and trades. Adding to these impending economic woes, Csallner was alarmed by the low fertility rates apparently exhibited by the nation's hereditarily best endowed groups. An obvious example of this line of argument is found in Csallner's perception of the clergy, teaching and other academic professions, which he believed to be inherently superior, given that for him (although obviously contradicting his view regarding degeneration being nurtured by small families), talents were always inherited, never acquired.[12] Even if these professions replenished themselves with surplus stock from lesser casts, such reinforcements were, by their very nature, inferior to those they replaced. But while this process marked a gradual and ongoing racial devaluation of the nation's elite, an increasing number of mixed marriages represented an absolute, racial hygienic aberration.

3) Mixed Marriages: The End of Racial Integrity

Csallner never minced words when it came to decrying mixed marriages. For him, they were both the logical consequence of a nation in complete ignorance of the struggle for survival it was fighting, as well as imminent cause of its ultimate destruction as a recognizable organic entity. Although Csallner published a mere handful of tracts on the degeneration caused by mixed marriages,[13] their elimination—

[12] Csallner wrote several "socio-biological" studies largely focusing on the fertility patterns exhibited by various professions in the towns of Schässburg, Bistritz, Mediasch and Hermannstadt. See, for example, Alfred Csallner, "Statistisches über den Bistritzer sächsischen Gewerbestand," in Csallner, *Kleine Aufsätze,* 1928); *Der Bistritzer deutsche Akademikerstand, gesellschaftsbiologisch betrachtet* (Hermannstadt: Honterus, 1934); *Der Mediascher deutsche Akademikerstand, gesellschaftsbiologisch betrachtet* (Mediasch: Feder, 1932).

[13] The bulk of Csallner's work on mixed marriages is found in a three-part article published in the *Auslandsdeutsche Volksforschung* in 1937/8. Here Csallner proceeds to examine, where possible, the marriage patterns of the Saxon, Romanian, Hungarian and Jewish populations and their degree of intermarriage in the towns of Sächsisch-Regen, Bistritz, Schässburg and Hermannstadt (mostly for the period 1895–1937): Alfred Csallner, "Die Mischehen in den siebenbürgisch-sächsischen Städten und Märkten: Teil I," *Auslandsdeutsche Volksforschung* I (1937): 225–55; "Die Mischehen in den siebenbürgisch-sächsischen Städten und Märkten: Teil II." *Auslandsdeutsche Volksforschung* 2 (1938): 14–36; and "Die Mischehen in den siebenbürgisch-

by excluding all who could enter into them from Church and national offices—was one of his most ardently held demands. Beside the "pollution" introduced by ethnic others, Csallner was adamant that such a "transfer" of racial qualities actually ennobled the Saxons' very competitors for resources and *Lebensraum*.

4) The *Lebensraum* Nexus: Re-Homogenize, Regenerate and Rejuvenate
Although Csallner never ascribed the loss of *Lebensraum* the same central role Siegmund had in diagnosing the Saxons' deepening degeneration, both were unequivocal in demanding the expulsion of the foreign "invaders" who had set about dismembering and colonizing it. And as with other parts of Csallner's catalogue of degenerative processes, he insisted that the Saxons were perpetuating the problem themselves. Stubbornness and misplaced pride (above and beyond the purely mathematical lack of human capital) stood in the way of ridding (especially) rural communities of the "foreign" labor, agricultural or otherwise, they employed. Through calculation, tabulation, and any other form of statistical evidence he could conjure, Csallner repeatedly pointed to the rising cumulative wages that Saxons paid these ethnic others, which they subsequently used to buy their way into Saxon towns and settlements. If this heralded a perpetual downward spiral, then the Saxon consumption of alcohol and tobacco only amplified the problem.[14] In turn, Csallner was deeply concerned with the amount of national wealth wasted on these societal scourges, compounded by the amount of land, farms and so on that could have been saved had these funds been deployed in the service of his eugenic cause rather than being squandered on paying "outsiders."

sächsischen Städten und Märkten: Teil III." *Auslandsdeutsche Volksforschung* 2 (1938): 227–60. These three articles were a source of great pride to Csallner who, in the 1975 version of his *Meine wissenschaftlichen Arbeiten*, claimed they had outlined the means of "strengthening the national sense of responsibility as practical measures to combat mixed marriages by influencing internal migration patterns and career choices" (Csallner, *Meine wissenschaftlichen Arbeiten* [1975], 18).

14 An example is afforded by the following article, in which Csallner demanded that all those claiming to be national leaders should lead by good example and abstain from drinking entirely: Alfred Csallner, "Offener Brief an alle, die Führer sein wollen in unserem Volke" [also: *Bistritzer Deutsche Zeitung* 09.01.1931; *Großkokler Bote* 11.01.1931; and *Mediascher Zeitung* 24.01.1931], *Selbsthilfe* (February 01, 1931): 2–4. Csallner also published a small number of pieces in the pages of the Order of Good Templar's journal *Der Aufbau*, such as: Alfred Csallner, "Der sächsische Pfarrer als Guttempler" [also: *Der Aufbau* 31 (1928): 1–3], in Alfred Csallner, *Kleine Aufsätze* (Hermannstadt: Selbstverlag, 1928).

Consequently, Csallner offered a wide range of measures to cleanse Saxon lands of the spiritual and physical evils destroying it, most of which were based on a remarkably authoritarian desire to contain and control virtually every aspect of Saxon socio-economic life—as dictated by the statistics he had come to sacralize as absolute truths. Apart from the imperative to enlighten and awaken the Saxon masses to their national, racial hygienic duties, Csallner eventually came to demand the exclusion of all who "refused" to bear their nation four children, the re-deployment of entire settlements so that they could better serve as "bastions," and the control of the supply and demand of Saxon labor in order to rid the economy of foreign apprentices, maids, farm hands and so on. All of these centralist programs for re-entrenching and regenerating the Saxon nation where to be grounded in an array of databases and scientific tables, an indicative example of which was the immensely ambitious "hereditary fitness index" Csallner hoped to have realized and managed by the Saxon Church.

That said, it should be little surprise to find that Csallner perpetually demanded the Church and its clergy awaken to their pivotal role as leaders in his eugenic revolution. Csallner's perception of the Saxon Church, its historical mission and contemporary duties were always somewhat removed from the detached religious brief of a theologically trained priest. This is apparent in his dislike of "seminarists," as compared with academically qualified theologians[15] or his increasingly vocal quest to refashion the clergy into 'eugenic missionaries' embodying, first, a form of centrally coordinated assessors of the flock's racial hygienic health, and second, a role as "torchbearers" of the eugenic movement by proliferating (and enforcing) its ideological aims. The Church itself was to become a *Staatsersatz* of sorts, a substitute for the impossibility of a politically independent and viable Saxon state. This objective constitutes a remarkable role reversal, in which racial hygienic ideologues had become the "prophets" whose tenets the Church, along with its priests and teachers, were duty-bound to spread and manage.[16]

15 See, for example: Alfred Csallner, *Die volksbiologische Forschung unter den Siebenbürger Sachsen und ihre Auswirkung auf das Leben diese Volksgruppe* (Leipzig: S. Hirzel, 1940), 99–102.

16 This demand is most passionately advocated in an article entitled "Torchbearers," in which one finds a remarkable passage postulating racial hygienists as "prophets": "The preachers of these new truths, no, of these old truths that we have only now begun to understand, they are prophets and called upon to refashion the way the future will feel and act" (Alfred Csallner, "Fackelträger" [also: *Kirchliche Blätter* (1928)]. Csallner, *Kleine Aufsätze*, 2.)

While not to deny his own piety, this explains why virtually all of Csallner's eugenic population policies were designed to be realized by the Protestant Church and its vast network. In 1928, for example, Csallner submitted complete proposals for a drastic revision of the Church's salary scale and child benefit payments in favor of large families.[17]

## Alfred Csallner's Institutions: An Overview

Csallner's organizations represented the institutional manifestations of a eugenic agenda that had coalesced into a discernible ideological matrix under Heinrich Siegmund's tutelage in the early 1920s. It was one invariably radicalized and politicized by Csallner's fervent enthusiasm for, and involvement with, the nascent Saxon "Self-Help" movement founded by Fritz Fabritius (1883–1957). Demanding direct and immediate action, Csallner had already founded his "Society of Child Enthusiasts" [*Verein der Kinderfrohen*, VdK] in 1927. The Society, though, failed to attract as numerous a membership as Csallner would have liked, the policy proposals submitted to the Church largely failed to achieve their desired outcomes and, as he later recalled, he "let it fall asleep" again. But Csallner was soon to find a sympathetic and willing partner in the Self-Help. In 1929, Csallner's Society joined Fritz Fabritius' growing fascist movement, officially due to the "extraordinary advantages"[18] membership brought in terms of the fiscal resources available to fund their work. Csallner became consumed by the fascist project to regenerate and refashion the Saxon nation—to design and build a eugenic fortress. Csallner was even listed as the movement's representative [*Vertrauensmann*] in Roseln, the village whose priest he had become in July 1929, having left his previous parish in Stolzenburg utterly embattled.[19]

17 Alfred Csallner, "Vorschläge zur Neuregelung der Familienzulagen" [also: *Kirchliche Blätter* (1928)], in Csallner, *Kleine Aufsätze*. See also: Alfred Csallner, Heinrich Brandsch, Herman Buesch, Ruediger E. Fritsch, and Hugo Sitzler, *Denkschrift des 'Vereins der Kinderfrohen'* [also: *Evangelischer Fürsorger* 5/6 (1928): 48–67] (Hermannstadt: Honterus, 1928).

18 Csallner, "'Selbsthilfe'," 10.

19 See the issue of *Selbsthilfe* (20 July 1929).

In 1932 Csallner approached Fabritius with the offer to set up and run a *Self-Help Race Office* [*Rassenamt der Selbtshilfe*].[20] His ravenous embrace of Saxon fascism was beyond doubt the single most radical reorientation (if certainly not revision) of Csallner's eugenic agenda. The Race Office offered Csallner the spiritual home his VdK had earlier failed to build, and also gave him a politically—rather than clerically—grounded framework through which to operate. And while virtually all of his eugenic policies involved the Church to some respect (from setting up work groups to micro management and enforcement), his belief that "ever more tasks that until now only our Church seemed able to tackle can now be addressed by our new national political organization"[21] underlines the extent to which a fascist Saxon political superstructure was to become his preferred organizational means. Despite being involved in various population policy debates, the rest of Csallner's brief but industrious career was dominated by what had humbly begun as the "Self-Help Race Office."

While Csallner's "Race Office" remained a rather virtual institution prior to 1935, it nevertheless succeeded in creating a political and institutional base. Csallner's enthusiasm for Fabritius' movement suffuses, if nothing else, his October 1932 announcement of the Office's creation in the "Self-Help:"

> The very best that we, as a nation, own, and which forms the very precondition for a continuously industrious nation worthy of life and survival, is our hereditary substance. Our racial worth has, naturally, declined since our immigration as a result of an undesirable degree of mixing with other nations; the emigration of our hereditarily best endowed; a declining fertility amongst our best classes; alcoholism and other processes that are currently accelerating our decline at an unprecedented pace. This is why we must act now to not only mobilize and preserve our hereditary fitness from further rot but, via the acquisition and securing of further *Lebensraum*, to encourage an above average

20 Alfred Csallner, *Das Landesamt für Statistik und Sippenwesen der Deutschen Volksgemeinschaft in Rumänien* [henceforth: *Das LSS*] (Hermannstadt: Manuscript, dated October 10, 1939 [NA Sibiu, Volksgemeinschaft, "Inventar Supplementar," D.XIV.19), 1: "In 1932 I approached the leader of the 'Self-Help,' our current national leader Fritz Fabritius, with the suggestion of creating a 'Self-Help Race Office' and offered to direct it in an honorary position myself. Fabritius accepted my proposal and entrusted the Race Office's management to me."

21 Alfred Csallner, *Meine Wissenschaftlichen Arbeiten*. NA Sibiu, Brukenthal, Z 1-7, Nr. 200, 1944 Nr. 15.

> level of population growth so as to turn an ongoing degeneration into regeneration. The latter goal will always be our '*Aufnordung*'. But considering the particular realities within and outside of our nation, it might currently be prudent to not necessarily underline this point and focus on the immediate goal of protecting and increasing that in our nation which is worthy.[22]

Not only does this proclamation, if not manifesto, provide a concise summary of Csallner's overall eugenic agenda—one explicitly driven by the fear of racial degradation caused by the most talented parts of society's failure to reproduce their superiority at the desired rate—it also underscores the necessity for the re-homogenizing racial hygienic measures that were so fundamental to fascist eugenics. Dismissing the Church as a route to appropriate population policies, Csallner proceeded to claim that "[t]he 'Self-Help' is the only one of our organizations to have the will and ability to begin the necessary work, and does so by entrusting it to a special Race Office, one that might eventually head race offices for the individual districts and local communities."[23] His boundless ambition notwithstanding, Csallner realized that the geopolitical climate of the early 1930s (essentially, prior to the ascendancy of the Third Reich) demanded a more cautious approach. Csallner thus set his new "Race Office" more realistic short-term targets, such the organization of lecture series and an increased proliferation of racial hygienic knowledge. Nonetheless, his primary aim was to advance a sweeping assessment the nation's racial health, as it was the Race Office's "main task to study settlement by settlement our racial composition, our physical and spiritual capabilities in general and those of our most valuable as well as inferior families in particular."[24] This ambitious, meticulous study of the overall Saxon constitution would, significantly, be carried out by "voluntary workgroups," the *Self-Help Working Teams* [*Selbsthilfe Arbeitsmannschaften*, or SAM]. Csallner believed the extensive participation of the Self-Help's youth movement' would, apart from providing the necessary manpower, offer the best opportunity to reach the masses,

22 Alfred Csallner, "Zur Errichtung des Rassenamtes der 'Selbsthilfe'," *Ostdeutsche Beobachter* (1 October 1932): 4.

23 Csallner, "Zur Errichtung des Rassenamtes," 4.

24 Csallner, "Zur Errichtung des Rassenamtes," 4.

to "sharpen their national sense of duty in general and their hereditary and racial conscience in particular, thereby averting many a disaster."[25]

Shortly after Saxon fascism's rise to political dominance, first in the Saxon "National Council" in 1933 and then in the "Association of Germans in Romania" in 1935, Csallner set about expanding his Race Office which, in late 1935, emerged drastically enhanced and re-branded as the "National Department for Statistics, Population Policy and Genealogy" [*Landesarbeitsstelle für Statistik, Bevölkerungspolitik und Sippenwesen* or LSBS]. In November 1933, Csallner produced an eight-page pamphlet insisting that the LSBS "needs to be an office that always knows exactly, to the tiniest detail, what the state of our *Lebensraum* and hereditary composition is, as well as know what possibilities to grasp when they offer themselves."[26] In outlining its mission statement and revamped organizational hierarchy, Csallner assigned the LSBS with its singularly most important task: a 26-questionnaire thick survey covering all German settlements in Romania.[27] That such an endeavor could, despite Frabritius' orders to all national and local offices to cooperate and assist in its successful completion, be brought to fruition without the state's interference was perhaps a naïve idea, particularly given the incredible scope of data Csallner hoped to collect.[28] As was to be expected, Csallner's census was quickly brought to a standstill by the inner political turmoil dividing Romania's Germans. Csallner was forced to suspend his study when faced with the arrest, interrogation and harassment "endangering" his students' field work, culminating in the police's confiscation of three meter's worth of material.[29] With this Saxon political dispute effectively stamped out by Berlin in 1938, Csallner reached the pinnacle of his political influence and literary output with the reorganization and expansion of his Department into the *National Office for Statistics and Genealogy* [*Landesamt für Statistik und Sippenwesen*, or LSS].

25 Csallner, "Zur Errichtung des Rassenamtes," 4.

26 Alfred Csallner, *Die Arbeitsstellen für Statistik, Bevölkerungspolitik und Sippenwesen* (Volksgemeinschaft der Deutschen in Rumänien, 9 November 1935).

27 Csallner, *Die Arbeitsstellen für SBS*, 3.

28 Fabritius to NAF – dated November 14, 1936: NA Sibiu, CGR, D.XIV.840, Nr. 5.

29 Csallner, *Meine wissenschaftlichen Arbeiten*, 18–9; and Alfred Csallner, "Die rumaeniendeutsche Bestandsaufnahmen von 1937/38 und 1940," *Sueddostdeutsche Vierteljahresblaetter* 3 (1997): 216–218.

## The "National Office for Statistics and Genealogy"

The LSS, from its humble origins as a *Race Office* in 1932 to the radical expansion of 1938, was thus the institutional manifestation of Csallner's perception of a national and racial dysgenic crisis and, ultimately, the means with which to counter it. The LSS marked a clear departure from working solely within the organizational framework afforded by the Church by creating an independent Saxon research institution both informing and increasingly making Saxon policy decisions at the reinvigorated and increasingly totalizing political level:

> The National Office for Statistics and Genealogy does not merely endeavor to evaluate all statistical matters concerning the German national community in Romania, to alert them to their national duties and to guard their appropriate application, but, what is more, has the duty to ensure that all other national and racial hygienic duties are realized and properly implemented by all other relevant departments [*Stellen*].[30]

The extent to which Csallner was to dominate the LSS' agenda can be illustrated by his own position: he was both the director of its "National Office" [*Landesamt*] in Hermannstadt and head of the Transylvanian "District Office" [*Gauamt*] "in order to be able to work in at least one district myself—to not only stimulate the work of others through central directives but to influence and drive forward by example."[31]

Of the handful of extant publications and manuscripts pertaining to the LSS' inner workings, surely the most significant is a 27-page report composed by Csallner in October 1939. This document—along with appended questionnaires, graphs and maps—constitutes the nexus of the LSS' agenda and activities over the latter half of the 1930s. Regrettably, Csallner provides only minor detail on its employees and their respective roles within the organization, nor does he elaborate on their respective research

30 Csallner, *Das LSS*, 1. See also the initial founding charter signed by Csallner, Fritz Fabritus and Alfred Bonfert in May 1939 (Alfred Csallner [Deutsche Volksgemeinschaft in Rumänien, signed by Csallner, Fritz Fabritius, Alfred Bonfert – dated 25 May 1939], *Organisationsrichtlinien des Landesamtes für Statistik und Sippenwesen* [Schässburg: Friedrich J. Horeth, 1939].)

31 Csallner, *Das LSS*, 1.

interests in any meaningful detail (despite including a fascinating four page lament of his fiscal woes and pleas for more funds). Instead, the report offers a detailed structural analysis of the LSS' central mission statement(s) in the wake of its recent expansion, in addition to an outline of the work it intended (but was largely unable) to carry out in the following years. Nonetheless, taking a closer look at the LSS' most important departments and the mission statements composed for them provides remarkable insights into Csallner's overall eugenic agenda.

The LSS, like its predecessor, was headed by a National Office whose director was appointed by the "National Head" [*Landesobmann*]. It was directly responsible to the "Commissioner for the Nation's Organic Constitution" [*Beauftragter für den Volksorganischen Aufbau*] Wilhelm Schunn, while enjoying an advisory seat on the Saxon "National Council" [*Volksrat*]. In turn, the *National Office* appointed the leaders of its various "District" [*Gau-*], "County" [*Kreis-*], and "Local Offices" [*Ortsstellen*] offices, with the latter to be established in either especially large or endangered settlements. The LSS consequently added a number of new departments (all of which included several individual sections) to the existing infrastructure created by its predecessor and, by 1939, consisted of six departments tasked with *Statistics* [*Statistik*], *Genealogy* [*Sippenwesen*], *National Betterment* [*Volkspflege*], *Lebensraum*, *Foreign Nations* [*Fremde Völker*] and *Enlightenment, Recruitment and Education* [*Aufklärung, Werbung und Erziehung*] (see fig. 2).

## 1. Statistics

The department of statistics was tasked "with collating, evaluating and providing all documents necessary to the national leadership and to work towards an increasingly perfected meshing together of all of these *völkische* organizations."[32] Its main function was to accumulate and evaluate all demographic data, to assess the extent of the Saxon dysgenic crisis and devise solutions for it.

Of its six subsections, the first and most industrious one, *Censuses* [*Bestandsaufnahme*], was given the immediate responsibility for completing and evaluating the data collected in Csallner's abortive 1937/8 census,

[32] Csallner, *Das LSS*, 2. See also: Csallner, *Organisationsrichtlinien des LSS* - 25.05.1939, 6.

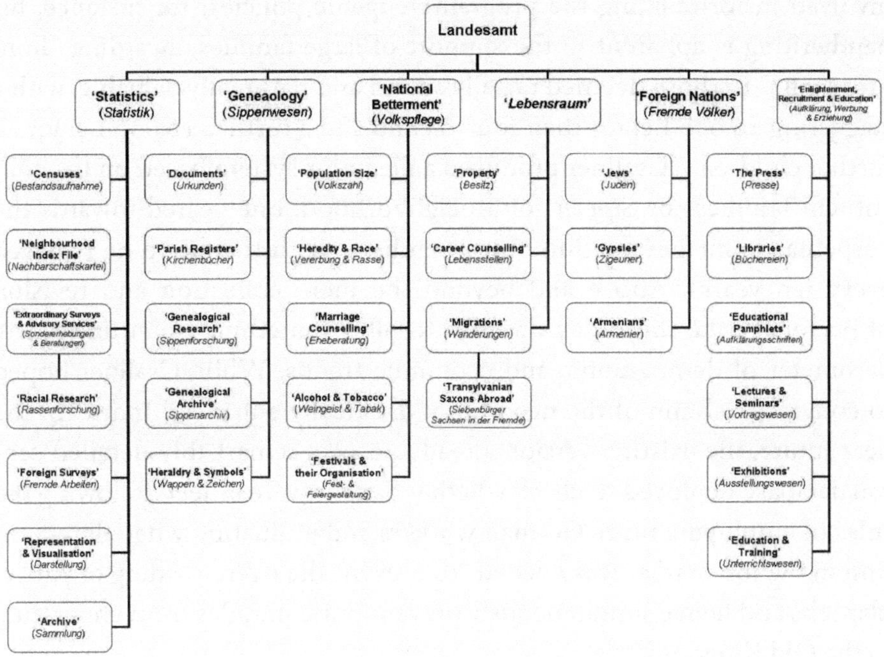


Fig. 2: LSS Organigram (October 1939)

as well as augmenting it with further surveys where needed. Under Csallner's direct supervision, Transylvanian regional data was to be the only set systematically evaluated before the outbreak of World War II. While the data for the Bukovina region had been fragmentary, its *Gauleiter* Dr Herbert Mayer, had subsequently conducted his own survey. And to Csallner's consternation, Bessarabia's *Gauleiter*, Prof. Arthur Fiechtner, had assumed a teaching position in Radautz in northern Bukovina, meaning he could no longer benefit from the VDA's provision of funds towards a 10 month stay in Hermannstadt to evaluate the remaining data.[33] Instead, Csallner was forced to rely on the foreign students who had initially aided him in 1937, sending the bulk of the material to Vienna to be analyzed there.[34]

A second section equally close to Csallner's eugenic heart was the *Neighborhood Index File* [*Nachbarschaftskartei*], which Wilhelm Schunn had asked the LSS to design and implement. Csallner had been intimately

33 Csallner, *Das LSS*, 3.

34 Csallner, *Das LSS*, 3.

involved in formulating the program's eugenic policies; for instance, his handwriting is apparent in the support of large families, awarding "honorary gifts" to those deemed racially sound and nationally valuable, with a staggering 20,000 Lei for their fourth child and a further 10,000 Lei for all further children.[35] Csallner produced a file index system based on the constituent families [or *Sippen*] of a neighborhood, one geared towards the perpetual, annual evaluation of its members (with the file to be renewed every ten years). Above and beyond the mere collection and revision of personal data, the Index was also to offer something of an up-to-date barometer of demographic and economic trends. While Csallner hoped to convince Schunn of the need for a "drastically enhanced Index" in the near future, the existing version already posed a remarkably-detailed central database deployed to check whether Saxons were indeed following the rules of employing more German workers and evaluating where intervention in the job market was needed "to prevent the overcrowding of particular jobs and hence simultaneously prevent our comrades from emigrating to the Old Kingdom."[36]

The degree of total statistical control Csallner had so consistently pursued is perhaps more apparent here than anywhere else: "In a few years time we will be able to evaluate the statistical data for all Germans annually, and can do so more thoroughly and frequently than, generally, most states can—and we will simultaneously create the most powerfully compelling means imaginable to quickly and purposefully intervene in its

35 While this particularly interesting eugenic policy has (to the author's knowledge) so far evaded scrutiny entirely, Schunn himself described its ideological aims, racial hygienic means and practical ends in great detail in the fascinating short booklet: Walter Schunn, and Oskar Pastior, *Die Ehrung des Kinderreichtums bei den Deutschen in Rumänien* (Hermannstadt: Krafft & Drotleff, 1940). It also offers a case study of one of Hermannstadt's 39 "Neighborhoods," namely the impoverished suburban "Am Taubenbrunnen" where the average income of the 24 families to have been awarded these "honorary gifts" revolved around the two to three thousand Lei per month mark, thus highlighting the significant economic impact for the local community provided by these "gifts." As for the overall origins, ideology and structures informing the "Neighborhood" system of local micro-management in interwar Romania, see: Wilhelm Schunn, *Die Nachbarschaften der Deutschen in Rumänien* (Hermannstadt: Krafft & Drotleff, 1937); *Schulungsbüchlein für Amtswalter des Nachbarschaftswesens nachdem Stande der Arbeit im Herbst 1939* (Hermannstadt: Krafft & Drotleff, 1939); Viktor Quandt, *Nachbarschaft – die überschaubare Einheit mit Herz: Ein kurzer Überblick über Sinn und Wandel des Nachbarschaftswesens der Siebenbürger Sachsen in Vergangenheit und Gegenwart* (Vienna: Kramer Verlag, o.J. [But after 1971]); Hans-Achim Schubert, "Wilhelm Schunn's ‚Völkische' Nachbarschaften – Wunsch und Wirklichkeit," in Walter König, ed., *Siebenbürgen zwischen den beiden Weltkriegen* (Cologne: Böhlau, 1994), 169–82; and Georg Weber, "Zum Struktur- und Funktionswandel der siebenbürgisch-sächsischen Nachbarschaft in Rumänien und Westdeutschland: Eingrenzung, Begriffe, Institutionen," *Ethnologia Europaea* 12 (1981): 131–161.

development."[37] Csallner's work clearly tended to sacralize statistical analysis and he strove consistently toward the compilation of increasingly vast databases. Statistics were to him the ultimate self-reflective narrative on societal (d)evolution; and, simultaneously, the objective determinant of centrally implemented eugenic population policies.

This desire to collect ever more data is also evident in the remit assigned to a third section, "Special Surveys and Advisory Services" [*Sondererhebungen und Beratungen*]. Here, seven indicative surveys were either being prepared or already underway in 1938. The first, on landownership in Saxon villages [*Hofbesitz in den sächsischen Dörfern*], was carried out by a group of students that had offered their help and two of Csallner's "work groups"—consisting of teachers and priests—focusing on the Bistritz county parish.[38] Another survey dealt with the related question of "Urban House Ownership" [*Städtischer Hausbesitz*] and was to be undertaken by the neighborhoods. A third, similar survey on mixed marriages (stemming from Csallner's own private work) was to inform his three part article on this particularly destructive specter. By 1940, Csallner had already begun preparatory work on extending his study to include the Bukovina (which he deemed particularly prone to inter-ethnic marriages). A fourth survey dealt with Migrations [*Wanderungen*], the product of which was an article on migratory trends prevalent in the village of Grossalisch[39]. A fifth census studied "Rural Youth" [*Ländische Jugend*] in conjunction with a sixth on the critical issue of employment patterns for Saxon versus non-Saxon "Farmhands" [*Knechte und Mägde*]. Not only was Csallner concerned with the increasing frequency with which Saxon youths (in particular young women) emigrated to the Romanian Old Kingdom, he also returned to his customary

36 Csallner, *Das LSS*, 4.

37 Csallner, *Das LSS*, 4.

38 Csallner, *Das LSS*, 5.

39 Alfred Csallner, *Die Erforschung der Abwanderung vom Dorf, dargestellt am Beispiel Grossalisch* [*Volk und Rasse – Volk und Raum* 11 (1940); also: *Evangelische Fürsorger* 7/10 (1940)] (Hermannstadt: Honterus, 1940). Apart from providing an illustrative example of Csallner's overall eugenic mentality and methodology, this article was to serve as a template and good practice guide for future surveys. But it also reflects Csallner's deep-seated dislike of the Romanian "Old Kingdom" and the supposedly degenerative impact it had upon Saxon migrants. He laments that "one [of five] had returned, but with a Hungarian wife and ruined by his stay abroad. Of a fourth it is said he was also likely to return. But those returning from the Old Kingdom usually bring—see the mentioned case—so much that is undesirable with them so that they are themselves corrupted, so that one is not necessarily sure whether to actually wish they had better not come back."

complaint that employing anyone other than a Saxon would, by definition, be detrimental to the national body. What is more, as most of the ethnically Romanian servants tended to live in Saxon settlements, this invariably led to an ongoing loss of *Lebensraum*, "with these hundreds of millions of Lei which we pay Romanians year on year—to our most dangerous adversaries in the struggle for our continued existence—we strengthen them precisely there where they have already breached our *Lebensraum* and are further snatching more of it away, piece by piece."[40] The last of the seven surveys entrusted to the Special Surveys and Advisory Services was on the question of "landed property (gained from and lost to Romanians)" [*Bodenflächen (gewonnen und verloren mit bezug auf Rumänen)*] planned for summer 1940. Beyond this ambitious workload, Csallner had also been entrusted with a highly politicized survey, one quantifying the interwar "losses" all German enclaves suffered at the hands of "purposeful over-taxation," the "agricultural reform's great land robbery," the decline in the number of German civil servants and so forth. Linking all of these surveys is the explicitly ideological goal to "verify the injustice that we have suffered in the past 20 years and continue to suffer, to efficiently ground our demands for compensation."[41]

In equal measure both enthralling and unnerving due to the current absence of any further documentary evidence on its activities, the Special Surveys and Advisory Services department also featured a *Sub-department for Racial Research* [*Unterabteilung für Rassenforschung*]. This was headed by Prof. Albert Hermann (1893–1975), who had published a landmark biometric study on the Saxon peasantry in southeastern Transylvania in 1937.[42] Together with Michael Fleischer (1912–1982)—who, for

---

40 Csallner, *Das LSS*, 7.

41 Csallner, *Das LSS*, 7. While this study appears to have remained incomplete, Csallner did produce an unpublished manuscript entitled *Damages Inflicted upon the German National Community in Romania by the State* [*Schädigungen der deutschen Volksgruppe in Rumänien durch den Staat*] in 1940.

42 Albert Hermann was one of the few indigenous racial anthropologists to have pursued biometric evaluations of Saxon settlements, and he did so with the active support and encouragement of Prof. Eugen Fischer. Regrettably, Hermann was to publish very little of his work apart from his incredibly useful 1937 book on *The German Peasants of the Burzenland* (Albert Hermann, *Die deutschen Bauern des Burzenlandes* [Jena: Fischer, 1937]). Written for a wider audience, it included a general overview of the region's history to then describe the results of fieldwork he conducted with the help of local teachers, priests and peasants in 1932. Csallner clearly admired Hermann's work, which he was delighted to point out marked the "first large racial anthropological study to be conducted by an *auslandsdeutscher* researcher, and is likely to be the most substantial racial anthropological study of a homogenous national community to date" (Csallner, *Die Volksbiologische Forschung*, 74).

his part, had carried out anthropometric studies of Saxons in the north-eastern regions of Transylvania (paid for by a research grant negotiated by Csallner)—Hermann's subsection was tasked with the management and expansion of biometric research across all Saxon communities (interestingly enough, it included local Romanian and Szekler populations) eventually to encompass every German ethnic enclave.[43]

Another subsection of the Department of Statistics dealt with "Foreign Surveys" [*Fremde Arbeiten*]; namely, "the task of evaluating the statistical and related works of others in so far as they are pertinent to the Germans of Romania, to adopt those portions of use, to correct errors made and challenge that which is damaging and dangerous."[44] In particular, this section was supposed to amend and correct the official data released by the state, such as the "fundamentally wrong" ethnographic map that had been based on the official 1930 census for Romania.

Many of maps, graphs and illustrations produced by the LSS were the work of Günther Herbert, a close friend of Csallner's and head of the fifth section, *Representation and Visualisation* [*Darstellung*]. The sixth and final department to be housed in *Statistics* was the LSS' Archive [*Sammlung*].

## 2. Genealogy

The LSS' second department was that of *Genealogy* [*Sippenwesen*] and was assigned the critically important task of "assessing the national community's genealogical [*sippenmässigen*] makeup, to teach it to think, see and act in the genealogically and hereditarily [*sippenhaften und sippengerechten*] proper way. To that end it has the particular duty to preserve and evaluate parish registers and to introduce and encourage the private pursuit of genealogical research and hereditary health."[45] The department was subdivided into five subsections, with the first, "Documents" [*Urkunden*], assigned the task of obtaining "register-excerpts and other documents, to certify *Ahnennachweise*, and is primarily used by those who have immigrated to Germany and their descendants."[46]

43 Csallner, *Das LSS*, 7–8.

44 Csallner, *Das LSS*, 8.

45 Csallner, *Das LSS*, 9. See also: Csallner, *Organisationsrichtlinien des LSS - 25.05.1939*, 7.

46 Csallner, *Das LSS*, 9.

Another section, headed by Father Gustav Arz, was dedicated to the study of "Parish Registers" [*Kirchenbücher*]. Inspired by a visit to the Kirchenstelle Alt-Berlin, and with Bishop Glondys having consented to his plan, the tedious work of cataloguing all parish registers was vital for Csallner's ambition to compile genealogical studies not merely of individuals or their families (for which he could always rely on either the "Hereditary Fitness Archive" [*Nachbarschaftskartei* or *Erbtüchtigkeitsarchiv*]), but of entire Saxon settlements. What is more, Csallner hoped it would also serve to revitalize the *Volkskörperforschung* [National Constitution Studies] that had lost much of its impetus since 1936, when Father Johann Bredt[47] succumbed to his illness. That said, the third section, *Genealogical Research* [*Sippenforschung*], intended to further introduce Saxons to the kind of "biologically framed" genealogical research that Csallner had been advocating since the early 1920s.

The LSS generally placed a heavy emphasis on the social utility of genealogical research insofar as this allowed a given individual to study his or her hereditary makeup—to identify that which Csallner had defined as good and also what was less so. Csallner had frequently defended the virtuous fruit the genealogical tree could offer in terms of deciding whom (if at all) one should marry and have children with,[48] to discover what talents had been hardwired and to awaken the nation to the blood-ties [*Blutsbande*] he had often ascribed to internal migrations. Genealogy was thus something of a racial anthropological microcosm, an easily accessed window onto the

47 The priest Johann Bredt, who was a close friend of Csallner's, had taught at middle schools for several years and was parish priest in Oberneudorf, Waltersdorf, Windau and Kleinbistritz before retiring to Bistritz in the early 1930s due to illness. Bredt's core contribution to the Saxon eugenic discourse was his pursuit of "national-body-studies" or "*Volkskörperforschung*." Bredt aspired to create a "transparent" nation not only through the study of family trees but by the anthropo-genealogical study of entire towns. In other words, he studied the historical evolution and spatial variations of particular settlements in order to produce a perpetual economic and demographic survey for any given point in time. Csallner was not the only one to see the biopolitical utility of a statistical system offering such "scientifically tailored" evaluations of any respective town's struggle against displacement and degeneration, and the Leipzig "*Stiftung für deutsche Volks- und Kulturbodenforschung*" actively encouraged him to publish a summary of his theoretical approach as *Volkskörperforschung* in 1930 (Johann Bredt, *Volkskörperforschung* [Breslau: Ferdinand Hirt, 1930]). This 55-page pamphlet, in addition to his landmark 1929 case study on *Windau* in *Past and Present* (Johann Bredt, *Windau in Gegenwart und Vergangenheit: Eine Ortsgeschichte auf Grundlage der Volkskpörperforschung* [Bistriz: Buchdruckerei Gustav Zikeli, 1929]) constitute the bulk of his short but intriguing career. A much-revised version was later published by Günter Litschel [Hilfkomitee der Siebenbuerger Sachsen (ed.)], *Windauer Heimatbuch* (Schweinfurt: Schweinfurter Druckerei und Verlagsanstalt GmbH, 1971).

48 See, for example: Alfred Csallner, "Von der Bedeutung der Gattenwahl für Wesen und Schicksal der Kinder" *Volk und Rasse - Volk und Raum* 1 (1934). See also: Aufbau (1934) (Hermannstadt: Honterus, 1934).

historical past as well as a Janus head predicting the future, since "hardly anything is so suitable to permanently enthuse so many people as to allow them to recognize the laws of heredity and the importance of hereditary traits by themselves, to teach them how to see and think biologically—to awaken in them the sense of community and a communal destiny, to their responsibilities towards each other."[49] That proliferating genealogical research amongst the Saxon masses was not merely an end unto itself is, again, apparent in Csallner's reification of statistics. This was complemented by the notion that a complete statistical excavation could create a database even modern states dare not contemplate. And the beating heart of this database, predicting and controlling virtually everything from cradle to grave, came in the form of the *hereditary health and fitness archive* [*Erbgesundheits- und Erbtüchtigkeitsarchiv*], which Csellner envisioned creating in league with the medical professionals and the Church.[50] This department also lobbied political and Church authorities alike to coerce their employees into submitting their own family trees — to ostensibly force the public at large to grapple with the concept of immutable heredity, at least temporarily, if not to spark permanent enthusiasm. Not surprisingly, then, a fourth section was to act in the genealogy department as the LSS' own *Genealogical Archive* [*Sippenarchiv*].

The final subsection was that of *Heraldry and Symbols* [*Wappen und Zeichen*]. This was headed by prominent heraldic expert Albert Arz von Straussenberg (1857–1935), who had also been intimately involved in the design of the various coats of arms sported by Schunn's *Nachbarschaften*. Straussenberg's engagement with the LSS had been rather accidental: he had initially approached the *Cultural Institute* in Hermannstadt with a proposal for setting up a *heraldry department*, upon which Csallner offered him free reign in his LSS. Considering that the latter had initially thought this subsection would do little more than collate information on already existing coats of arms and act as an advisory body to those endeavoring to create new ones, Straussenberg's *Heraldry and Symbols* was to gain considerably more significance in the course of time. Bearing this out, consider Csallner's rather astonishing statement on the subject:

49 Csallner, *Die Arbeitsstellen für SBS*, 3.

50 Csallner, *Die Arbeitsstellen für SBS*, 3–4.

> With the creation of the department 'Heraldry and Symbols' I believe to come a good bit closer to realizing an old dream of mine: I would like to assist in the creation of a cast of German noble leaders [*Führeradel*] that will constitute – while not too far removed from the rest of the community – a closed breeding community [*Zuchtgemeinschaft*] raised to noble thinking and noble composure in the clear awareness of their particular duties towards the national body, to be and do all that it can in the coming centuries along the lines of what Hans F.K. Günther demanded of a *Führeradel* in his *Führeradel und Sippenpflege*.[51]

Apart from this idiosyncratic embracing of Günther (who, otherwise, features rather rarely in his texts), Csallner hoped to introduce an honorary coat of arms for particularly valuable families, awarded—and potentially revoked—by the national leader (on grounds of fertility). While this appeared to be a particularly valuable means of encouraging (and, importantly, maintaining) higher fertility rates amongst the socio-biological casts Csallner rated most highly, he failed to specify the "high demands" set out for potential candidates.[52]

## 3. National Betterment

The department of *National Betterment* [Volkspflege] was perhaps the LSS' most significant, having been given the task of surveying the past, present and, particularly, the future of the German hereditary stock. The department's stated aim was:

> to evaluate and further expand upon the results of studies conducted by the main departments 'Statistics,' 'Genealogy,' 'Foreign Nations,' as well as those carried out by other agencies with regards to the evolution and nature of the Germans in Romania.[ ... ] Particular attention and research is to be paid to the dangers inherent to fertility decline, emigration and mixed marriages and to work towards overcoming these threats.[53]

In continuously close cooperation with the *National Office for National Health* [*Landesamt für Volksgesundheit*], this was to be undertaken in order to "awaken ever wider sections of the population to their responsibility for

51 Csallner, *Das LSS*, 11.

52 Csallner, *Das LSS*, 11–12.

53 Csallner, *Das LSS*, 12.

the nation's future composition and fate, to awaken and entrench in them the pride and willingness to bear these responsibilities for the recovery and regeneration of the German people."[54] Despite the strong degree of jurisdictional overlap apparent in the mission statements of virtually all of the LSS' head departments, Csallner explicitly tasked "National Betterment" and its five subsections with his eugenic ideology's principal agenda: to lead the battle against the degeneration inherent to small families, to combat the existential threat posited by differential fertility, to arrest the Saxons' eugenic decline by ending mixed marriages and to eliminate the socio-economic scourges of alcohol and tobacco.

The subsection "Population Size" [*Volkszahl*] correspondingly assumed exceptional importance and "has primarily to lead the battle against declining fertility rates and work towards a strong German population increase."[55] This section was supposed to work closely with Schunn's *Nachbarschaften*, whose practice of awarding "honorary gifts" to large families already bore Csallner's handwriting. Added to this incentive, Csallner introduced a book of honor and matching honorary emblems [*Ehrenzeichen*] to be handed to recipients in the festive *Nachbarschaften* meetings. Csallner also resubmitted his 1927 outline for an equality fund [*Ausgleichskasse*][56] to the Church via his LSS, insisting that "[b]ack then I found no appreciation [for my plan]. Now the time has come, as members of the 'movement for renewal' form the decisive majorities in the Church's corporate bodies and the *Landeskonsistorium* in particular, for the next *Landeskirchenversammlung* to finally accept my own or a similarly thorough solution."[57]

With an increasingly draconian mindset, Csallner also submitted a considerably more radical resolution to both the Church and political establishment alike, calling for the exclusion of all who could, but failed, to bear

54 Csallner, *Organisationsrichtlinien des LSS - 25.05.1939*, 7–8.

55 Csallner, *Das LSS*, 12.

56 This had initially posed as the flagship policy advanced by the VdK in a "*Denkschrift*" with the rather unwieldy title "Draft law on the creation of a fiscal equality between employees with few or no children and those with many within the national Protestant Church A.B. in Romania towards the protection and furthering of large families". In essence, the fund had recast the carrot as a stick, by ensuring that all Church employees were taxed beyond any financial advantage gained by having less than four children, it was suggested the Church finance this fund by taking 30% of the wage of single, divorced or widowed employees, 20% of that of married employees without children and 10% of that of married employees with only one child. (Csallner, et al., *Denkschrift des 'Vereins der Kinderfrohen'*).

57 Csallner, *Das LSS*, 12.

more than four children from all Church and political offices. In other words, Csallner aimed to exile such underachievers from the Saxon community at large:

> I am expecting particularly favorable results from a plan that I shall propose to the national leader, and which I will ask the *Landeskonsistorium* to present as a draft law to the next *Landeskirchenversammlung* [... namely that] those who, although they could, do not bring themselves to have at least four children—and are thereby refusing to give the national community that most valuable of all gifts—should henceforth be excluded from all national and Church honorary positions, thereby taking away from them any means to partake in national and Church life.[58]

But for Csallner, as one life was certainly not equal to another in terms of inherited talent, it fell to a second subsection, *Heredity and Race* [*Vererbung und Rasse*], "to ensure that beyond their mere number their quality is not forgotten."[59] This group thus dealt directly with two of Csallner's greatest concerns: the continual loss of racial nobility due to particularly low birth rates amongst the hereditarily most valuable sections of society, and the rapidly increasing occurrence of mixed marriages. In turn, this section "has to lead the fight against mixed marriages, it has to call upon the pride for one's own nation one's duty towards it while strengthening and directing the nation's resistance to mixed marriages, but it also has to discover all the reasons that lead to mixed marriages [such as wage earning Saxon women setting such a high standard that they deter German suitors] and to work towards the dissolution of these causes."[60]

The third section of the *National Betterment* department, perhaps unsurprisingly, was *Marriage Counseling* [*Eheberatung*]. Here the intention was not to merely discourage mixed marriages but to actively promote the introduction of marriage eligibility licenses [*Ehetauglichkeitszeugnisse*]: "I also believe that following the appropriate enlightenment and courtship our Protestant Church will, in the foreseeable future, only carry out Church weddings if marriage eligibility licenses are produced."[61] More drastic still,

58 Csallner, *Das LSS*, 13.
59 Csallner, *Das LSS*, 13.
60 Csallner, *Das LSS*, 13.
61 Csallner, *Das LSS*, 14.

Csallner called upon the Church—in the pages of Siegmund's *Evangelischer Fürsorger* no less—to realize where its national duty lay, namely that it insist brides and grooms furnish racial anthropological evaluations (conducted by German doctors wherever possible), in addition to checking their own hereditary archives prior to sanctioning any wedding.[62]

A fourth subsection added in September 1939 led the battle against society's twin ills *Alcohol and Tobacco* [*Weingeist und Tabak*]. That Csallner would have created such a department is to be expected considering Siegmund's ideological influence and their collaboration in the "International Order of Good Templars," while its placement within the organization reflects Csallner's understanding of their negative impact lying less with the nation's actual health and much more with its wallet. He explains that:

> [n]ot only have I, for many years now, industriously labored in this field and want to continue the work that has grown close to my heart, but the greatest threats inherent to alcohol and tobacco consumption are not even health related, but constitute an eerie and irresponsible waste of national wealth in addition to severely maiming our moral [*sittlich*] strength to resist in the battle for our continued existence.[63]

The final section of the "National Betterment" department, *Festivals and their Organization* [*Fest- und Feiergestaltung*], was created to ensure that all festivities took the basic tenets of national betterment [*Volkspflege*] into account, that is, by celebrating the large family and its beneficial, revitalizing, impulses.

## 4. Lebensraum

The fourth of the LSS' six departments, *Lebensraum* consisted of only three subsections, all of which dealt with the founding leitmotif of Saxon eugenics: evaluating the extent of territorial losses to ethnic "others" and assessing the most fruitful means of overcoming them. The first subsection, *Property* [*Besitz*], was given the task of exploring rural settlements. This entailed quantifying the amount and size of abandoned properties (due to, for example, emigration or inheritance), the employment of "foreign" agricultural

62 Alfred Csallner, "Familienbuch – Erbarchiv," *Evangelischer Fürsorger* 1 (1939): 2.
63 Csallner, *Das LSS*, 14.

laborers and handymen, the extent to which land holdings (and their individual plot size) were shrinking and creating local 'work groups' by furnishing the necessary conceptual and methodological support for investigating these trends in greater depth. Given the extent of perceived local dysgenic processes, this department sought to prevent any further degeneration by manipulating internal colonization/migrations by flagging areas in need of Saxon human capital and by enshrining the importance of employing and selling only to one's ethnic kin. A typical example of what form this sort of encouragement was to take is afforded by one of the LSS most direct appeals, in this case to the "Saxons of Arkeden!" A case study of striking simplicity, the poster featured two pictures, one of the local church-castle and the latter of a street. Their respective captions read: "Your fathers have built this church-castle so the community may brave the storms of history and persist unto today" and "What do you do to protect yourselves from those who threaten you? You let Gypsy hawks buy up this entire street!" (See fig. 3.)

The second subsection, *Career Counseling* [*Lebensstellen*] had a rather difficult task given that the very idea of a *Lebensstelle*—life-long, permanent employment such as in a family-run pharmacy—was progressively losing its economic (if not symbolic) significance. Nonetheless, feeling ever more embattled by the perceived onslaught of ethnic rivals into the job market, the notion that "Career Counseling" needed to enlighten and encourage the masses, to "not only defend and further the existing ones, but struggle to regain lost *Lebensstellen* while creating new, German ones,"[64] gains its paramount importance as the form of an, albeit more nationalist than utilitarian, form of career counseling.

*Lebensraum*'s final subsection covering *Migrations* [*Wanderungen*] had two overarching goals: to explain how internal migrations created and perpetually recreated an organic nation bound by blood and guided by a communal destiny [*Blut- und Schicksalsgemeinschaft*] and, second, to extol this destiny. Deriving from this agenda was the twofold policy to directly influence migration patterns by encouraging migrants to resettle in German rather than "ethnically foreign" areas and to lead the battle against emigra-

64 Csallner, *Das LSS*, 15.

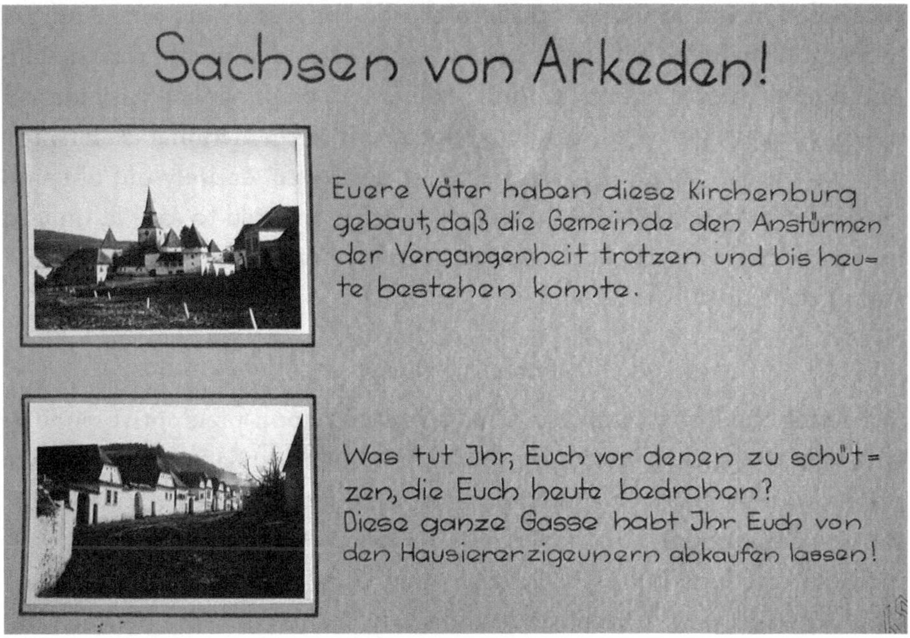


**Fig. 3: "Saxons of Arkeden!"**

tion more widely—in particular to both "contain emigration to the Romanian 'Old Kingdom'"[65] as well as to the Third Reich.

This section also featured a rather intriguing further subdivision, if only in its *Siebenbürgen Gauamt*, for *Transylvanian Saxons Abroad* [*Siebenbürger Sachsen in der Fremde*]. Largely self-explanatory, this was set up to oversee one of Csallner's favorite pet projects: to amass the addresses of as many emigrated Saxons as possible and to send them letters from home [*Heimatbriefe*] encouraging these prodigal sons to return home or to at least aid in the maintenance of their Germanness whilst in the foreign wilderness.[66] Ideally, Csallner wanted to publish four twelve-page issues per annum, in collaboration with the VDA Kassel. Correspondingly, he had written to Saxon societies in Graz, Budapest and Vienna in late 1938, requesting that their local membership lists be sent to him.[67] In fact, Csallner had actually

65 Csallner, *Das LSS*, 16.

66 Csallner, *Das LSS*, 17.

67 Csallner an die Gauobmänner der der Vereine der Siebenbürger Sachsen in Graz, Ofen und Wien – dated 7 November 1938 [NA Sibiu, CGR, D.XIV.688, Nr. 13].

succeeded in amassing files on about 10,000 émigrés by the time Andreas Schmidt dismantled the LSS in 1943. More curious, though, was the fate of this nascent card index. We find Csallner complaining that Schmidt had not, as would have been expected, relocated it to the national leadership's new base in Kronstadt. Instead he had it destroyed, entirely out of spite: "he had it pulped, regardless of how my daughter tried to save it, only to hurt me: 'If one were to need such a thing in ten or twenty years, one could simply do it again.'"[68]

## 5. Foreign Nations

Reflecting the heavy emphasis Csallner placed upon preemptive population policies (such as managing the labor market in relation to migratory patterns) in the battle to prevent further territorial losses to ethnic others, the fifth department, *Foreign Nations* [*Fremde Völker*], monitored and evaluated interactions between Germans and their neighbors. In Csallner's words, its aim was to "to observe and study the relationship between the German ethnic group[s] and the other nations, and to transmit their conclusions to the other head departments [...] to mediate and pay special attention to, in particular, matters concerning culture, the economy and biological issues."[69]

Foreign Nations was created to report on these other nations, their internal discourses and how these could pose threats further down the line. Founded in early October 1939, this department was one of the LSS' newest innovations and remained largely uncompleted. Methodologically, its subsections were ethnically structured at the district level (while the *Landesamt* itself would only create a new section if the same 'ethnic other' was found in at least two regions). Although sections overseeing developments within Jewish, Gypsy and Armenian communities where certainly operational by 1939, it is presently impossible to discern further details on how many such offices there eventually were, let alone how they operated on the ground.

A major exception was, however, the department's activities in the *Gauamt Siebenbürgen*, which held sections on the Romanians, Szeklers, Gypsies

68 Csallner, *Meine wissenschaftlichen Arbeiten* [1975], 17.
69 Csallner, *Das LSS*, 17.

and Jews and to which Csallner had assigned a wide array of surveys. Csallner was concerned with the potentially disastrous impact of falling birth rates and rising debt levels amongst Romanians and launched a study to investigate "the incredible biological damage incurred by an internally unprepared Romanian nation as a result of its sudden political and social rise since 1918."[70] A second survey dealt with the "structural changes amongst the Szeklers that have led to a dangerous degree of infiltration of our cities." Turning his attention to Gypsies, another study assessed what Csallner deemed to be their ruinous [*verderblich*] influence on Saxon morality in the Reps region in particular, and evaluate the damage suffered by the Saxon *Lebensraum* at their hands (especially with regards to the already encountered example of Arkeden). Finally, a fourth study was to investigate the ongoing Jewish "extortion" [*Auswucherung*] of Saxons in Bistritz and Sächsisch-Regen.[71]

## 6. Enlightenment, Recruitment and Education

The sixth LSS department, Enlightenment, Recruitment and Education symbolized a further centralization of Csallner's eugenic propaganda. If almost all of his proposals necessarily rested upon a centralized organization to enforce them, this final department oversaw the Saxons' ideological indoctrination by "coordinating and furthering the informative, recruiting, and educational projects of the other departments, and ensuring their frictionless and increasingly complementary augmentation of each other's information, recruiting, and educational work."[72] To this end, six individual subsections oversaw and influenced the work of the LSS. The first, Press [*Presse*], worked in close cooperation with the *National Department for Press and Propaganda* [*Landesamt für Presse und Propaganda*] while Libraries [*Büchereien*] managed and expanded the LSS' house library and coordinated 'wandering libraries.' Another section produced and proliferated Educational Pamphlets [*Aufklärungsschriften*][73] and a fourth subsection

70 Csallner, *Das LSS*, 17.

71 Csallner, *Das LSS*, 17.

72 Csallner, *Das LSS*, 18. See also: Csallner, *Organisationsrichtlinien des LSS - 25.05.1939*, 8–9.

73 An interesting instance showing how Csallner continued to sustain and further deepen his relationship with the *Reichsbund der Kinderreichen Deutschlands*, which had given him the initial impetus to found his short-lived VdK, is provided by the following example: they sent him 200 copies of the *Völkischer Wille* [*National Will*], which Csallner, under the greatest of fiscal lamentations, claimed he was unable to post to his co-workers as he could not afford the stamps (Csallner, *Das LSS*, 23).

trained speakers and organized Lectures [*Vortragswesen*]. The fifth subsection acted as the curator of wandering Exhibitions [*Ausstellungswesen*] and finally, Education [*Unterrichtswesen*] was vitally entrusted with the twin responsibilities of recognizing and rectifying any damage inflicted by the unacceptable Romanian school curriculum, one so unequivocally unsuitable for raising "German children." Csallner had, in fact, personally written additional teaching materials based on his eugenic research. It seems these textbooks had been field tested and were even approved by an unnamed school in Schässburg.[74] But if raising a new generation in the spirit of this biological world view was a long term project, Csallner's last LSS department had the most immediate mission of ensuring that priests and teachers were taught the key tenets of "heredity, race and national betterment."

On a final note, Csallner was consistently adamant about the need for further funds towards translating these grand plans into practice. It thus comes as little surprise that his November 1939 report on the LSS' restructuring concluded with a four page lament on the precariousness of his fiscal position. Indignant about his perceived monetary maltreatment, Csallner proceeded to underline—in far from modest terms—his personal importance to the Saxon eugenic discourse and to the LSS as its nurse and caretaker. Categorical on the ways in which he had driven forward not only racial hygienic research, so crucial to any program of national regeneration (spiritual or otherwise), Csallner drew attention to a new generation of laymen-cum-racial-anthropologists, explaining that: "I can calmly claim that as yet no one has made as great an effort to focus our science on these true-to-life and nationally important tasks—while supplementing our current lack of proper researchers with an increasing number of laymen—as I am doing, and as I have done long before I was entrusted with the care for our country's German science."[75] Grounded in this self-assessment, Csallner threatened to resign were he not awarded a significantly better salary. Yet what Csallner could not know was that almost precisely a year later his career, as one of the most prominent of Saxon eugenicists, would come to a frustratingly abrupt end in the wake of Andreas Schmidt's return from Berlin.

74 Csallner, *Das LSS*, 18.

75 Csallner, *Das LSS*, 25.

Andreas Schmidt's appointment as National Leader in 1940 marked a critical shift in Saxon socio-political reality, one that did not favor Alfred Csallner. Initially, though, he was delighted to see his LSS further expanded into the ISB or *Institute for Statistics and Population Policy of the German National Community in Romania* [*Institut für Statistik und Bevölkerungspolitik der Deutschen Volksgruppe in Rumänien*].[76] Strife was not long in coming, however, and Csallner soon fell out with Schmidt. Subsequently sent to Germany, Csallner's institute was dissolved with a sizable portion of his work destroyed by the time he returned to Transylvania in 1943. A year later, Csallner was arrested and interred in the Romanian prisons *Târgu Jiu* and *Turnu Măgurele*.[77] Upon release in 1946, Csallner returned to his priestly robes and picked up the pen again—this time to write novels and short stories.[78] Following two further stays in Romanian prisons accompanied by a string of interrogations and house searches, Csallner eventually emigrated to Germany in 1974, where his publications won prizes such as the Cultural Award [*Kulturpreis*] of the *Landsmannschaft der Siebenbürger Sachsen* in 1982. Csallner died ten years later, in 1992, at the age of 97.

## Conclusion

The 1930s were a seminal period in Saxon history, and far more research is needed on virtually every facet of the genesis, evolution and impact of Saxon eugenics during this period. This includes, but is certainly not limited to, the substantial influence Austria and Germany had on the Saxon eugenic discourse ideologically, fiscally and institutionally; the interaction between Saxon racial hygiene and its Romanian counterpart; the question of how this program compares, for example, to that advanced by other German minorities across Europe; and, fundamentally, how Saxon eugenics challenges our current understanding of the viability of eugenic discourses promoted by ethnic minorities within the conditions of an empowering nation-state.

76 Csallner, *Meine wissenschaftlichen Arbeiten* [1975], 7.

77 Csallner, *Das LSS*, 8.

78 Such as his autobiographical narrative *Rottenholz und die Rottenholzer Grosseltern* (Traunreut: Selbstverlag, 1976), and his collection of (extremely interesting and indicative) short stories *Der Baruch und andere Erzählungen aus Siebenbürgen* (St. Michael: Bläschke, 1980).

Although the excavation and analysis of Saxon eugenics' aims and means, and its interaction with the local, national and international theaters is still in its infancy, it will undoubtedly prove a fruitful new field of study, not only due to its specific importance to Saxon historiography but because of the numerous ways in which the history of Saxon eugenics may contribute to our understanding of the German, Austrian and even Romanian societies and organizations that it collaborated with. While the maturation and partial empowerment of Saxon eugenics was grounded in the growth of an indigenous movement for renewal, it is difficult to imagine what it would have looked like without the prominent backing received from Germany in particular. Not only did virtually all Saxon eugenicists study abroad, most of its most prominent figures did so in Berlin under Eugen Fischer's direct tutelage (for example, Albert Hermann and Eckhard Hügel [1908–1977]). The question of what impact Fischer had on Saxon racial anthropology is a pressing one. Similarly, Csallner would hardly have been able to expand his LSS so dramatically or conduct many of his studies without the financial support he received from the VDA and the *Deutsche Forschungsgemeinschaft,* amongst others. These international links are hence not only crucial to our ability to define the trajectory Saxon eugenics itself took but to locate it within the far greater network of European eugenic movements that shaped the interwar period.

That said, this study has pursued the far humbler aim of sketching the contours of interwar Saxon eugenics in terms of its dogmatic conceptualization and political instrumentalization of a biologically redefined Saxon identity, notably the manner in which it was shaped and progressively institutionalized by Alfred Csallner. Critically, this was a eugenic discourse that found itself increasingly radicalized through contact with an indigenous fascist movement led by Fritz Fabritius. Often acting in tandem, both pursued the common goal of reversing a perceived dysgenic crisis, a demographic meltdown in terms of both "quantity" and "quality," and the loss of *Lebensraum* at the hands of increasingly expansionist and threatening Romanian neighbors. This battle though, could only be won through the spiritual and racial rejuvenation of the Saxon nation, the pro-active reclamation of lost territory and the unwavering quest for ethnic exclusivity.

# FIGHTING THE WHITE PLAGUE

## DEMOGRAPHY AND ABORTION IN THE INDEPENDENT STATE OF CROATIA

*Rory Yeomans*

When Josip Blažek, student commissar and Ustasha death squad member, fell in battle fighting against the Partisans in 1943, he was declared a martyr. In the days afterwards, the obituaries that followed noted his fanaticism, his devout Catholicism, his loyalty until death both for the fascist Ustasha Movement and his comrades, as well as his love for the newly-created Independent State of Croatia. Student leaders declared eight days of mourning. Many obituaries recalled Blažek's student activism in the dying days of the Yugoslav state, especially his concern for the social welfare and hygiene of his fellow students. As the young assistant of a highly respected professor of social hygiene at the University of Zagreb who had been instrumental in improving the health, diet and living conditions of impoverished students, Blažek's interwar activities would surely have made him a model Ustasha student—if not for the fact that the professor he had worked under was Andrija Štampar (1888–1958), a liberal, an advocate of eugenics and, worst of all, a committed Yugoslav. Štampar was one of the first public figures to be arrested after the Ustasha came to power in April 1941. His former assistant was more fortunate. Blažek's career prospered in the new Croatian state and he ultimately became commissar of the Department for the Social Protection of University and High School Students. Notwithstanding this success, when Blažek died he was eulogized as someone who always remained a student, "modest in his demands" and seeking neither position nor privilege. At the same time, although imprisoning Štampar, the fascist state that Blažek served so fanatically had appropriated many of his ideas and theories.[1]

The relationship between Blažek and Štampar serves as a microcosm of the wider story of an ultra-nationalist regime in Europe and the way in which it used eugenic theories, racial hygiene and pro-natalist principles as well as a religious conception of life to regenerate the state, racially cleanse the nation and morally purify society, the latter by encouraging women to return to the home and fulfill their "natural" function as wives and mothers. In 1941 Yugoslavia was invaded, occupied and divided by Nazi Germany and Fascist Italy. After less than a month of fighting, the Independent State of Croatia (*Nezavisna Država Hrvatska* – NDH) was established in April, comprising Bosnia and Herzegovina as well as most of present-day Croatia. The new state was divided into two zones of occupation, one controlled by the German army and the other by Italian army units and local Italian Fascists. However, official power in the state resided nominally with an extreme nationalist and terrorist organization, the Croatian Ustasha Movement. Founded in the early 1930s out of militant youth groups and radical student societies, the founding members of the movement lived in exile in training camps in Hungary, Italy and Germany, from where they periodically infiltrated Yugoslavia to carry out terrorist attacks. However, the Ustasha also had significant support within Croatia and Bosnia, especially among the lower clergy, seminary and university students and high school youth as well as Catholic and rightist intellectuals. Following the establishment of the Independent State of Croatia in 1941, Ante Pavelić (1889–1959), the leader of the Movement, was declared the supreme leader or Poglavnik of the state.

The Ustashas' long-stated aim was the establishment of a Greater Croatia purified, by force if necessary, of all "alien" populations, especially Serbs, Jews and Gypsies. They rejected many—although by no means all—aspects of modern urban life, including cosmopolitanism, bourgeois liberalism and female emancipation and idealized the traditional and patriarchal life of the peasant *zadruga*, which the Ustashas and their mainly young Catholic intellectual acolytes claimed was the true expression of Croatian

[1] "Ustaša-sveučilištarac Josip Blažek položio je svjoj život kao ustaški zastavnik u borbi protiv odmetnika," *Nova Hrvatska* (25 March 1943): 7. All translations from the original are my own unless otherwise cited.

nationalism. Although the Ustasha Movement was not doctrinally religious (like the Romanian Iron Guard) or led by priests (like the Slovakian Hlinka Guard), this Fascist movement was intensely mystical, ritualistic and powerfully influenced by the rites and beliefs of Catholicism. This resulted in a world-view which was chiliastic and apocalyptic, encapsulated in its mystical cult of death.[2] Once in power, the Ustasha regime embarked on a campaign of extermination against its "racial" enemies, a campaign carried out with such brutality and ruthlessness that it shocked even hardened Third Reich officials. This was preceded by a public campaign of vilification in which Serbs and Jews were portrayed by regime officials and the press as the most dangerous enemies of the Croatian nation, needing to be eliminated if the Croatian nation was to survive. By the time the Independent State of Croatia collapsed in 1945, the Ustasha regime was responsible for the deaths of over half a million people, with both the state and the movement becoming synonymous with sadism, fanaticism and genocide.[3]

But the ideology of the Ustasha Movement was inherently contradictory. Despite the regime's violence and emphasis on racial and moral purification, the Ustashas' relationship to eugenics and racial hygiene was complicated. On the one hand, when it concerned the persecution, expulsion and mass murder of dangerous "aliens," the regime showed no compunction in utilizing the most extreme interpretations of racial hygiene and negative eugenics in order to purify the nation. On the other, the predominantly Catholic roots of many regime functionaries and the strong support the regime received from young Catholic intellectuals meant it officially opposed the "scientific" principles of negative eugenics in the case of euthanasia and abortion. Although the Ustashas aimed to improve the Croatian nation through racial purification, the exterminatory excesses of negative eugenics were to be applied only to biological enemies of the Croatian

2 Regarding the Ustasha Movement's cult of death see Rory Yeomans, *Visions of Annihilation: the Ustasha Regime and the Cultural Politics of Mass Murder, 1941–1945* (University of Pittsburgh Press, forthcoming) and Rory Yeomans, "Cults of Death and Fantasies of Annihilation: the Croatian Ustasha Movement in Power, 1941–1945," *Central Europe*, 3, 2 (2005): 121–42.

3 Unfortunately, there are no completely satisfactory English-language histories of the Ustasha Movement and the Independent State of Croatia. The most comprehensive political history of the Ustasha regime and the Independent State of Croatia can be found in Jozo Tomasevich, *War and Revolution in Yugoslavia: Collaboration and Occupation* (Stanford: Stanford University Press, 2001). See also the collection of essays in Sabrina Ramet, ed., *The Independent State of Croatia, 1941–1945* (London: Routledge, 2008).

nation. In the case of racially worthy members of the nation—defined as anyone who was not Serb, Jewish or Gypsy—the Ustashas railed against negative eugenics. Therefore, the blind, deaf, mentally and physically disabled were to be protected so long as they were Croatian; those that were not could perish.

The Ustashas' rejection of abortion and euthanasia was partly inspired by the Catholic influence on its ideology. However, it certainly was not evidence of any kind of respect for human life. As shown by the campaign of mass murder carried out against Serbs, Jews and Gypsies, the regime had a profound disregard for human life, including infant life. The establishment of concentration camps at Jasenovac, Jadovno and Stara Gradiška in the summer of 1941 was testament to this.[4] All three camps contained children's sections in which infants were abused, poisoned and routinely murdered by the camp guards. In fact, what the regime really feared in the practice of euthanasia and especially abortion was the threat posed to the NDH's campaign to alter Croatia's demographic balance by raising the birth-rate of racially valuable stock. One method to achieve this demographic transformation was through the mass murder of racially inferior communities. The other method, however, involved the imposition of pro-natalist policies in which motherhood, fertility and family life were idealized, the "selfish" single life assailed and abortion outlawed.

## Eugenics and Emancipation: Women and Demography in Interwar Yugoslavia

Modern, urban, emancipated: this was interwar Yugoslavia in the western imagination. As the satirist Branislav Nušić (1864–1938) commented in 1925: "After the war everything got shorter: hair, skirts and marriages."[5] But popular perception did not entirely reflect reality. Yugoslavia was not

4 There are few scholarly studies of the Ustasha system of terror, mass murder and concentration camps in the English language. The exception is Tomislav Dulić's expertly nuanced *Utopias of Nation: Mass Killing in Bosnia and Herzegovina, 1941–1942* (Uppsala: Uppsala Universitet, 2005). The most comprehensive documentary collections in Serbo-Croatian are Vladimir Dedijer, *Jasenovac i Vatikan: dokumenti* (Belgrade: Rad, 1987) and Vladimir Dedijer and Antun Miletić, eds., *Protjerivanje Srba sa ognjište, 1941–1944: svedočanstva* (Belgrade: Prosvjeta, 1989).

5 Branislav Nušić, "Pre i posla rata," *Politika* (6 January 1924): 3.

as liberal or modern as many hoped it would be and its citizens, especially women, were not as free as might have appeared. In fact, Yugoslav women labored under a number of restrictions. Since there were no unified legal codes in interwar Yugoslavia, the "Yugoslav woman" simply did not exist legally. Instead, women were at the mercy of existing laws governing the region in which they lived. Women in Serbia, for example, lived under the 1844 Constitution legislating that women were not entitled to have private property and had no rights in relation to their fathers and husbands. In Croatia and Bosnia, meanwhile, women lived under the 1811 Austrian Civil Code which, though more liberal than Serbia's, were hardly the personification of female emancipation. Women were also restricted in terms of employment: they could not work as school inspectors or directors; in 1934 a *numerus clausus* was introduced into the telephone and telegram service, which limited the number of women working there and in 1929 a new regulation was introduced that stipulated that a female teacher could not be married. Women were also disadvantaged in many other ways: they could not vote and had higher rates of illiteracy. Women also tended to receive lower wages and were often forced to work longer hours, sometimes at night. In the 1930s, a growing army of unemployed in Yugoslavia meant women's wages and jobs were increasingly vulnerable to economic pressure. Perhaps most germane of all, women did not have legal reproductive rights: birth control was hard to obtain and abortion outlawed for most of the interwar period.[6]

During the 1920s and 1930s, a number of laws were passed that aimed at improving the lives of working women. From the beginning of the 1930s the government prohibited women in industry from working at night. According to the Law for the Protection of Children and Mothers, women had the right to paid maternity leave from up to six weeks prior to the birth of the baby and for the first six weeks thereafter. However, the sum they received was often so small as to mean many pregnant female workers toiled almost right up until they began their labor. Irrespective of how diligently these laws were followed by private industry, for the first time, the government required industry to provide nursery facilities for employees as well as medical insurance. Working women with children were also offered breaks at least once

6 Jovanka Kecman, *Žene Jugoslavije u radničkom pokretu i ženskim organizacijama, 1918–1941* (Belgrade: Institut za savremenu istoriju, 1978), 24–30.

every five hours.[7] Even anti-government observers like the Communist journalist Vera Novak were forced to concede that by the 1930s, Yugoslav women were entering the job market in large numbers, although in her analysis, this was due to the "rationalization" of industry and dire working conditions that ensured women, especially younger women, were cheaper to employ.[8]

Novak also wrote with disdain about the government's military training for young professional women through the Sokol gymnastics organization.[9] For her, this could never be a harbinger of female emancipation, only another sign of their bourgeois oppression. Among Yugoslav ideologues in the 1920s and 1930s, however, improvements in the lives of women were of central importance. According to the journalist Laza Popović, the new Yugoslav woman would be the partner of the new Yugoslav man: together they would build the new Yugoslavia. Envisaging the equality of women, Popović declared that women should be vocal in expressing their opinions rather than being "inert" or "a dead mass." Many Yugoslav ideologues further linked the emancipation of Yugoslav women to the unveiling, liberation and education especially of Muslim women, who received government support through the secular Muslim society Gajret.[10] Some western visitors were impressed with the progress Muslim women were making. Writing in 1928, the travel journalist J. W. Wiles observed that, even in Bosnia, there had been a change in both the attitude and conduct of young women. Following the introduction of compulsory education for all children, Muslim women's societies had sprung up and, as a result, "hundreds of Moslem women are employed outside their homes; in the various professions; in factories. They can be seen going to their work, some veiled, some with transparent veil, some only with the symbol of the veil, others complacently lacking even the suggestion of a symbol." As far as Wiles could see, the atmosphere in Yugoslavia was becoming "electrified with feminine enterprise and progress."[11]

7 Kecman, *Žene Jugoslavije*, 56–64.

8 Vera Novak, "The Position of Working Women in Yugoslavia," *International Press Correspondence*, 10, 10 (1930): 175–76.

9 Vera Novak, "Military Training of Women in Yugoslavia," *International Press Correspondence*, 33, 10 (1930): 623–24.

10 L [Lazar Popović]., "Jugoslovenska žena," *Nova Evropa* 4, 1 (1922): 1–3. See also Ahmed Muradbegović, "Problem muslimanske žene," *Nova Evropa* 4, 1 (1922): 3–9.

11 J.W. Wiles, "Moslem Women in Yugoslavia," *The Moslem World* 18, 1 (1928): 65.

Reflecting this advancement, writers, artists and novelists cultivated an image of independent Yugoslav womanhood as a defining symbol of the Yugoslav state. In one of Branislav Nušić's most famous plays, set in 1920s Belgrade among a fashionable bourgeois family, the major part of the humor revolved around the subversion of the conservative mother's expectations of how her daughter-in-law should behave. Needless to say, wearing pyjama pants, smoking cigarettes and having a career are not what she envisages for the mother of her grandchildren.[12] The popular novelist Milica Jakovljević (1897–1952) wrote a number of articles for the popular press in the 1920s on young, new women she encountered in Belgrade: bob-haired, cigarette-smoking, Charleston-dancing heroines more interested in cars, jazz music and athletics than in finding a husband.[13] However, the extent of female emancipation in interwar Yugoslavia should not be exaggerated. The kinds of women portrayed in Jakovljević and Nušić's portraits were wry and larger-than-life stereotypes not intended to be taken entirely seriously. Most Yugoslav women did not lead the carefree existence envisaged by these literary and journalistic creations. Still, neither should it be denied that Yugoslav women, as elsewhere in Europe and the United States, were generally freer and better able to take advantage of socio-political opportunities than they ever had been before. No matter how imperfectly the Yugoslav state had realized the emancipation of women, it nonetheless stood in stark contrast to the attitude of many Croatian nationalists who in their cultural and intellectual journals made it clear that female emancipation and feminism had, even with these modest steps, gone too far.[14]

Where discussion of women's roles was concerned, talk of demography and birth-rates usually followed. And even here, there was plenty of evidence to suggest that social attitudes to abortion and birth control as well as the place of women in society—even in patriarchal villages—were evolving. In the late 1930s the Croatian anthropologist Vera St. Erlich wrote a series of articles for the Yugoslav press based on questionnaires she had sent to teachers, physicians and others working in small villages.

12 Branislav Nušić, *Beograd nekad i sad* (Belgrade: Narodna štamparija, 1924).

13 See, for example, Milica Jakovljević, "Portreti beograđanski – Kiki sa beogradskih ulica," *Politika* (18 April 1927); and "Četvrtna brzina u životu Beograđanke," *Politika* (11 May 1927).

14 Zdenka Jušić-Šeunik, "Ženskoj omladini," *Mlada Hrvatska* 1, (1 January 1936): 3, and "Ženski pokret i studentice," *Alma mater* (11 April 1935): 83.

The aim of her research was to record the culture and practices of rural Yugoslavia. Her research showed that, while Yugoslavia had the highest birth-rate in Europe, the widespread use of birth control alongside a high mortality rate was helping keep population levels down. She found that in nearly half of all Croatian villages abortion was practiced on a wide scale, with three-quarters of villages using birth control. In regions where abortion rates were low, this was more often attributable to a lack of knowledge than moral reservations. In Slavonia, for example, St. Erlich estimated that a staggering three-quarters of all women had had abortions.[15] In a 1940 study, the feminist and socialist researcher Nada Sremec came to nearly identical conclusions, emphasizing the role that economics and social pressure, including familial pressure, played in limiting the number of births.[16]

During the 1930s, Yugoslav demographers and researchers undertook a number of ethnographic surveys looking in a more systematic and scientific manner at birth-rates and mortality figures in various parts of Yugoslavia. Undoubtedly, these studies reflected widely held concerns about the demographic future of the nation. In perhaps the most well-known study of population changes and patterns in Croatia, the Slovenian medical demographer Bojan Pirc (1901–1991) carried out a survey looking at demographic changes in Otok near Vinkovci in eastern Slavonia. In order to ascertain the wider pattern of births and deaths in his selected region he categorized marriages according to whether both partners were old settlers in the region, mixed or were both new settlers. Pirc found that the lowest fertility rate was found amongst regionally endogamous marriages, with a higher rate in regionally exogamous marriages and the highest birth-rate when both spouses came from outside the region. Pirc noted the following motivating factors behind these low birth-rates: the parcelization of land should several children be born, conflicts in joint families in which one of the wives had more children than the others and thus threatened the equal distribution of property and, finally, the women's desire to hand down their dowry "intact" to a single daughter. Pirc also noted

15 Vera St. Erlich, *Family in Transition: A Study of 300 Yugoslav Villages* (New Jersey: Princeton University Press, 1966), esp. 287–305.

16 Nada Sremec, *Nismo mi krive. Kako živi narod. III. knjiga: Slavonska žena* (Zagreb: Gospodarska sloga, 1940).

that men and women appeared to be in almost total agreement about the need to limit births and, drawing attention to the high rate of child mortality, suggested that this pointed to the incidence of child neglect as a deliberate means of limiting family size. He ascribed birth-rates and wider demographic changes to "a new sexual morality" characterized by liberal individuality.[17] In a 1937 study, the Croatian demographer Fedor Mikić produced similar findings. He found that in Croatia birth-rates were relatively low, infant mortality relatively high (especially among infants born out of wedlock) and, compared to other industrialized and modernized European countries like Holland and Germany, there was a far higher level of stillbirths, infant deaths, deaths of women and men of childbearing age and deaths in childbirth.[18]

Pirc and Mikić's studies highlighted a more general concern among health experts and racial anthropologists. The perceived cataclysmic demographic future for Yugoslavia prompted a scientific desire to study the native population at close range. Consequently, scientists traveled to remote villages to observe their inhabitants. During these research projects they studied nose sizes, took blood samples and made notes on the menstruation patterns of young girls. Their findings were compared with those of similar groups in Yugoslav cities. Health professionals recognized that many demographic problems resulted from poor diets, arduous lives and deplorable living conditions where highly infectious diseases could flourish. For Yugoslav health experts, an effective means of combating the high levels of infant mortality, low birth-rates and general ill health was through a concentration on social issues, in particular in improving the living conditions of the impoverished masses. Since any coordinated national health program would need to begin by addressing the health of mothers and children, health experts maintained that highly infectious diseases such as malaria, tuberculosis and malnutrition were best com-

---

17 Bojan Pirc, *Opadanje stanovništva u Slavoniju: socijalna-medicinska studija o prilikom rađanja i smrtnosti u pet slavonskih srezova* (Belgrade: Štamparija Centralnog higijenskog zavoda, 1931); E.A. Hammel, "Economics 1, Culture 0: Fertility, Change and Differences in the Northeast Balkans, 1700–1900," in Susan Greenhaigh, ed., *Situating Fertility: Anthropology and Demographic Enquiry* (Cambridge: Cambridge University Press, 1995): 238–257.

18 Fedor Mikić, *Pomor dojenčadi u Zagrebu sa statističkog, socijalnog i kliničkog stanovništva referat na 1. jugoslav. pedij. kongresa na Bledu* (Belgrade: Centralna higijenska zavod, 1936).

bated through the establishment of hygiene centers and health clinics in every region of the country that were accessible to all.[19]

The preoccupation of physicians and scientists with demographic developments—as well as their connection to hygiene, sanitary conditions and the relative lack of pediatric facilities—was to be expected in a state still recovering from the decimations of World War I.[20] Yet proposals to reduce the rate of infant mortality inevitably evolved from ideas about improving the availability of health facilities toward rather more explicitly eugenic suggestions about replenishing the national stock. In 1934, at the annual Congress of the Yugoslav Physicians' Society in Bled, Slovenia, these concerns were expressed explicitly, with numerous speakers combining calls for liberal health and social reform with openly eugenic suggestions. M. Zelić, for one, argued that in order to increase the birth-rate the government should penalize "selfish" affluent single women who "refused" to have children by forcing them to volunteer as nurses in rural communities. This would impress the national and eugenic importance of having children upon them. At the same time, he also argued for a comprehensive sex education program to be initiated in high schools and for the legal recognition of children born out of wedlock, ideas which, although eugenic in their motivation, were also championed by many progressive campaigners.[21] A year later, Svetislav Stefanović (1877–1944), president of the Yugoslav Physicians' Society, echoed this pro-natalist call. Writing in the journal *Evgenika*, he argued that abortion and sterilization should be permitted in the case of criminals, alcoholics, psychopaths and the congenitally blind and deaf in the interests of "medical, racial and spiritual hygiene." In all other cases, however, abortion should be opposed since in "racial-hygienic

19 See, for example, Matija Ambrožić, "Zavod za zdravstvenu zaštita matera i djecu," *Socijalno-medicinski pregled*, 4, 1–3 (January–June 1933): 68–79; Bogoljub Konstatinović and Bojan Pirc, "Bolnice i bolnička služba u Jugoslaviji," *Socijalno-medicinski pregled* 4, 8 (1936): 260–78; Branislava Kalić, "O školskoj higijeni kod nas i na strani," *Glasnik Centralnog higijenskog zavoda*, 1, 1–3 (1926): 61–73.

20 See, for example, Josip Rasuhin, "Vitalna statistika," in Andrija Štampar (ed.) *Socijalna medicina*, vol. 1 (Zagreb: Institut za socijalnu medicinu, 1925), 34–61; Natalija Nikolajević-Davidović, "Zaštita zdravlja školske dece," *Socijalna medicina*, 95–122; and Matija Ambrožić, "Zaštita dojenčadi i male dece," *Socijalna medicina*, 61–94; and Branislava Kalić, "O telesnom razvoju školske dece prema ekonomskom stanu roditelja," *Socijalno-medicinski pregled*, 10, 4 (1938): 41–52.

21 M. Zelić, "Srestva protivu začeća u cilju regulisana rađanja," in Svetislav Stefanović, *Naučni rad xvi godišnjeg skupa Jugoslovenskog lekarskog društva, 8–10 Septembra 1934 na Bledu* (Belgrade: Jugoslovensko lekarsko društvo, 1935), 282–323. See Z. Rosenberg, "Pravo vanbračnog deteta," *Nova Evropa*, 16, 6 (1926): 150–163.

and racial-biological terms" it destroyed the health of the mother, damaged the nation and generally had a "degenerative effect" on the race.[22]

Such views were not marginal. One of the most enthusiastic propagators of the idea that eugenics could be used to transform Yugoslav society was the medical pioneer, Minister of Health and founder of the School of Public Health in Zagreb, Andrija Štampar. Štampar complained that few countries, including Yugoslavia, had adopted an organized approach to the implementation of eugenic policies. He argued that eugenic practices should form the basis for the health policy of the state. This approach would not only help to increase the birth-rate but the quality of the Yugoslav race as a whole. He ascribed low birth-rates to the "rationalization" of sexual activity in which it had become linked to pleasure and separated from reproduction. In his opinion, this reflected a more general societal trend, where individual interest had superseded eugenic necessity. In the current climate, people could secure a comfortable life without great effort, meaning that more and more people were choosing not to get married and have children. For Štampar, one easy way to increase the national birth-rate would be through state encouragement: every "healthy" couple should have at least three children, with any number after that supported by the state.[23]

Štampar was not a supporter of simple natalist policies and he rejected the idea of raising the birth-rate by simply encouraging everyone to have large numbers of children. He was an advocate of qualitative eugenics, his concern for selective breeding and the future of the Yugoslav race demonstrating how his ideas about social medicine had been influenced by race-hygienic thinking. To prevent the "degenerate" and those of "lesser worth" from procreating, for example, Štampar argued that the state would have to implement policies to discourage indiscriminate reproduction. In a theoretical sense, the scientific community would have to gain knowledge about the "biological makeup" of Yugoslavia through biometrics and mass examinations. In the meantime, there were other, more practical measures that could be taken. One of these was the regulation of marriages by counselors and qualified physicians, who could determine whether couples were

22 Svetislav Stefanović, "Rasna higijena, abortus i zaštita matere," *Evgenika*, 1, 5 (1935): 1.

23 Andrija Štampar, "Politika o populaciji," *Socijalna medicina*, 19–28.

suitably healthy to be married. Those deemed unsuitable for such eugenic marriages included, according to Štampar, "all those with sexual diseases in the infectious stage, leprosy, chronic mental illnesses, epileptics and the retarded, those who are paralyzed, have dementia, manic-depressive illnesses, imbeciles, degenerative hysteria, chronic poisoning or chronic addiction to morphine, alcohol and cocaine."[24]

While Štampar opposed the more drastic policies of compulsorily castration and sterilization of criminals or those deemed unsuitable for procreation as pursued in some states in the United States, he did suggest that groups shown to be dysgenic could be isolated. This would be a humane solution to the problem and would liberate society from "a whole list of degenerates." In Štampar's opinion, this "general isolation" would better serve the "purification of human society from those elements unsuitable for breeding." As a preventative measure, alcoholism, sexually transmitted diseases, tropical diseases and tobacco would all have to be eradicated in the interests of the health of the population. At the same time, Štampar argued that the general public needed to be educated about the importance racial hygiene through national programs of eugenic enlightenment. In many of his ideas, he drew inspiration from the proposals of leading European proponents of eugenics, such as Francis Galton and Christian von Ehrenfels. For instance, changes to the established sexual "order" might be contemplated, he wrote, so that particularly physically and eugenically superior individuals could have the opportunity to produce ever larger numbers of offspring. Štampar also suggested reforming the education system so that it teach knowledge about scientific, social and sexual ethics, the prevention of war, which "destroys the biologically best and eugenically most worthy individuals" in addition to rewarding particularly able citizens and students with eugenic awards and diplomas. Finally, in its eugenics campaign, Štampar noted, the state could draw on the old folk poetry and songs that had been a feature of the South Slavs prior to the liberation and unification of Yugoslav lands.[25]

Yet, in an example of the curious "liberalism" which characterized interwar Yugoslav eugenics, there was another, more socially progressive aspect to Štampar's beliefs. This was expressed in his advocacy of social medi-

24 Štampar, "Politika o populaciji," 30–31.
25 Štampar, "Politika o populaciji," 30–31.

cine and hygiene—the positive eugenic idea that the health of the nation was inextricably linked to the social conditions in which the masses lived. He argued that disastrous economic conditions, hunger, poverty, a lack of decent housing, war and the prevalence of highly infectious diseases—often caused by poor sanitary conditions and disproportionately affecting the young and the healthy—were important factors conditioning the health and racial quality of the nation. Even if a mass program of selective breeding could be introduced, children would still physically deteriorate if the buildings in which they learned were unheated, unsanitary and primitive or if the shacks in which they lived were sweltering in the summer and freezing in the winter or, worse still, infested with malaria.[26]

This picture of deprivation and desperation, Štampar submitted, must be replaced with clean, modern village homes guided by social hygienic principles in the countryside, hygienic working conditions in the factory, the workshop and on the farm and the establishment of garden cities to aid the material and physical health of urban populations. He also proposed that the massive levels of venereal disease be addressed through the introduction of a "rational" sex education program in schools and suggested that humane policies to address the social and economic causes of prostitution would be more effective than draconian laws criminalizing prostitutes. The scourge of alcoholism, meanwhile, could be addressed through the substitution of non-alcoholic fruit juices for hard liquor alongside the imposition of high taxes on tobacco producers and breweries. He also believed poverty and inequality could be tackled through the imposition of high taxes on opulent properties and luxury goods with the revenue used for the socialization of health services. Nevertheless, even here his negative eugenic principles revealed themselves: while calling for taxes to be levied on everyone to finance a National Health Service in Yugoslavia, Štampar suggested that taxes should be levied at a higher rate for the unmarried and childless.[27]

As director of the Institute of Social Medicine at the Ministry of Public Health, Štampar—like so many Yugoslav physicians, health experts and

26 Štampar, *Deset godine unapređivanja narodnog zdravlja* (Zagreb: Narodne novine, 1934); idem, "O socijalnoj terapiji," *Glasnik ministarstva narodnoga zdravlja*, 2 (1920): 1–17. This "liberal" form of eugenics was not unique to Yugoslavia. It was also an integral feature of the modernization process in some other Southeastern European countries such as Romania. See, for example, Maria Bucur, *Eugenics and Modernisation in Inter-War Romania* (Pittsburgh: University of Pittsburgh Press, 2002), esp. 78–121.

27 Štampar, *Deset godine, unapređivanja narodoga zdravlja*; idem, "O socijalnoj terapiji," 1–17.

eugenic advocates—was keenly aware of the need to improve provisions for mothers and babies in order to better the health of the nation. In the introductory section of a book written in 1925 to mark the first five years of the Yugoslav National Health Service, he drew attention to the numerous advances made in a short space of time: the creation of hundreds of socio-medical institutions, including school polyclinics, maternity, infant and child health centers, malaria control and anti-venereal centers founded across Yugoslavia and the thousands of children treated by these facilities each year. Nevertheless, he acknowledged that unless physicians spent more time with ordinary citizens, they could not hope to be responsive to their needs or educate them in the importance of hygiene. He advocated that medical students should spend time working in villages in order to better appreciate the health needs of the masses. National propaganda campaigns for social hygiene should also be launched while social hygiene needed to be adopted as a core part of the school curriculum.[28] By 1940, Štampar, now professor of public health at the University of Zagreb, concentrated on projects to improve living conditions and medical access for the University's impoverished students—in the process earning the ire of nationalist students for his Yugoslavist attitude to the question of student health.[29] Yet this was the same year he authored a book drawing attention to the declining Croatian population and its future implications.[30] Ultimately, despite his many contradictions, as a eugenicist Štampar was representative of wider intellectual, scientific and national conflicts in interwar Yugoslavia, being both a Croat racial hygienist who advocated "negative" eugenic ideas for the betterment of modern society and a "positive" eugenic proponent of South Slavic demographic improvement. Despite his Yugoslavism and liberal political beliefs, many of Štampar's ideas were appropriated—albeit in a much distorted and often even inverted form—by the Ustasha regime. Quite unintentional and profoundly ironic, in their preoccupation with the demographic decline of the nation, their criticism of young single women and their eugenic quest for the betterment of the race long before the Independent State of Croa-

28 The basic principles of social medicine are explained in Andrija Štampar, "O socijalnoj medicini," *Socijalna medicina*, 5–18.

29 See, for example, Janko Škrbin, "Socijalno pitanje, sveučilišta, dr. A. Štampar i mi...," *Uskok* 1, 1 (1940): 1, 2.

30 Andrija Štampar, *Problem o populaciji u Hrvatskoj* (Zagreb: Hrvatski liječnički zbor, 1940).

tia came into being, leading scientists and physicians in interwar Yugoslavia had already laid much of the ideological groundwork for the Ustasha regime's campaigns of natalism, re-population and racial purification.

## Purifying the Nation: A New Nationalist and Religious Morality

Traveling through the Slavonian countryside in the autumn of 1941, the young writer Ivo Balentović (1913–2001) observed that the role of women was sorely in need of betterment. He argued that the life of the average peasant woman in Croatia was utterly miserable. Economic anarchy, backwardness, unhygienic living conditions and a massively high rate of illiteracy were much more likely to be the lot of the average peasant woman than a "civilized" life. In order to rectify this state of affairs, he provided a list of eugenically-progressive proposals: the Ministry of Health had to provide more physicians and midwives as well as more effective communication links to the cities, new schools and hospitals would have to be built and education programs specifically to combat female illiteracy and ignorance about hygiene also needed to be introduced. Balentović found conservative attitudes to the role of women largely responsible for the backwardness of economic and social conditions in the country, especially in Muslim regions of the NDH such as in Bosnia. Like earlier Yugoslav ideologues, he criticized the "veiling of Muslim women, their exclusion from public life and their partial enslavement in the home" in addition to old-fashioned closed mindedness and jealousy at the root of much of the primitiveness and unhappiness of contemporary Muslim woman in Croatia. Removing this old-fashioned attitude and consigning anachronistic religious doctrines to history would do much to make the life of the average Muslim peasant woman healthier and happier. This had been illustrated by Turkey's banning of the veil and the subsequent improvement in the quality of women's lives there. Modern Turkish women were to be found in all kinds of professions, Balentović claimed, as well as attending universities.[31]

31 Ivo Balentović, "O teškom životu naše seljačke žene," *Hrvatska smotra* 9, 9 (1941): 533, 536.

However, Balentović's comments were not motivated by a concern for women's rights in and of itself since, for him, the progress of women was inextricably linked to their ability to have more and healthier babies. This was the reason behind the need to improve the level of female literacy, since women were the natural educators of the young. Likewise, in highlighting the high rate of mortality among peasant women, his main concern was that in dying young, their unborn children were lost to the nation. Moreover, women, he maintained, were entirely responsible for combating the unhygienic living conditions in which so many peasants lived and which he characterized as "filth in the kitchen, dirt in the bed" and linked this to the fact that the highest rate of child mortality was to be found in rural Bosnia. He also pointed out that twenty thousand "young, healthy" females died in interwar Yugoslavia annually before the age of fourteen in the main from some, short unexplained illness prior to giving birth, or instead as a result of the "artificial termination of a pregnancy." In contemporary rural Croatia, he noted, wives and mothers were accorded very little respect and, in some regions, this led to a kind of "rationalization of sexual morals" characterized by "a limitation on the number of births, an ever greater number of abortions and a fearfully high rate of mortality among young women." The situation in Croatian cities was just as calamitous. There, women were surrounded by various luxuries encouraging them—especially those from the affluent middle-classes—to "escape from their maternal and domestic duties."[32]

Balentović's interest in the fertility of Croatian peasant women predated the establishment of the Independent State of Croatia in 1941. Like many nationalist writers, he identified the low birth-rate among Croatians as one of the most serious problems facing the nation. Writing in April 1935, he expressed concern about the low birth-rate in Slavonia and wondered whether the lifestyle of contemporary Croatians—in particular their relative wealth, prosperity and "sexual morals"—had been responsible. Citing Bojan Pirc's study from the previous year, Balentović claimed that as soon as women were given individual freedoms they selfishly opted to spend their free time on themselves: "They indulged in dreaming and yearning and dressed themselves in silk and gold. There was a desire for physical

32 Balentović, "O teškom životu naše seljačke žene," 534.

beauty and the emphasis of this beauty resulted in the rationalization of sexual life and a subsequent reduction in the number of children." Perhaps, he added with distaste, it was not coincidental that, at this time, there had also emerged a poetry of a sexual and erotic nature that celebrated "the free lives of women." Rather than simply ascribing the pro-natalist crisis of the nation to a moral collapse, however, Balentović pointed to a range of environmental factors: terrible social conditions in the countryside, including the dire economic situation, deplorable levels of hygiene in many rural households, the climate, unhygienic food, a lack of cleanliness in local water, the widespread prevalence of tuberculosis and "an excessive reliance" on alcohol. In this situation it was hardly surprising that there was such a high incidence of mortality amongst babies.[33]

The problems caused by this "white plague," as well as the fact that Croatian families were still not having enough children, became an important theme for newspapers in the Independent State of Croatia, especially in the early months of the new regime. According to *Hrvatska krajina* [Croatian frontier] in Banja Luka, a low birth-rate was as deadly an enemy of Croatia as Serbs or Jews. Over the past twenty years, the newspaper asserted, the white plague had manifested itself with the result that many regions had been depopulated of Croats and natality eviscerated. This plague appeared in both cities and villages throughout the NDH, partly as a result of "misery and poverty" and partly due to "empty, unruly and malign fashion." Jewish physicians in particular had materially benefited from the destruction of thousands of urgently needed unborn Croatian citizens, in addition to the health and in some cases even the lives of mothers. This had been orchestrated at the hands of a corrupt government in Belgrade, which had sought the "systematic destruction" of future Croatian generations in order to "destroy the Croatian nation and make it incapable of resistance and, along with many other perfidious methods, counted on passivity towards the destruction of its progeny."[34]

In the opinion of another writer in the same newspaper, the "artificial, partial or complete sterility of woman" was "enemy number one of the Independent State of Croatia" and it needed to be made a capital crime immediately.

33 Ivo Balentović, "Bijela kuga u Slavoniji," *Hrvatska smotra*, 3–4 (1935): 217–223.
34 "Nestat će bijela kuga," *Hrvatska krajina* (15 August 1941): 1.

For centuries the Croatian nation had fought and defeated many fierce enemies, yet, ironically, wealth and fashion had brought the greatest enemy—abortion and birth control—to both peasants and the urban middle-classes. The desire for comfort and enjoyment had fatally reduced the number of births, with the dying out of numerous families and whole villages subsequently being settled by non-Croats. Responsibility for the devastation of these villages lay with selfish women "who did not want to have any children, or, at the least, did not want to have more than one or two children." Women could not be held solely to blame for this situation, however, as it was a well-known fact that "some dishonest physicians—above all Jews—raked in enormous fortunes carrying out for a steep price the artificial termination of pregnancies and, I believe, still carry out these artificial abortions today."[35]

The writer went much further than simply advocating the outlawing of abortion, however. True, the introduction of such a law would be a "potent" sign of a new Croatia, but the law needed to go further. Not only should the death penalty apply to all those who carried out abortions, irrespective of the circumstances, but any advertisement promoting birth control should be banned. At the same time, the outlawing of abortion could serve a useful function in leveling social inequalities since the property of those carrying out abortions could revert to the state after their execution. Equally, radical changes in the law of inheritance would ensure that the property and wealth of families who did not have children—or, even worse, did not want to have children—reverted to the state, being placed in a special fund to assist poorer families with large numbers of children. The wealth and property of families with only one or two children should also be inherited by the state for the same purposes.[36]

Whether this kind of article was a genuine reflection of the attitude of not only a small group of nationalist and Catholic intellectuals but a section of the wider population is not clear. Nonetheless, while the emphasis on the failure of women to give birth to a large enough number of children was motivated by elite concerns about demographic survival of Croatia—magnified by the need for a large population in a time of war, conflict, hunger and disease—it probably also reflected more widely held anxieties in

35 "Neprijatelj Nezavisne Države Hrvatske broj 1," *Hrvatska krajina* (25 May 1941): 1.
36 "Neprijatelj Nezavisne Države Hrvatske broj 1," 1.

the Independent State of Croatia about a rapidly changing society and women's place in it. The Ustasha regime left no doubt that one of its primary social aspirations was to return women to the home and men to the factories and fields. As the Ustasha functionary Zdenka Žanko explained in 1942, women had been encouraged to spend their days in idle chat in cafés in interwar Yugoslavia rather than in having children and, if they did have children, limiting themselves to one child. However, in the new Croatia the "national, religious and moral betrayal" of childless women would be eradicated. The regime would instead create a healthy moral society based on the principle of motherly devotion and fatherly responsibility. This meant removing women from the public sphere and giving their jobs to men so that they could have the opportunity to support their own families in such a way that they were able to restore their pride.[37]

According to Žanko, one of the most damaging aspects of life in interwar Yugoslavia was the way in which traditional Croatian values of female gentility and maternalism had been marginalized by a morally decadent and callous regime. A "cruel" Yugoslav regime had deliberately promoted the white plague among the Croatian masses, with the nefarious intention of reducing their overall number. In addition, public opinion and contemporary trends dictated that the only way a woman could be considered to be fashionable was if she had, at most, one child: "Things got so bad that women began to kill their own descendants, consciously and calculatingly killed and slaughtered them and no one could do anything about it."[38] Similarly, Ela Maroš, another ideologue, wrote that the new Croatian woman—in contrast to Yugoslavia which had taught her decadence, cosmopolitanism and vanity—would live for her family rather than herself in the Independent State of Croatia. An obedient wife to her husband, she would be the "soul of the house" and "modest in dress, dignified but proud," she would be "a bodily angel, a moral assistant in her husband's struggles in life, a gold-winged bird."[39] Nonetheless, this rhetoric did not hold much appeal for at least some actively revolutionary younger women. Bosiljka Perše, for one, a student of pharmacy, categorically rejected the idea that a women's place was "in the house, in the kitchen, with the children, in clean-

37 Zdenka Žanko, "Majka u svietlu ustaštva," *Ustaškinja* 1, 8 (1942): 7.

38 Žanko, "Majka u svietlu ustaštva," 7.

39 Ela Maroš, "Uloga Hrvatske žene u Hrvatskoj," *Ustaškinja* 3, 8 (1944): 32–37.

ing, at the neighbors," and demanded the same right for women as for men to participate in the military defense of their homeland.[40]

Perhaps as a result of this cultural resistance to casting women purely in terms of their biological function as mothers, many Ustasha youth and female journals also sought to portray more egalitarian and liberated images of womanhood, lauding Croatian women's innate heroism, warrior nature and strength. They praised the liberation of Muslim women from the "primitiveness" of the veil, which the Ustasha regime identified as one of its most important cultural and social achievements.[41] Even as devout a Catholic as the prominent Ustasha activist Mira Dugački (1917–2008) felt the need to make concessions to this spirit of assertiveness among younger women. True, she railed against the "poison" of foreign ideologies which had resulted in the destruction of many Croatian families, "many murdered unborn children, many corrupted lives of young women" and informed young women that, in the new Croatia, the family and home would be their kingdom, with motherhood their first and "most sacred" duty. But she also acknowledged the desire for greater independence by many young women in the movement, emphasizing that the new Croatian woman would combine the progress of modern times with the long held traditions of her nation. She could pass through "fields and factories and offices and schools of all sorts" and "like every female worker and peasant she will know about her nation as well as world affairs." Although her main task involved the protection of the family hearth, this "New Woman" would work with her husband as a comrade and not a slave.[42] As well as a New Woman, the Ustasha regime also heralded the coming of a New Man. Writing in 1943, Mijo Bzik, (1907–1945) official historian of the Ustasha Movement, depicted the New Man brought into being by the Ustashas' national revolution as a spiritually cleansed, clean-living warrior who abstained from all swearing and drunkenness and whose whole life was founded on the family.[43]

40 Bosiljka Perše, "Žene ustaše," *Ustaška mladež* (3 August 1941): 4.

41 See, for example, "Ustaška radnica," *Hrvatski narod* (1 June 1942); Hn-H, "Duh Kule Fazlagića," *Osvit* (13 June 1943); "Pismo o borbenim ženama," *Ustaškinja* 4, 1 (10 April 1941): 13; Vika Bišćan, "Hrvatska djevojka u ustaškoj Hrvatskoj," *Ustaša* (10 August 1941): 8 and HURA, "RUŽA," *Ustaška mladež* (7 September 1941): 6.

42 Mira Dugački, "Ženska mladež u novoj Hrvatskoj," *Ustaška mladež* (3 August 1941): 15, 3.

43 Bzik, *Ustaški pogledi*, 23. For a fuller discussion of the idea of the new Ustasha man and woman and its public reception and interpretation among both the Movement's activists and wider population, see Rory Yeomans, "Militant Women, Warrior Men and Revolutionary Personae: The New Ustasha Man and Woman in the Independent State of Croatia, 1941–1945," *Slavonic and East European Review* 83, 3 (2005): 345–87.

Traditional Catholic morality played an important role in the ideology of the Ustasha Movement. Many Ustasha functionaries had been educated at Franciscan schools, Catholic priests, including the Archbishop of Sarajevo, had been among the most fervent supporters of the movement, and seminary students enthusiastically joined the regime's paramilitary death squads following the establishment of the NDH. Catholic intellectuals also played a key role in the formulation of social policy in the state. Similar to many extreme Croatian nationalists, leading figures in the Catholic Church in Croatia had long been concerned about a perceived decline in the Croatian birth-rate. And insofar as traditional Catholic morality became a defining ideological feature of the Ustasha Movement, by the beginning of the 1940s, extreme nationalism had become a defining feature of much of the Croatian Church, in particular among the lower clergy and Franciscan orders. Some, such as Ivo Guberina (1897–1945), went so far as to accuse the Yugoslav regime of deliberating trying to destroy the Croatians, using miscegeny to dilute their Catholic faith in favor of the Serbian Orthodox Church. Looking back at the interwar period in 1943, he recalled that Catholic areas of Croatia had been "systematically colonized" by the Serbs. Worst of all, widespread miscegenation at the expense of the Catholic Church was "skillfully encouraged" by sending young Catholic girls' school teachers to Serb areas, while Serbian officers were dispatched to Catholic areas to conduct marriages according to the matrimonial rites of the Orthodox Church. By 1940 there had been over thirty thousand of these "mixed" marriages, with the result that over the previous generation, the Catholic Church had lost almost two hundred thousand members of its congregation through "apostasy and mixed marriages."[44]

No matter how much the Croatian Catholic Church would have liked to blame "foreigners" for the birth-rate decline, many clerical writers and religious figures highlighted what they perceived to be an equally important factor for the white plague: the deterioration of personal morality among Croatians themselves. For a number of Catholic writers, the widespread use of contraceptives was to blame. In this interpretation, such "religious-moral decadence" could only be reversed by shaming married couples into

44 Ivo Guberina, "La formazzione cattolica della Croazia," Krunoslav Draganović, *Croazia sacra* (Rome: Tipografia angostiniana, 1943), 15–21.

avoiding the use of birth control.[45] The Croatian primate, Alojzije Stepinac (1898–1960), spoke for many Catholic writers when, in a speech to Croatian physicians, he placed the blame for the white plague upon the frequency of abortion, which had become so great it was fast assuming the character of "national suicide."[46] Catholic intellectuals worried that the white plague was "rampaging" in many parts of Croatia and that as a result, whole villages had been emptied, to then be settled by "foreigners," who quickly became the majority. Indeed, so serious had the white plague become that the very survival of the nation was threatened. In consequence, Croatia's once large living space was shrinking and the nation was becoming ever weaker. At the same time, aggressive aliens—in particular, Serbs and Jews—were becoming ever more dominant and powerful. There existed a "great lake" of Serb settlements in the heart of Croatia where in man districts they comprised an absolute majority. Croatian women were failing in their national duty since they were not having babies in sufficient numbers, aided by the "hellish" intentions of "nationally unconscious" physicians. Were this trend to continue, the nation would die out since the nation that allowed unscrupulous physicians to kill its young clearly did not want to live. Croatia could only withstand these pressures if physicians carried out their "medical-patriotic duty" and encouraged a greater level of pro-natalism. For the nation to be truly "protected from suicide," however, the government needed to show no mercy to those "unscrupulous speculating" physicians, above all, those of "foreign" origin, who "for a little money poison the Croatian nation."[47]

Mile Budak (1889–1945), the novelist and Ustasha Minister for Education, Religion and Cults, famously claimed that religion was the basis for the Ustasha Movement, even going so far as to compare its members to medieval Crusaders. This might have been rhetoric but the Movement's official statute emphasized the way in which traditional Catholic values and ideas about moral purification were inextricably linked to the Ustashas' wider program to racially regenerate the nation. The statutes defined the Ustasha regime as a "respectable movement" which would purify the nation from "godlessness, cursing, swearing, drunkenness, immorality, disharmony, lies and profanity," protect "the sanctity of the family" and defend "the honor and repu-

45 J. Majerović, "Borba svećenika protiv bijele kuge u ispovijedaonici," *Katolički list* (19 September 1940), 450–51.

46 B-c, "Za ljepšu budućnost Hrvatske – protiv bijele kuge," *Katolički list* (15 February 1940): 73.

47 B-c, "Za ljepšu budućnost Hrvatske – protiv bijele kuge," 73.

tation of maidens."[48] The regime also launched a number of moral "crusades" against swearing, drunkenness and prostitution, many of which were explicitly framed in terms of national survival.[49] Drunkenness, for example, should be banned, reasoned one young regime supporter, because it shamed the nation and led to the destruction of the family which, in turn, had a negative effect on the birth-rate and demographic prospects of the nation.[50]

Nonetheless, the problem of birth control and abortion consumed more of the regime's energies than any other moral issue and was closest linked by its ideologues to concerns about national survival. A particularly striking example of the way in which moral purification and racial regeneration were associated in the minds of many Ustashas was illustrated in a broadside launched by Ivan Topljak, an Ustasha youth leader in Bjelovar in June 1941. He accused women who had abortions of trying "to destroy Croatia in their womb" and reduce Croatian living space. Such women were criminals "on the soil of Croatia." He was similarly contemptuous of "selfish" couples deciding to only have one child: they should present the new Croatia with as many children as possible since having a small number of children would condemn future Croatians to an ever-decreasing racial living space. His views of medical professionals carrying out abortions were even more vitriolic. Physicians and midwives who carried out abortions "for a few pieces of silver" were "murderers of the Croatian nation," slaughtering babies "with the stroke of a pen." For them, death was the mildest sentence that could be imposed.[51]

## The Ustasha Law on Abortion and its Implementation

According to the propaganda of the Ustasha regime, sexual and moral vice had flourished in interwar, cosmopolitan Yugoslavia. Prostitution was tolerated and the authorities had turned a blind eye to abortion. Like much Usta-

48 "Propisnik o zadaći, ustrojstvu, radu i smjernicama Ustaše – hrvatskog oslobodilačkog pokreta," Matija Belić and Josip Junašević, eds., *Zbornik zakona i naredaba Nezavisne Države Hrvatske* (Zagreb: Državo vijeće, 1942), 1031.

49 K. Gutić, "Borba protiv pijančevanja," *Bilogora* (18 August 1941): 3.

50 Gutić, "Borba protiv pijančevanja," 3.

51 Ivan Topljak, "Stare rane treba liječiti," *Nezavisna Hrvatska* (14 June 1941): 1.

sha propaganda, this was a half-truth at best. In fact, moral strictures had been strong enough in interwar Yugoslavian cities for police to occasionally arrest courting couples kissing in the public parks. The new laws outlawing abortion in the Independent State of Croatia were simply an amendment—albeit in a drastic and radicalized manner—of existing abortion laws. In interwar Yugoslavia abortion had been illegal and women seeking abortions—as well as physicians performing them—were liable to relatively long sentences. Admittedly, the penalties for abortion had been relaxed following a constitutional amendment in January 1929 and, in practice, harsh sanctions appear to have only rarely been carried out. The law was amended in 1933, legalizing abortion for the first time—albeit only in cases where not having an abortion would impact in a socially negative way on the expectant mother. Nevertheless, for most women abortion remained illegal, difficult to obtain and, even in cases where it was permitted, very tightly controlled.[52]

The abortion laws introduced by the Ustasha regime, in contrast, were frighteningly draconian. According to the first such abortion law published on 10 June 1941, the Legal Order for the Outlawing and Punishment of Artificial Abortions and the Termination of Pregnancies, any woman even attempting to end her pregnancy by artificial means could be sentenced to between five and ten years in prison, and a woman who had been successful in ending her pregnancy by such means could be sentenced to life imprisonment. However, the punishments that desperate women faced were relatively mild in comparison to those that physicians and midwives performing the operations could expect. For medical professionals found guilty of performing an abortion, the sentence was life imprisonment or, more usually, death. Even the descendants of the convicted physician or midwife could be punished, with all property or personal wealth being confiscated by the state. This money would then be used to provide social care for expectant mothers.[53]

The only exception to this law was in cases where the life of the mother was thought to be endangered should the pregnancy continue and if there

52 Thomas A. Emmert, "*Ženski Pokret:* The Feminist Movement in Serbia in the 1920s," in Sabrina P. Ramet, ed., *Gender Politics in the Western Balkans: Women and Society in Yugoslavia and the Yugoslav Successor States* (Pennsylvania: Pennsylvania State University Press, 1999), 37.

53 "Zakonska odredba o zabrani i kažnajavanju uzrokovanog pometnuća i o prekidanju trudnoće," *Narodne novine* (10 June 1941): 1.

were no other medical alternatives. In such cases, the physician was not permitted to act alone and any decision to perform an abortion would have to be taken in a public hospital. Such a procedure also needed authorization by a medical committee comprising the director of the hospital, an expert in the field of the illness or disease that the expectant mother was suffering from as well as the gynecologist or surgeon performing the operation. In the unlikely event that an operation was deemed necessary, details of the abortion had to be made known to the Ministry of Health within twenty-four hours. This was to take the form of a lengthy and detailed report, providing extensive information about the circumstances of both the operation and the patient. Physicians were expected to provide details of genital examinations carried out prior to the abortion, commenting on the general condition of the mother, what specific dangers existed to the life and health of the pregnant woman should the operation not be performed and, finally, the reasons why these health risks could not be addressed in any other way. Failure to abide by these strictures carried a sentence of between ten and twenty years in prison.[54] Abortions could only be carried out in public hospitals; those carried out in other locations were to be considered illegal and subject to the full weight of the law. A subsequent law confirmed that only in extreme cases, where the life of the mother was immediately threatened, could an abortion be performed by a gynecologist without waiting for the committee's final decision.[55]

Clearly, the regime did not want women dying as a result of being denied the right to a necessary abortion, for while an abortion represented the death of one future Croatian life, the death of a young, healthy woman who had been refused an abortion could represent the loss of many future Croatians. On the other hand, the ideology of the regime made it imperative to make abortion as difficult as possible. In the early months of the new state, it appears public hospitals did obey the laws diligently. Certainly, there were a few reports in the press of these laws being flouted. However, there is also evidence that some women and gynecologists took advantage of the exemptions provided by the law to terminate unwanted pregnan-

54 "Zakonska odredba o zabrani i kažnajavanju uzrokovanog pometnuća," 1.

55 "Zakonska odredba o promjeni i nadopuni zakonske odredbe o zabrani i kažnajavanju uzrokovanog pometnuća i o prekidanju trudnoće broj LXXXIII-149-Z. p.-1941. od 10. lipnja 1941," *Narodne novine* (28 August 1941): 2.

cies. To be sure, official hospital records told a different story, obediently reporting that abortions had been carried out for clear medical reasons, reluctantly agreed to by both expectant mother and gynecologist, usually in cases where the viability of the pregnancy itself was also in some doubt. In some instances this must have been an accurate reflection of what had actually happened. In addition, not every appointment for a termination resulted in the operation being carried out, although whether this decision was based on genuine pediatric considerations or the hospital's fear of the possible ramifications of a termination is hard to say. That hospital authorities were required to produce meticulous reports even in cases where a woman had not gone through with the operation illustrated both the seriousness with which the regime took its anti-abortion laws and the nervous atmosphere of the times: livelihoods and lives were at stake. Nevertheless, this was what was recorded in the case of Slava Pavičić, a twenty-seven-year-old worker's wife, who entered Sisak state hospital on 13 September 1941. Feeling ill throughout her pregnancy, she had begun to bleed heavily the day before being admitted. Following an examination of the patient, the gynecologist concluded that there were no immediate grounds for terminating the pregnancy: there was a possibility that the fetus could be saved and the patient could still give birth without complications.[56]

A similar decision was reached in the case of Zlata Karaić, a married housewife admitted to the state hospital in Stolac in Bosnia on 4 June 1941 after beginning to bleed heavily and suffering from severe stomach pains. Here, the anxiety of the hospital might well have played a part. As the hospital director nervously pointed out, the Ministry of Health had sent them no paperwork, meaning that any termination would technically be in breach of abortion laws.[57] Even in cases where there was ostensibly a clear medical need for an abortion, the facts were not as clear cut as they might have appeared. Marica Turina, for instance, who was admitted to the Institute for Tuberculosis suffering from an infection of the lungs, was quickly given an abortion on the grounds that carrying her baby to term would greatly endanger the life of this mother of three children. While this might have

56 Hrvatski Državni Arhiv (HDA), Nezavisna Država Hrvatska (NDH), Ured Ministarstva (UM), fond Ministarstvo zdravlje, "Zapisnik," 14 September 1941, br. 226/2978/41, s.1.

57 HDA, NDH, UM, fond Ministarstva zdravlje, "Zapisnik," 4 August 1941, br. 226/818/1941, s. 1-2.

been true, given that campaigners for the legalization of abortion frequently cited mothers suffering from tuberculosis in support of their cause—and in light of the fact that she already had three children and was pregnant again in a time of war and dislocation—Turina's case raises the possibility that some gynecologists and abortion committees connived in a woman's willful decision to terminate a pregnancy. At any rate, suspiciously large numbers of pregnant women were admitted to hospitals and specialist institutes in the Independent State of Croatia between 1941 and 1945, suffering from tuberculosis and in need of terminations.[58]

The problems facing Lenka Zeisler, also a patient at the Institute for Tuberculosis, were somewhat more complicated. The seriousness of her tuberculosis and pleurisy were not in doubt, nor was the likelihood that continuing the pregnancy would make her condition graver since surgeons would be prevented from carrying out invasive therapies. At the same time, the fact that she was a nineteen-year-old, unmarried shop assistant raised serious moral dilemmas for both the Institute and the Ministry of Health itself. Officially, the Ministry was meant to uphold the regime's belief in the sanctity of the Croatian fetus. But did the state want to be seen as giving license to immoral behavior, such as conceiving children out of wedlock? It appears in fact that sometimes they did. Indeed, the fact that some Ustasha activists and natal experts seriously advocated the legal recognition of illegitimate children suggested that the regime's rejection of abortion was dictated as much by a desire to raise the birth-rate as by ideas of moral rebirth. In the struggle between demography and morality, demography would take precedence.[59]

## Propaganda, Popular Culture and the Consequences of Abortion

In order to impress the disastrous consequences of having an abortion upon women, various forms of propaganda were used, ranging from polemical articles in the Movement's journals to poems and short stories. This partic-

58 HDA, NDH, UM, fond Ministarstva zdravlje, "Zapisnik," undated but almost certainly 1941, br. 226/919/1941, s. 1-2.

59 HDA, NDH, UM, fond Ministarstva zdravlje, "Zapisnik," 11 August 1941, br. 226/908/1941, s. 1-2.

ularly appears to have been the case from 1942 onwards, when it is likely that initial adherence to the law on the part of gynecologists and women was disintegrating. That reports of abortion law violations in the popular press were infrequent suggests very few cases were ever actually prosecuted. It is therefore likely that, as time went on, legal sanctions lost their deterrent power. Evidence that abortion laws were increasingly being ignored is suggested by an article written in 1943 by the medical expert Eduard Miloslavić (1884–1952) warning women of the terrible fate awaiting them as a result of "that unnatural, harmful and destructive phenomenon in our homeland that has a damaging influence on the organism of the nation and leads to the unavoidable destruction of the population."[60] For Miloslavić, procreation was simply part of women's nature, and those without the urge to breed were unnatural and doomed to sad, empty lives. Women, he wrote, had a natural instinct for motherhood; anything else was "degenerate thinking." Young couples without children or who sought an abortion because the pregnancy had come at an inopportune moment were doomed to emotional unhappiness—the wife above all. Worse, this first abortion could lead quite easily to any number of medical complications, including internal injuries to the sexual organs of the young mother, inducing changes of "fatal significance." Not least, it could mean that the young mother would be infertile and incapable of bearing children in future. Since the soul and body of the "female organism" was programmed to give birth and bring up children as well as to be the support for her husband, what would become of such an infertile woman? Disaster surely lay ahead for her.[61]

Miloslavić claimed that women submitting to the surgeon's knife were at risk of destroying their marriages too. Women who had consented to abortions ended up as partners in marriages in which there was "unhappiness, quarrels, mutual distrust and eventually the break up of the marriage." There was also a corresponding effect on the physical condition of the couple. Both partners developed a depressive psychological state of mind, and in the woman's case, "frigidity and hysterical complexes" ensued. The development of sexual hysteria had been noted in some unmarried women, while others had "lost all inhibitions of normal sexual lives and sank into

60 Prof. Dr. E.L. Miloslavić, "Kobno spriečavanje poroda," *Ustaški godišnjak* 2 (1943): 222.

61 Miloslavić, "Kobno spriečavanje poroda," 228.

prostitution." In short, with abortion, the woman "degenerated both mentally and physically."[62] These warnings were accompanied by color diagrams, illustrating the effect of a botched abortion on both the healthy fetus—which was not surprisingly depicted as a fully-formed baby—and the internal organs, including the reproductive organs of the unfortunate young mother. Other illustrations showed the physician's scalpel missing the fetus completely and piercing the wall of the mother's uterus or causing massive internal hemorrhaging. Diagrams also depicted the grisly sight of a fetus following an attempted abortion, with the head of one defenseless fetus decapitated from its tiny body and another tangled as it was ripped from the uterus. Such was the butchery of the abortionist. Who could possibly defend such radical evil?

Abortion also meant the destruction of the nation and the extinction of the race as far as Miloslavić was concerned. Its growing acceptance, even in the village, put the whole nation at risk. In the past, he wrote, the Catholic faith and morals had protected the peasants from such "criminal acts and suicidal intentions." But now "pseudo-sociologists with their eugenic propaganda and ideas about the need for the rationalization of sexual life and birth-rates" and "physician-abortionists with their social indications in the profit of their wallets" were spreading among the people "deleterious ideas" about the necessity of artificially removing fetuses or preventing pregnancies. The dismissive reference to eugenics reflected the general rejection of "negative" eugenic methods by the Ustasha, as these would have the effect of reducing the number of Croatians born every year and, ultimately, the size of the Croatian nation. The reference to the "rationalization" of sexual life, meanwhile, was surely an allusion to the work of Štampar, an early victim of the regime.[63]

What it did not express, however, was an outright rejection of eugenic principles. The Ustasha regime simply invented its own form of eugenics. The social and racial ideologues of the movement did not object to the idea of a better race, one which would be healthier and stronger and fitter for breeding. Miloslavić, like Štampar before him, argued for laws in Croatia stipulating that, before getting married, a couple should provide medical

62 Miloslavić, "Kobno spriečavanje poroda," 229.
63 Miloslavić, "Kobno spriečavanje poroda," 230–32.

certificates attesting to their physical and mental health. This represented "practical" eugenics to Miloslavić, as opposed to the "destructive" tendencies of more "perfectionist" eugenics. His version of eugenic biopolitics advocated that young men and women ready to enter into marriage needed to avoid all "poisonous pleasures" such as alcohol and cigarettes that might impact on their sexual organs—and thus, their ability to procreate—lest this contribute to the development of unhealthy and abnormal descendants. They should also endeavor to protect themselves, through sexual abstinence and purity, from the risk of venereal diseases, specifically, the risk of infection by syphilis since this could adversely affect future generations of their family or even precipitate mental degeneration among their own children. To Miloslavić it was the role of eugenicists to teach the younger generation to look after their bodies in order to increase their fertility and produce "future healthy biological generations."[64]

Far from opposing eugenic beliefs, Miloslavić argued that the Catholic faith embraced them. Naturally, Catholicism could not easily base its faith upon the "superficial assumptions and the dangerous experiments of supposed scientists" but would resolutely defend itself against all attacks against its principles and morals. "Positive" eugenics, which would increase the health and size of the Croatian population while upholding Catholic morality, was to be especially encouraged.[65] "The eugenics of the individual and eugenics in the family are the eugenics of the nation," he proclaimed. Abortion, however, could play no part in Croatian eugenics. On the contrary, physicians and sociologists who thought it could were the greatest enemies of the nation, criminals whose work should be publicly stigmatized and "energetically prevented."[66] And reflecting his concern that abortion laws were no longer being strictly obeyed, at the end of his article he implored readers to join him in his struggle against "the murderers of human fetuses" in order to prevent the commission of their destructive work among the people. "We must direct all our energy," he concluded, "towards these criminals and then their destructive work—the crime of

64 Miloslavić, "Kobno spriečavanje poroda," 232.

65 Many other medical experts and educationalists also believed that eugenic principles, including sexual education, could be taught within the framework of Catholic morality. See, for example, Josip Hajda, "Mogućnosti bolje metode u nastave biološke skupine predmeta," *Napredak* 85 (1944): 139–49.

66 Miloslavić, "Kobno spriečavanje poroda," 233.

abortion—will disappear. We should not talk anymore about this crime—we should seek out the criminals!"[67]

Relatively few people in the Movement bought the theoretical regime-conform journals in which articles like Miloslavić's were written. It was thus likely to have been read only by fellow intellectuals already sharing his opinion. Fortunately for the anti-abortion crusade, the state press, able to reach a far wider readership, aggressively promoted the new abortion laws, linking abortion with national betrayal and the injustices of liberal capitalism, using militant language ordinary Croats could easily understand. If abortion was enemy number one of the Croatian nation, then, according to one Ustasha newspaper in Banja Luka, the promotion of abortion by the Yugoslav regime to bring the Croatian nation to the brink of extinction meant that any woman seeking an abortion or any Croatian involved in the performing of an abortion was also an enemy of the state.[68] This newspaper welcomed the introduction of pro-natalist policies by the Ustasha regime, especially the encouragement of large families that the abortion law implied, not just for reasons of demography but also social justice. Gone were the bad old days when a landlord could deny a roof over the head of a Croatian family with several children because he knew that in the NDH, doing so could end in a sentence to hard labor. With additional payments for the birth of each new child, families with a large number of children now knew that they could continue reproducing without fear of being labeled "social parasites" and without anxiety about what the future might bring.[69]

In fact, the situation had now reversed itself in the NDH. In the traumatic Yugoslav era, the families of poor but honest Croatian workers and peasants were forced to be ashamed of their children and had "fallen under the influence of various Jewish physicians and their assistants who had, using their medical licenses, destroyed the progeny and the living strength of the Croatian nation." This had come to an end in 1941. Irrespective of whether they lived in the city or the village, those "who from recalcitrance still choose not to take part in the renewal of Croatia so that the homeland

67 Miloslavić, "Kobno spriečavanje poroda," 233.
68 "Nestat će bijela kuga," 1.
69 "Nestat će bijela kuga," 1.

receives a healthy and needed progeny" would have to be "marked out by the force of the law as unproductive members of society." This would apply especially not only to those "modern women who refuse to bear children so they can remain 'eternally young and beautiful'" but also to "women from the rich and fertile plains of Slavonia," where the white plague had spread "like nowhere else" and as a consequence, the land had fallen into foreign hands. In Slavonia, matters had degenerated so far that there were many villages without a single native Croatian. The Croatian presence had completely disappeared while the villages had been colonized by foreigners. The situation had come to such a sorry state, the newspaper concluded, that it recommended altering the inheritance laws in order to recognize illegitimate children, if such a measure could raise the number of live births.[70]

The regime also appropriated popular culture in its struggle against abortion. Numerous poems and stories were written about the disastrous consequences of abortion in an attempt to shape public opinion. In his poem "A Woman, A Dove and Abortion," the young poet Đuro Šnajder (1919–1993) recounted the story of a woman who had undergone an abortion, vividly evoking the devastating psychological and physical effects this had had on her, her uterus irreparably damaged by the butchery of the abortionist's knife. Šnajder contrasted the bloodstained violence of abortion with the pristine tranquil flight of the dove: the "murmur of his unfurling wings" with the "death from my wings." The carnage of the abortionist was described in gruesome detail, the "bloody crude rags," the unborn child waiting for the "silver needle that will drown it," the hideous "milky hands" of the murdering midwife, the "sweaty white sheets" and carbolic acid, the womb boiling over "in a volcano of black clotted blood" and searing pain.[71]

The poem was published in 1943 in the cultural journal for young Ustasha intellectuals, presumably to warn young people about the dangers of abortion. But given its select readership, this was another example of preaching to the converted. Moreover, the fact that it had to be published at all—and in a journal for revolutionary active youth—indicated the limits of the effectiveness of the anti-abortion campaign. Popular literature provided another means of channeling the anti-abortion message of

70 "Nestat će bijela kuga," 1.

71 Đuro Šnajder, "Žena, golub i abortus," *Plava revija* 3, 1–2 (1943): 43.

the regime. Mara Švel-Gamiršek's 1942 short story "Ranka's Mistake", for instance, offered a cautionary tale about the physical and emotional consequences for a young female medical student who has an illegal abortion after a failed love affair with a dissolute young man. However, given the mauling this collection of short stories received from cultural critics close to the regime, it is again not clear how many people would have read Švel-Gamiršek's work, never mind considered its message.[72] A far more effective method for promoting an anti-abortion and pro-natalist message, the regime believed, was through a week dedicated to the eulogization of motherhood and maternity.

## Celebrating the "Week of the Croatian Mother and Child"

On 31 May 1943, the second annual Week of the Croatian Mother and Child was launched. In the presence of Ante Pavelić's wife and dozens of Croatian mothers assembled in the Croatian State Theater, the commander of the Ustasha Women's Vine Irina Javor gave an opening speech declaring that the purpose of that week was to "protect the Croatian mother and child" by offering all possible state assistance to both. The idea of a week devoted to Croat mothers and children had been motivated by a desire to follow the instructions of the Poglavnik, who had ordered the social protection of motherhood as one of the most pressing tasks of the Ustasha state. Immediately following liberation, she explained, the Poglavnik had ordered the construction of homes for workers throughout Croatia and provided maternity products for even the poorest mothers. Javor's comments were roundly applauded and were followed by a public concert which, in addition to opera arias and poems on the theme of motherhood, featured a fertility dance performed by the Croatian prima ballerina Ana Roje (1909–1991) intended to reflect the central theme of the Croatian mother and child.[73]

72 Šnajder, "Žena, golub i abortus," 43; Mara Švel-Gamiršek, "Rankin grijeh," *Potreti nepoznatnih žena* (Zagreb: Matica Hrvatska, 1942), 222–43. They were derided for being outdated, sentimental and for contributing nothing to the establishment of a national revolutionary literature. See, for example, Stanko Gasparović, "Portreti 'slavonskih' žena u pripovijestima Mare Švel," *Hrvatska revija*, 15, 11 (1942): 621–24.

73 "Svečanom priredbom u kazalište otvoren je hrvatski tjedan majke i djeteta," *Nova Hrvatska* (1 June 1943): 7.

Ostensibly, the "Week of the Croatian Mother and Child" celebrated the contribution mothers and their offspring were making to the new Croatian state. Equally, it provided the regime with an opportunity to advance pronatalist ideas for the nation and to enlighten ordinary citizens. The week-long program of enlightenment and education took many forms. Adjutants of the Ustasha Vine Organizations, for example, visited local schools to both celebrate the Croatian mother and educate school children and teachers on her national importance. Children were also asked to write a composition on the Croatian mother, with the winning essay receiving a prize.[74] Entertainment formed an important part of the festivities, with one of the largest concerts in aid of the "Week of the Croatian Mother and Child" given on the 4 June each year at the premises of the Higher Teachers' School in Zagreb. This featured actors reciting traditional and contemporary poems extolling the virtues of the Croatian mother and opera singers performing arias in which figures from Croatia's glorious history extolled motherhood. Speeches by state officials followed. In 1943 the keynote speech was given by Vilim Peroš (1906–1945), editor of the newspaper *Nova Hrvatska* [New Croatia]. Entitled "The Mother in Ustasha Croatia," the speech chronicled the sacrifices that mothers had made for the Croatian homeland and the "boundless" love shown to their children. Motherly love, Peroš stated, was mystical and could not really be explained; it was natural to all women. He also restated the movement's belief in the role of fertility to regenerate the nation:

> The Ustasha state sees in the family the foundation of all national and state renewal. The Independent State of Croatia, under the leadership of the Poglavnik, aims to use all methods to facilitate new marriages so that more Croatian families can be created and so that the power of the whole Croatian nation can become stronger, because the power of the family and the state are in a reciprocal relationship.[75]

The Week involved a special series of radio lectures devoted to the subject of motherhood and child rearing by leading theoreticians, experts and regime ideologues offering their advice and opinion.[76] Indeed, an impor-

74 "Proslava tjedne hrvatske majke i djeteta u školama," *Nova Hrvatska* (4 June 1943): 7.

75 "Liepa svečanost u čast majke u visokoj pedagožkoj školi," *Nova Hrvatska* (4 June 1943): 8

76 "U nedjelju 30. svibanja počinje hrvatski tjedan majke i djeteta svečanom akademijom u Hrvatskom državnom kazalištu," *Nova Hrvatska* (28 May 1943): 7.

tant aspect of the Week was the opportunity it offered the regime to instill healthy motherly practices in Croatian women. Some of these proposals were the kind of common-sense advice that would be useful for any mother. For instance, physicians and health experts advised mothers to breast feed their babies instead of bottle-feeding them and leaflets published by the Ministry of Health warned women of the dangers they could be placing their babies in were they not fed with "mother's milk." These leaflets also warned mothers that in the ensuing summer months they should make sure that their homes were adequately ventilated and that the baby got sufficient fresh air. Mothers were also advised to keep their homes clean and to dispose of leftover food in order to prevent the risk of contamination and disease.[77]

However, the week also gave regime functionaries the chance to publicize the transformation the regime had made in the lives of mothers. In Banja Luka, for example, the director of the Institute of Hygiene Dr. Ivan Grujić presented a lecture on the sanitary health of mothers and children, pointing to the large number of infant and children's clinics, kitchens and homes that the regime had constructed to improve the health of the youngest members of society. This was followed by the presentation of a film describing the activities of the School of Public Health in Zagreb, whose founder and director had been Andrija Štampar.[78] The Mayor of Zagreb, Ivan Werner, highlighted the wide range of state facilities available for mothers and children. Listing the existing facilities in the capital for the protection of the mother and child, Werner mentioned that Zagreb now hosted children's shelters, school physicians, dentists and ambulances in addition to milk kitchens, ensuring that all school children received at least some milk every day.[79]

Despite these advances, state officials stressed the work that still remained to be done in the struggle against abortion and infant mortality. According to Luka Bežić, an official in the Department for the Protection of the Mother and Child, while there had been some improvements, especially with the introduction of maternity classes for mothers and young

77 "Savjeti mladim majkama za zdravlje djece," *Hrvatski narod* (3 June 1943): 4.

78 "Za zdravlje naše mladeži," *Nova Hrvatska* (4 June 1943): 5.

79 "Zaključen 'Hrvatski tjedan majke i djeteta,'" *Hrvatski narod* (9 June 1943): 6.

girls, notions of health and hygiene were insufficiently understood in the village. In order to raise the collective standard of health and hygiene it would be necessary to ensure that propaganda about "infant culture" was spread effectively, and that every organization aided in the "growth of pregnancies in healthy families at the local level." He argued that it was particularly important to concentrate on illegitimate children, research showing that they were the most threatened and their level of disease being the greatest. Moreover, from the health and social point of view, they were equal to children born in wedlock.[80]

The Ustasha regime liked to believe its eugenic and racial hygiene policies operated at a higher moral and spiritual level than the kind of eugenics and racial hygiene practiced in liberal societies. The NDH, it frequently asserted, did not employ the destruction of human life, namely abortion and euthanasia, in search of racial betterment. In their Independent State of Croatia, all citizens, irrespective of their physical and mental conditions, would have a place. In a very limited sense this was correct, but on the whole it was self-serving fiction. Respect for human life certainly did not extend to children the regime considered to be racially inferior—and thousands of Serb, Jewish and Gypsy children died either at the hands of its death squads or in concentration camps from disease, malnutrition, deliberate starvation, poisoning and outright murder. In what amounted to a program of euthanasia and the destruction of newborn life, the cruel sadistic manner in which Serb, Jewish and Gypsy children were treated stood in stark contrast to the officially idealized image of the lives of mothers and babies in the Ustasha state portrayed during the "Week of the Croatian Mother and Child."

It is true that the regime established numerous schools for blind, deaf and disabled children as well as for orphans. By 1943, twenty such schools had been established and trades schools had also been founded so that children judged to be vulnerable could receive some training to survive in the outside world. But there was a sharp distinction in the way these groups of children were treated. Able-bodied orphaned children—many of them from war-torn parts of the country—were placed with "healthy and pro-

80 "Zdravstvena zaštita dojenčadi i male djece," *Hrvatski narod* (9 June 1943): 7.

gressive families" in a process the Ustashas called "colonization." However, the regime felt it was impossible to place children with incurable illnesses and "defects" such as blindness or severe disabilities with families. Such children were instead placed in institutional homes. They might learn "to work and again become useful members of the human community" but they were also effectively segregated from the rest of Croatian society.[81]

The protection of vulnerable Croatian children was hardly as altruistic as it appeared since most, if not all, of the initiatives rested on the premise that the birth-rate needed to be increased and that the women's role could be reduced to the production of children. When during the massive Ustasha youth rally in Zagreb in June 1942, female members of the Ustasha youth had marched in front of the Poglavnik waving dolls, carrying kitchen utensils and pennants, pushing baby carriages and carrying placards stating: "We want to be little mothers!," their demonstration was both a manifestation of the fanaticism of which the movement's youthful activists were capable and a telling reflection of the regime's view of women.[82] Although the NDH's attitude towards female activists was complex and often contradictory, officials never ceased reminding women that their primary role in life was to have babies, babies ultimately belonging to the nation.[83] Thus, the care homes established in summer 1941 to look after vulnerable teenage female workers and apprentices in Zagreb attempted to improve the "subordinate" and "miserable" conditions of working-class girls, but it also sought to transform these abandoned and "socially marginalized" young girls into "honorable young women" and Croatian mothers of the future. Motherhood in the service of the demographic struggle, as numerous newspaper articles made clear, was the ultimate destiny of these girls. Clearly, Croatia needed "healthy" female apprentices and "tough and resilient" female workers, but it needed healthy mothers who would provide the next "healthy generation" more urgently.[84]

81 See for example, "Smještanje sirotčadi kod zdravih i naprednih obitelji," *Nova Hrvatska* (5 June 1943): 5; "Ministarstvo udružbe," in Ivo Prpić, ed., *Spomen-knjiga prve obljetnice Nezavisne Države Hrvatske 10.4.41-10.4.42* (Zagreb: Izvještajni i promičbeni ured, 1942), 336–8. However, it is not clear how many of these healthy but vulnerable children were Croatian since orphaned Serb children who had been rescued from Ustasha concentration camps were also sometimes adopted by Croatian families.

82 "Velebni mimohod ustaške mladeži pred Poglavnikom," *Hrvatski narod* (9 June 1942): 1.

83 "Svaka Hrvatica mora biti u prvom redu majka, prava hrvatska majka," *Nova Hrvatska* (4 June 1943): 8.

84 "Ustaška skrb za siromašne naučnice i radnice," *Hrvatski narod* (25 June 1942): 7.

The frequent articles heralding the dramatic increases in birth-rates since April 1941 among Croatians due to improved diets and living conditions that the NDH had provided for pregnant women also made this point eloquently.[85] If any doubt remained concerning the regime's determination to increase the birth-rate, then the radio broadcast given by Josip Rasuhin (1882–1973), an assistant in Štampar's Institute for Social Medicine prior to becoming an official in the Ministry of Health, stated the intentions of the regime unambiguously: in his talk of 6 June 1943 entitled "Our Demographic Politics" he declared that the key to the survival of the Croatian nation was for each married couple to have "at least" three children.[86] Nor could there be any doubt about the designated role of women. When the Poglavnik met a delegation of peasant women from the regions being honored in the "Week of the Croatian Mother and Child" for the large number of children they had given birth to, the leader of the women's delegation, Sofija Brajša, a mother of twelve, declared that the primary and "natural" duty of a woman was procreation.[87]

Not all messages were so blatantly stated. Many articles that called for women to return to the home in the demographic interests of the nation synthesized traditional Catholic interpretations of the mother's role with the modernizing language of eugenics. An article of May 1943, for example, portrayed the nation in explicitly biological terms, reminding its readers that a healthy state required a healthy and happy family since the family was "the most intimate cell" of the national organism. From this cell the individual emerged, and therefore this individual belonged to the state. However, family life would be useless unless the mother and father were capable of supporting their children and "economic and societal conditions for the fulfillment of the national, economic and social tasks of the family" existed. This could only be achieved if the state ensured all prospective marital couples were physically and mentally healthy. Ultimately, though, the truly pronatalist state saw its role in ensuring the material prosperity of the couple, in ensuring that the man earned enough so that the wife need not leave the home to pursue her own "selfish" interests. Without the constant presence

85 See, for example, "Porast porođaja u Osijeku," *Hrvatski list* (22 May 1942): 8 and "Nagrađene majke brojne djece," *Hrvatski narod* (10 June 1942): 7.

86 Josip Rasuhin, "Hrvatski narod ne će nikada izumrijeti!" *Hrvatski narod* (9 June 1942): 5.

87 "Hrvatska država bit će sigurno državu hrvatskih majki i hrvatske djece," 5.

of the mother, neither the family nor nation could prosper. In the building of a harmonious home, the nation would simultaneously be regenerated.[88]

The way in which the regime sought to fuse modern social ideas with the language and ideology of national regeneration can be illustrated in a radio broadcast to mark the first "Week of the Croatian Mother and Child" in June 1942. In his speech on "Social Provisions for Mothers and Children," Dr. Kamilo Bresler, a high-ranking official in the Department for Family Care and Children, argued that the regime's health and social protection policies were now more needed than ever to compensate for the neglect that Croatian children had endured in interwar Yugoslavia—not to mention the fact that Croatia had struggled to protect its "national individuality and bare existence" while other countries had been developing social welfare policies. Conditions for Croatian children in the interwar period had been terrible, he claimed. An atmosphere of family collapse, declining birth-rates and increasing numbers of abortions had forced most countries to introduce a range of measures to protect the mother and child. In contrast, in Croatia between 1929 and 1939, thousands of children had died each year due to disease, unsanitary conditions and malnutrition, even in Zagreb. The figures he cited were shocking: over three-quarters of children with diseases associated with poverty such as tuberculosis died in the first three years of life, one quarter of children were dying before the age of eight and 30% of Croatian children were dead before their eighteenth birthday. This stood in stark contrast to the situation in other modern European states such as Belgium and Germany with their developed social welfare systems. According to Bresler, nearly sixty out of every one hundred Croatian deaths were those of children. "Every second grave that we dug was a child's grave," he recalled solemnly.[89]

Given that "a nation without children is a nation without a future," the first principle of the new "national social conscience" and spirit should aim at the protection of the Croatian mother and child from conception to adolescence. The variety of modern social welfare initiatives that the Independent State of Croatia had put in place—from maternity clinics, milk sta-

88 "Obitelj je najbitniji činbenik u preporodu naroda," *Nova Hrvatska* (28 May 1943): 7.

89 These figures are unlikely to be right. Although Mikić's data only extends to 1936, even for infants below the age of one, the highest regional mortality figure he gives is 20%. Moreover, the last census taken in 1939 showed that the population was actually increasing.

tions, school kitchens and schools for children with disabilities to summer camps, libraries and employment laws to protect apprentices—was ultimately designed to assist the mother and child. They were proof of the "decisive will and clear conscience of Ustasha Croatia" and the Poglavnik to protect the mother and child as part of a wider program for "the social renewal of the Croatian nation." The next challenge for the new state would be to bring these social reforms to a greater number of mothers and children. The responsibility for diligent mothers as "the source of future generations" and the education of children as "the carriers of the national future" needed to be shared equally between the family and the state, which, "in the drive for survival and the future of the nation," had to develop a universal system of social provision for the mother and child.[90]

These sentiments were echoed a year later by Josip Rasuhin in his own radio broadcast. The Croatian nation would not disappear "neither morally nor biologically," he asserted, because the creation of a social welfare system based on the very principles Bresler had outlined would "protect the national organism from degeneration."[91] Despite the language of national regeneration, these ideas were in many ways simply a recapitulation of the progressive eugenic principles developed by a liberal Yugoslav eugenicist for a secular Yugoslav state. This was an irony predictably and entirely lost on a regime under whose rule a young activist like Josip Blažek could combine his social welfare advocacy on behalf of the state's university students and high school youth with service in one of its most ruthless death squads.

## Conclusion

Eugenics in the Independent State of Croatia was unique in a number of ways and for a variety of reasons. Not only was the Ustashas' Independent State of Croatia the first Nazi satellite state to inaugurate a campaign

90 Kamilo Bresler, "Društvovna zaštita ustaške majke i djeteta," *Hrvatski narod* (4 June 1942): 6. Bresler was not a supporter of the Ustasha regime and was dismissed from his position in 1943. With Dijana Budisavljević and a group of Croatian Red Cross nurses, many of them active in illegal Communist cells, Bresler was instrumental in the rescue of thousands of Serb children incarcerated in the Jasenovac-Stara Gradiška camp complex. After the war he became a respected and influential pediatric expert in the People's Republic of Croatia.

91 Rasuhin, "Hrvatski narod ne će nikada izumrijeti!" *Hrvatski narod* (9 June 1942): 5.

of mass murder during World War II; it was possibly the only Nazi satellite state in which official hostility to negative eugenics, especially abortion and euthanasia, was directly contradicted by the zeal with which such ideas were embraced in the service of racial purification and mass murder. At the same time as deaf, dumb, blind and mentally disabled children were being given the opportunity for education and training as worthy members of Croatian society, hundreds of thousands of citizens, many of them children, were being indiscriminately murdered by the Ustasha for the very same purpose: the survival of the nation.

Both regime functionaries and the NDH's intellectual supporters feared that the nation faced demographic extinction. To remedy this, the size of the Croatian population needed to be increased. This was to be achieved in two ways. The numbers of fast-breeding "aliens," particularly Serbs and Jews, would have to be permanently and drastically reduced. Through a program of mass murder, forced assimilation and expulsion over a generation they would cease to exist in Croatia. At the same time, the birth-rate of Croatians would have to dramatically increase. Modern notions of female independence would need to be replaced by a cult of motherhood and a return to the "natural" procreative female role. Abortion, along with all other sexual and moral vices infringing on a high national birth-rate would need to be outlawed and severely punished. Since the regime associated these vices with interwar Yugoslavia, this law would serve a dual function in both the racial regeneration and moral purification of the Croatian nation.

This campaign of pro-natalism failed. While in some regions the national birth-rate did increase, with men away at the front and inspired by the Ustashas' own revolutionary rhetoric of the militant female warrior, women entered the workforce in unprecedented numbers, thus placing their supposedly natural role in abeyance. Moreover, the terror and brutality of the campaign of mass murder orchestrated against communities deemed to be racially inferior increasingly turned many ordinary Croatians against the regime. While there was some intellectual and popular support for the pro-natalist initiatives of the Ustasha regime—above all from Catholic intellectuals, clerical organizations and the Church hierarchy itself—among ordinary citizens and women in particular it remained marginal. Relatively few Croatians were prepared to be mobilized behind pro-natalist fantasies and racial utopias. Ultimately, many Croatians identified the regime and its eugenic ideas for artificially raising the birth-rate rather than alien populations or a declining birth-rate to be the true disease ailing the nation. Racial persecution, not racial contamination and enforced motherhood, not the right to abortion, were Croatia's real enemies. For most Croatians, the black plague of the Ustasha regime, rather than the white plague of low birth-rates, was the greatest threat to the survival of the nation.[92]

92 Although there are no concrete statistics on the number of women entering the workforce between 1941 and 1945, even regime propaganda admitted to the increasing feminization of the professions in the absence of men. See, for example, *Kroz ustašku Hrvatsku: progovorio je podravski seljak i radnik,* (Zagreb: Tipografija, 1942), esp. 26–28.

# PART IV

# NEW RESEARCH AGENDAS

# REMAPPING THE HISTORIOGRAPHY OF MODERNIZATION AND STATE-BUILDING IN SOUTHEASTERN EUROPE THROUGH HEALTH, HYGIENE AND EUGENICS

*Maria Bucur*

## Current Trends in the Historiography of Southeastern Europe

The historiography of Southeastern Europe in the modern period is ripe for reconsidering the paradigm of nation–building from increasingly nuanced and challenging vantage points. No longer are historians satisfied to read the personal papers of the "great men" of letters and politics or to look at legislative changes and diplomatic documents.[1] From scholars interested in the modernization of the military to those looking at socio-political change on the ground, historians are developing richer analyses of the transformation of the Balkans from a relatively vapid (or, by contrast, wildly mythologized) area on the map of modern Europe into a vibrant locale of change, modernization and contestation.[2]

---

1 Some important recent scholarship that touches at least in part on Southeastern Europe includes: Pieter Judson, *Guardians of the Nation: Activists on the Language Frontiers of Imperial Austria* (Cambridge, MA: Harvard University Press, 2007); Maria Bucur and Nancy Wingfield, eds., *Staging the Past: The Politics of Commemoration in Habsburg Central Europe, 1848 to the Present* (West Lafayette: Purdue University Press, 2001); Laurence Cole, ed., *Different Paths to the Nation: Regional and National Identities in Central Europe and Italy, 1830–70* (Basingstoke and New York: Palgrave Macmillan, 2007); Mary Neuburger, *The Orient Within: Muslim Minorities and the Negotiation of Nationhood in Modern Bulgaria* (Ithaca, NY: Cornell University Press, 2004); Krassimira Daskalova, *Literacy, Books, Readers, and Reading in Bulgaria on the Road to Modernity* (Sofia: LIK, 1999); and Keith Brown, *The Past in Question: Modern Macedonia and the Uncertainties of Nation* (Princeton: Princeton University Press, 2003).

2 Maria Todorova, *Imagining the Balkans* (New York: Oxford University Press, 1997); Maria Todorova, ed., *Balkan Identities: Nation and Memory* (London: Hurst, 2003); Maria Todorova, *Balkan Family Structure and the European Pattern: Demographic Developments in Ottoman Bulgaria*, 2nd rev. ed. (Budapest:

The shift has been one of locus of research to some extent, looking for historical evidence in archives that are not as close to the center of institutional power as used to be the case. Regional archives, archival holdings in ethnographic museums and institutes as well as other historical traces—such as monuments, cemeteries and personal artifacts of average people—have become a more common ground for constructing new narratives about the past.[3] By the same token, research into areas that previously seemed the purview of 'specialized' legal or medical historians has become more widespread. Periodical publications by epidemiologists, collections of papers presented at international conferences of eugenicists or proceedings of national gatherings of doctors have become sites of research by social, political and cultural historians of Southeastern Europe.

It would be fair to say, overall, that Southeastern European historiography has also become more interested in theoretical exploration of what such empirical evidence means both in the local and national context as well as the wider regional and continental one. Questions over the continued relevance and limitations of the Orientalist paradigm for Southeastern Europe have been confronted by several historians and other historically-minded scholars of the region for a decade now.[4] Some, like Milica Bakic-Hayden, find Orientalism of a "nesting" kind a useful paradigm for understanding the socio-cultural realities of Southeastern Europe. Others, such as Maria Todorova, have expressed a more critical stance towards the term, instead preferring to focus on Balkanism as something specific to the area,

Central European University Press, 2006); Larry Wolff, *Venice and the Slavs: The Discovery of Dalmatia in the Age of Enlightenment* (Stanford: Stanford University Press, 2001); Katherine J. Fleming, *The Muslim Bonaparte. Diplomacy and Orientalism in Ali Pasha's Greece* (Princeton: Princeton University Press, 1999); and Karl Kaser, Slobodan Naumović and Miroslav Jovanović, eds., *Between the Archives and the Field. A Dialogue on Historical Anthropology of the Balkans. Studies on South East Europe* (Münster: Lit Verlag, 2004).

3 Constantin Iordachi, *Citizenship, Nation and State-Building: The Integration of Northern Dobrogea into Romania, 1878–1913, Carl Back Papers in Russian and East European Studies No. 1607* (Pittsburgh, PA: University of Pittsburgh Press, 2002); and "The Ottoman Empire: Syncretic Nationalism and Citizenship in the Balkans," in Timothy Baycroft, Mark Hewitson, eds., *What is a Nation? Europe, 1789–1914* (Oxford: Oxford University Press, 2006), 120–151; and also the individual contributions in Todorova, ed., *Balkan Identities*.

4 Todorova, *Imagining the Balkans*; Fleming, *The Muslim Bonaparte*; Katherine E. Fleming, "Orientalism, the Balkans, and Balkan Historiography," *The American Historical Review* 105, 4 (2000): 1218–1233; Elissa Helms, "East and West Kiss: Gender, Orientalism, and Balkanism in Muslim-Majority Bosnia-Herzegovina," *Slavic Review* 67, 1 (2008): 88–119; Milica Bakić-Hayden, "Nesting Orientalisms: The Case of Former Yugoslavia," *Slavic Review* 54, 4 (1995): 917–931; Maria Todorova, "Balkanizem in postkolonializem: O lepoti pogleda z letala," *Zgodovinski časopis* 61, 1–2 (2007): 141–55; Maria Todorova, "Does Russian Orientalism Have a Russian Soul?," *Kritika: Explorations in Russian and Eurasian History* 1, 4 (2000): 717–727.

with essential differences from the colonial and post-colonial debates that both framed and are generally identified with Edward Said's original concept of Orientalism.[5] There is little consensus over this issue, although it is clear that those working on Southeastern Europe have strong opinions on the matter. The Europeanness of Southeastern Europe thus remains a contested matter, now even more so through the division effected by the European Union in the region between Romania, Bulgaria and Greece on the one hand, and the post-Yugoslav countries, Albania and Turkey on the other.[6] Given the longstanding traditions of historiography in Southeastern Europe, it is unsurprising that contemporary political divisions continue to play an important role in how historians construct the centrality and marginality of what constitutes modern Europe. Though the crass Huntingtonian division of Europe between Christianity and Islam has been strongly criticized over the past decade,[7] assessments of the Ottoman legacies vis-à-vis the idea of Europe, especially with regard to the place of Islam in Europe, remain a significant point of scholarly debate. Some view the Ottoman legacies as either part of the pre-modern past or a marginal element of the region during the modern period, while others consider these legacies to be both structurally and culturally central to how Southeastern Europeans, Muslims and non-Muslims alike, developed modern institutions and practices. Moreover, this debate is nowhere as pertinent and complex as in Southeastern Europe, whose checkered history of involvement with the Ottoman Empire and Islam invariably render any region-wide generalizations impossible. For while the Ottoman legacy indeed became the chief, pre-modern legacy against which the modern Greek state was constructed starting in the early nineteenth century, the story of Yugoslavia after World War I is much more complicated, given the presence of Bosnia at its core.

Another related and important trend in the historiography of Southeastern Europe is the fractious relationship between religion and modernization. An important scholar in this area was Peter Sugar, who published *Nationalism and Religion in the Balkans since the Nineteenth Century* in

5 See Bakić-Hayden, "Nesting Orientalisms," 917–93 and Maria Todorova, "The Balkans: From Discovery to Invention," *Slavic Review* 53, 2 (1994): 453–482.

6 Maria Todorova, "Spacing Europe: What is a historical region?," *East Central Europe/ECE* 32, 1–2 (2006): 7–55.

7 Samuel P. Huntington, *The Clash of Civilizations and the Remaking of World Order* (New York: Simon and Shuster, 1998).

1996.[8] A number of other scholars have started to closely examine not only the impact of religious institutions and practices on the development of nationalism in Southeastern Europe but more broadly the impact of those forces in the process of transforming pre-modern Balkan politics into modern ones.[9] Far fewer historians have probed into the seemingly paradoxical ties between, on the one hand, the religious institutions and deep religiosity still prevalent in Southeastern Europe in the first half of the twentieth century and, on the other hand, the scientifically articulated projects for social engineering that were developing at the same time, embodying major vectors of secularization.[10] Rather than situating the Balkans at the fringe of European developments in the twentieth century, this complicated relationship more accurately reflects the embattled relationship between secularization and the need for spiritual belief that persists as a leitmotif in the history of the modern West, a phenomenon that has only recently garnered renewed attention by Anglophone scholars, for instance.[11]

## Health, Hygiene, Eugenics and the Nation-State

The current volume is an important and rich addition to this historiography. The thirteen chapters represent an impressive array of studies offering new insights on how Southeastern European countries were transformed

8 Peter F. Sugar, *Nationalism and Religion in the Balkans since the Nineteenth Century* (Seattle: University of Washington Press, 1996); earlier discussions of this theme include George Arnakis, "The Role of Religion in the Development of Balkan Nationalism" in Charles and Barbara Jelavich, *The Balkans in Transition (Essays on the Development of Balkan Life and Politics since the Eighteenth Century)* (Berkeley: University of California Press, 1963); Charles Frazee, *The Orthodox Church and independent Greece, 1821–1852* (Cambridge: Cambridge University Press, 1969); and Paschalis M. Kitromilides, *Enlightenment, Nationalism, Orthodoxy: Studies in the Culture and Political Thought of South-eastern Europe* (Aldershot: Variorum, 1994).

9 See, for instance, Amila Buturović, "Christianity and Islam in the Balkans from the Fifteenth to the Twentieth Centuries," in Jacob Neusner, ed., *Religious Foundations of Western Civilization: Judaism, Christianity, and Islam* (Nashville: Abingdon Press, 2006); Vjekoslav Perica, *Balkan Idols. Religion and Nationalism in the Yugoslav States* (Oxford: Oxford University Press, 2002); and Bojan Alexov, "Nationalism in Construction: The Memorial Church of St. Sava on Vračar Hill in Belgrade," *Balkanologie* 7, 2 (2003): 47–72.

10 See, for instance, Maria Bucur, *Eugenics and Modernization in Interwar Romania* (Pittsburgh: University of Pittsburgh Press, 2002); and Marius Turda, "'A New Religion': Eugenics and Racial Scientism in Pre-World War Hungary," *Totalitarian Movements and Political Religions* 7, 3 (2006): 303–325.

11 See, for instance, Dennis L. Durst, "Evangelical Engagements With Eugenics, 1900–1940," *Ethics and Medicine*, vol. 18, no. 2, available at http://www.ethicsandmedicine.com/18/2/18-2-durst.htm (Last accessed 1 July 2008); and Christine Rosen, *Preaching Eugenics. Religious Leaders and the American Eugenics Movement* (Oxford: Oxford University Press, 2004).

in the late nineteenth and first half of the twentieth centuries from kinship-based, overwhelmingly rural, intensely religious and pre-modern societies into populations functioning within the parameters of modern states (in a Foucauldian sense)—albeit still largely rural, with strong kinship bonds and religious practices.[12] Focusing on health, hygiene and eugenic legislation and practices, the authors shed new light on how not only lawyers and politicians but also hygienists and epidemiologists crafted the modern institutions and practices propelling Southeastern Europe from a backwater of Europe into an area that strongly identified with the idea of European civilization in terms of citizenship rights, state institutions and cultural practices. These contributions complicate and enrich how we approach the nature, complexity and reach of modernization in Southeastern Europe through, and sometimes parallel to, state-building processes during this period.

The assumption for much of previous scholarship was that the energies of state builders in the late nineteenth and early twentieth century were concentrated almost exclusively upon constructing central state institutions (legislative, juridical and administrative), establishing a strong military and a cultural elite as well as some broader educational institutions.[13] Economic policy was the object of some limited academic efforts, especially in the area of banking, taxation and tariffs.[14] Yet other areas of public policy in the categories of social engineering, biopolitics or, less ominously sounding, social welfare, have generally not been the focus of much regional scholarship.

This may be viewed as a justifiable absence given the fact that even modern constitutions and political parties were quite new to the region in the late nineteenth century. Such a position, however, also implicitly assumes that there is indeed a "normal" hierarchy or chronology to the process of

12 By this I mean in particular the disciplining and controlling functions of the state to identify citizens, connect them institutionally with forms of control (i.e., duties towards the state and the 'nation' or local community) and empowerment (i.e., the right to take advantage of the resources the state was developing as part of new discourses linked to its civilizing mission in education, employment, defense, public health and other public services).

13 See, for instance, Barbara and Charles Jelavich, *The Establishment of the Balkan National States, 1804–1920* (Seattle, WA: University of Washington Press, 1977).

14 A good example for the trends described above is Frederick Kellogg, *The Road to Romanian Independence* (West Lafayette, IN: Purdue University Press, 1995).

modernization. The Balkans present direct challenges to such assumptions, as Maria Todorova, Katherine Fleming, and others have already shown.[15] Here the process of state-building took place at the same time as a number of professionally-educated elites were able to travel in various directions (St Petersburg, Berlin, Paris, London, Budapest or Vienna), undertaking training with the most outstanding European scholars in law, engineering, philosophy, economics, biology, physics, medicine, architecture and other disciplines.[16] The individual talents and professional abilities of these new elites were not necessarily matched by their collectively grandiose ambitions to become the safe keepers holding the keys to their countries' respective futures upon their return. Yet at home in Southeastern Europe the results of these transnational trends in the development of professional elites across Europe were felt immediately. In Greece, for instance, the medical profession was organized through a national network of doctors prior to the establishment any Ministry of Health, a network which, in fact, paved the way toward the framing of public health policy.[17] Likewise, in Bulgaria, Yugoslavia and Romania, physicians educated abroad pushed aggressively the creation of institutions to bring to life their ideas about the need to biopolitically shape the well-being of their societies.

Looking at the public policies of the Bulgarian, Greek, Yugoslav and Romanian states after World War I means, thus, to think of policy making as a two-way street: most leaders of the political parties in these countries had, at best, vague ideas about what it meant to deal with malaria, high infant mortality rates, alcoholism or syphilis, to give just a few relevant examples. Yet ambitious physicians among the region's educated elites had very specific ideas about these problems and their ostensible solutions, and had also mastered a generally nationalist discourse painting their country in stark colors of biological decay and degeneration. These physicians were also able to successfully depict themselves as saviors of their people, rather than mere technocrats with a science degree. Thus, the public health pol-

---

15 Todorova, *Imagining the Balkans* and Fleming, *The Muslim Bonaparte.*

16 Todorova, *Imagining the Balkans*; Elena Siupiur, *Intelectuali, elite, clase politice în Sud-Estul european: secolul XIX* (Bucharest: Domino R, 2004); Stathis Gourgouris, *Dream Nation: Enlightenment, Colonization, and the Institution of Modern Greece* (Stanford, CA: Stanford University Press, 1996); and Georgeta Nazarska, "The Bulgarian Association of University Women, 1924–1950," *Aspasia* 1 (2007): 153–175.

17 See the chapter by Katerina Gardikas in this volume.

icies enacted across Southeastern Europe after World War I could be said to be the fruit of both political will on the part of specific politicians and a process of educating these political leaders by physicians and other professionals around them. The latter played the essential role of communicating ideas about public health in a language that was politically and popularly appealing.

What was politically appealing —especially after 1918— was premised itself upon Woodrow Wilson's vision for the postwar international system, whereby 'nation' and 'self-determination' were key words for both ethnic majorities and minorities. For those belonging to a religious minority, as was the case with Jews and Muslims in Southeastern Europe (with the exception of Albania, where Muslims were a majority), their position was also politically reconfigured along Wilsonian lines, both internationally and internally. Thus, Jewish and Muslim communities in these areas learned to identify themselves as "national" minorities even more than had been the case before World War I; this represented their only recourse to some form of international protection, under the auspices, for example, of League of Nations' Minorities Treaty, against internal abuses. The implications of how "self-determination" and the drastic population policies (from forced relocation to genocide) that took place during and after the Great War have been the focus of increasing scholarly attention on the part of some historians of international relations and genocide. In a recent article in the *American Historical Review*, Eric Weitz offers a compelling analysis of the complex relationship between nationalism, internationalism and the biopolitics of the post-Versailles era.[18] He argues that the significant shift that took place in Europe after World War I has to been seen in the broader context of population policies and nationalism, as essentially undergirding the incentive of many states after 1919 in participating in any international system. Though only implicitly, his article in fact suggests integrating the insights of the scholarship on body politics (e.g., the history of eugenics) into understanding how nationalism becomes articulated in international

18 Eric Weitz, "From the Vienna to the Paris System: International Politics and the Entangled Histories of Human Rights, Forced Deportations, and Civilizing Missions," *The American Historical Review* 113, 5 (2008): 1313–1343. Other important recent scholarship on the relationship between eugenics discourses and the development of human rights discourses in an international context includes Kenneth Cmiel, "The Recent History of Human Rights," *The American Historical Review* 109, 1 (2004): 117–135.

relations during this period from the late nineteenth century into the interwar period.

Therefore it is not surprising to see that the categories of inclusion and exclusion, of perceived threats to the health of one's society and specific solutions to these threats were constructed along the lines of ethnic and/or religious identity instead of, for instance, class. In the 1920s, Bolshevism was largely viewed as a threat to the Treaty of Versailles rather than an ally in managing European stability.[19] The revisionism that flared up in the 1930s only reinforced these earlier discourses of ethno-religious exclusion and inclusion.

While these are not original ideas, what is relatively new in looking at the growing polarizing and biologizing (ethnocentric, xenophobic, sexist and racist) discourses used in political debates and policy-making in interwar Southeastern Europe is the role played by medical elites in these processes.[20] This does not mean one can draw a direct line connecting the words of eugenicists like Iuliu Moldovan, for instance, with the ideology of ethnic exclusion and anti-Semitism of the fascist Iron Guard in Romania.[21] But it does mean that policies depicting threats to the nation—in terms that were exclusionary on the basis of purportedly inborn biological traits—were becoming mainstream in many areas of public speech and action, so much so that those critiquing them were frequently seen as out of step with science and progress. The pervasive influence of negative eugenic approaches to social well-being may be best exemplified by the way in which religious leaders in Orthodox, Catholic and Protestant Churches often used such discourses, even though one might argue that any kind of social Darwinist thought would clash with the basic tenets of Christianity followed by these Churches.[22]

In their juxtaposition, the papers collected in this volume argue for rethinking who and what mattered in Southeastern Europe in terms of

19 See Ivan Berend, *Decades of Crisis* (Berkeley, CA: University of California Press, 2001).

20 See Marius Turda and Paul Weindling, eds., *Blood and Homeland: Eugenics and Racial Nationalism in Central and Southeast Europe, 1900–1940* (Budapest: Central European University Press, 2007).

21 On Moldovan and his links to the radical right in Romania see Bucur, *Eugenics and Modernization*; and Marius Turda, "The Nation as Object: Race, Blood and Biopolitics in Interwar Romania," *Slavic Review* 66, 3 (2007): 413–441.

22 Bucur, *Eugenics and Modernization*.

intellectual discourses about modernization and especially actual changes on the ground. To be sure, poets and journalists wrote a great deal more about the woes of Balkan "backwardness" or, conversely, the great yet tragically unappreciated contributions of their respective cultures to European civilization. However, their articulations had, at best, an indirect impact on how statesmen crafted new institutions, budgets and policies. Poetic-literary musings had even less impact on how the growing cadres of technocratic elites of the late nineteenth century and particularly the interwar period involved themselves in the business of producing modern, smooth running state institutions and policies. Nevertheless, scholars working on the national "awakening" of the Southeastern European peoples have had a penchant for focusing on precisely these literary constructions by the intelligentsia, largely neglecting the kinds of discourses and actions that other important segments of the educated public were engaged in, not least the growing strata of professional technocrats.[23]

Highlighting the work of physicians and eugenicists in the new states of Southeastern Europe thus adds important depth and nuance to understanding how these educated strata sought to transform their societies—making themselves into important vectors of betterment for their societies in the process, both discursively and also practically. The contributions in this volume focus on these aspects, albeit with different degrees of emphasis on discourse versus practice. Ultimately, however, the direct impact biologists, physicians and other public health personnel had upon the populations they sought to reach, help and control is important only to some extent. For most of the case studies presented here, it would be difficult to argue for any long-term, measurable public health consequences in the areas where policies were implemented, for the simple reason that war (the Balkan Wars, World War I and World War II) or financial troubles (the war debt and the Great Depression) cut short most of the policies initiated by public hygiene and eugenics advocates. Yet the new ways in which these reformers framed the state's responsibilities towards citizens and of citizens

[23] There are some exceptions to this, such as Maria Kovacs, *Illiberal Professions and Illiberal Politics: Hungary from the Habsburgs to the Holocaust* (Washington: Woodrow Wilson Center Press, 1994); Bucur, *Eugenics and Modernization*; Gary Cohen, *Education and Middle Class Society in Imperial Austria, 1848–1918* (West Lafayette: Purdue University Pres, 1996); and Tara Zahra, *Kidnapped Souls: National Indifference and the Battle for Children in the Bohemian Lands, 1900–1948* (Ithaca: Cornell University Press, 2008).

towards their own communities (their spouses, offspring and wider kin, as well as beyond) significantly changed the self-definition of these Southeastern European states, adding to their important traits as modern political entities the notion that they needed to cater to the physical well-being of their citizens, however defined (individually, communally, inter-generationally, racially and so on).

Finally, these contributions bring before international audiences insightful information (from details of public health legislation to statistics about medical services and epidemics) that has been largely unavailable previously in this format. The detailed descriptions of many policies are of importance not only for historians of medicine and science but should become as important a tool for social, cultural and political historians as are general population statistics or legislative measures pertaining to property laws for instance.

As substantial as the general achievements of this volume are, it is also useful to examine and compare individual pieces in this collection regarding regional similarities and differences, both in terms of their relationship to the older historiography on public health and eugenics in the broader modern European context as well as in relation to each other. In an attempt to synthesize the findings of the individual contributions, I will focus on: (a) theoretical frameworks and definitions; (b) historiographic dialogues and debates; and (c) empirical comparative insights.

## Theoretical Frameworks and Definitions

The authors in this volume anchor analyses of their respective case studies with diverse theoretical and historiographic frameworks, largely implicitly, but on occasion with explicit reference to relevant historiography. Several authors, from Marius Turda to Brigitte Fuchs, employ the Foucauldian term "biopolitics" to signal the disciplining nature of the discourses about medical education and policy-making in Southeastern Europe. The word seems particularly fitting for the case studies here, which collectively offer evidence for revolutionary ways in which both the individual body and groups of bodies (defined by race, religion, ethnicity or gender) were manipulated for various political ends. These national examples also point towards the ways in which the body and biology became a recurrent framework through which political issues were articulated. Indeed, it would be extremely valuable for the wider historical community studying modern politics in Southeastern Europe to push even further the boundaries of how biologizing politics and politicizing biology are mutually constitutive. In the case of Greece, for instance, the flourishing of eugenics after World War I can hardly be explained without looking at the specific experiences of politicians with the biologizing discourse of physicians trying to create public health policies (from workplace protection to premarital certificates and a national anti-malaria policy) or in turn, the experiences of physicians with the ineffective policies of the interwar Greek governments. The synthesis of eugenicist biopolitics after World War I was, in part, the fruit of these co-evolving frustrations.

The terminology common to all authors here concerning public health, hygiene and eugenics are on the whole descriptive. With the exception of Gergana Mercheva—whose piece both describes and deconstructs the definition of such terms in order to historicize and integrate them more broadly in the theoretical scholarship about modernization and social identity—contributors to this volume prefer extensive description of the ways in which public health measures were defined and articulated in the period and location under investigation. This dominant approach is entirely appropriate, but makes it rather difficult to step back and focus comparatively on the relationship between different usages of the same term. Elements described by Rory Yeomans as falling short of eugenic thinking regarding Croatian public health, for instance, are identified in the Bulgarian and Greek cases as articulations of eugenicist visions for public health. How ought one read these differences? Are they disparities based on different authors' assumptions about what constitutes an 'authentic' eugenicist discourse? Alternatively, does this distinction offer evidence of how greatly local cultural and political context mattered for understanding the disparate ways in which public health reforms chose to both represent these challenges to their communities and envision comprehensive solutions? The jury is still out on these issues, given the rather limited evidence offered here, but I would venture to suggest that both answers can be correct, at least to some extent. It is clear from the institutional, social, political and economic details offered by the contributing authors that there were vital contextual differences between the cases investigated here. By the same token, the chapters are framed around different types of questions and methodologies, varying from the narrow investigation of one person's ideas to broad narratives over several decades of institutional developments and activities.

Though somewhat frustrating for the reader, these differences strike me as having great potential for more research in this area. Starting from markedly different vantage points is a great way for contemplating the cross-pollination of, for instance, varieties of regional micro-foci as evident in the piece by Tudor Georgescu, with the telescopic vision of Željko Dugac. There is room for both kinds of approaches so long as each is aware of the strengths and weaknesses implied therein while being explicitly mindful of other ways to approach similar questions.

## Historiographic Dialogues and Debates

If an active interest in theorization is only partly visible in most chapters, they are also uneven in considering the existing historiography on themes they deal with. Overall, a sense of being in the presence of pioneering work prevails, even though several authors have published together as a group elsewhere.[24] But while true regarding many of the national studies herein, there are plenty of precedents for analyzing similar historical themes from the late nineteenth century and into the middle of the twentieth century in Europe, from racial politics to birth control, gender roles and labor protection. Some authors pay especial attention to the historiography of eugenics in other locales (especially Germany, France and the United States) during the same period, in particular when analyzing the impact of German, French or American ideas about public health on the country under investigation.

Yet it would be a great gain to the historiography on eugenics to push this analysis still further, offering some discussion on how Bulgaria, Yugoslavia, Romania and Greece can be construed as important cases for understanding the ways in which the historiography of biopolitics needs to embrace

24 Turda and Weindling, eds., *Blood and Homeland.*

these cases as paradigmatically important—not just mere additions to a rich scholarship. The relationship between post-imperial modernization, anti-Ottomanism, pan-Europeanism and public health in both Yugoslavia (and particularly in Croatia) and Bulgaria offer vivid, contrasting examples of the way in which post-colonialism, as an assertion of indigenous legacies, has been a theme identified in most of the relevant scholarship. The racist elements of internal colonization in Croatia as well as the coercive measures implemented by public health officials in interwar Bulgaria remind us that the politics of post-colonial regimes are by no means liberating ones for their entire populations, nor are they always liberal or democratic. Realities on the ground in both cases seem to be less of an exception to the rule than a reminder to look critically through the rhetoric of ideologies of liberation, empowerment and progress as often having an organicist, biopolitical core.

Another possible route for further developing the historiographic inroads opened by these authors is to more closely focus on comparing the relationship between claims (discourse) and actions (legislation and policies) by public health enthusiasts (whether government officials or members of civic, professional or religious groups). In particular, how should historians approach biologizing discourses about the "health" of the nation, if such discourses do not result in controlling compulsory actions? And if there is evidence of such compulsory actions, as was the case with Muslim populations in Bulgaria, how does such evidence shape historians' views on the intellectual discourse concerning perceived threats to Bulgarian society and its future? Christian Promitzer reflects on this issue with great honesty, leaving his own analysis somewhat open to different interpretations of the extent to which Bulgarian public health policies in the interwar period were driven by racist notions of nation and social well-being.

A significant, albeit largely implicit, common thread in almost all of the pieces in this collection is the relationship between public health reform and changing gender roles. In nearly every chapter (less so in those on malaria in Greece and workplace public health concerns), women represent either the real or the potential objects for the bulk of ideas for reform and control described, from sterilization to delousing. Several authors bring the gendered nature of the health policies they describe into their discussion. In particular, Fuchs, Promitzer and Yeomans pause to assess specific ways in which women became central objects of policy-making and how gender norms were decisive in constructing the particulars of public health policies. For instance, the campaign for delousing among Muslims in both Bosnia and Bulgaria took gendered taboos among Muslims into account, especially regarding physical interaction between women in Islamic communities and strangers from outside the Muslim world. Since Muslim women were not permitted physical contact with non-Muslim males, this interdiction paved the way for the employment of Christian female physicians in order to implement the said policies. Fuchs spends considerable time in analyzing this important element and Promitzer also concludes his piece by considering the gendered elements of the delousing campaigns in Bulgaria.

A broader consideration of how gender roles were normatively and experientially constructed and negotiated during this period suggests that the biopolitics of public health reform were fundamentally gendered.[25] Physicians built their notions of community, health, crisis responsibility and agency from deeply ingrained ideas about the roles women and men could and *should* have in their communities. Politicizing reproduction as a process to be overseen in some form by the state (through birth control, pre-nuptial certificates, preventing miscegenation, puericulture, etc.) meant very different things to men and women in Southeastern Europe at this time. These categories were further complicated by assumptions about ethnicity, religion, race and class. But normative assumptions about gender roles also helped construct what it meant to be, for instance, a Muslim, Greek, "Aryan" or peasant. Eugenicists were not uniformly consistent in

---

[25] Some recent relevant work includes Melissa Feinberg, *Elusive Equality. Gender, Citizenship, and the Limits of Democracy in Czechoslovakia, 1918–1950* (Pittsburgh: University of Pittsburgh Press, 2006); Bucur, *Eugenics and Modernization*; Atina Grossman, *Reforming Sex. The German Movement for Birth Control and Abortion Reform, 1920–1950* (New York: Oxford University Press, 1995); Renate Bridenthal, Atina Grossmann, and Marion Kaplan, eds., *When Biology Became Destiny: Women in Weimar and Nazi Germany* (New York: Monthly Press Review, 1984); Michelle Mouton, *From Nurturing the Nation to Purifying the Volk: Weimar and Nazi Family Policy, 1918–1945* (Cambridge: Cambridge University Press, 2007). Closer to the Southeastern European context in the late nineteenth century and first half of the twentieth century see also Nazarska, "Bulgarian Association"; Maria Bucur, "Between Liberal and Republican Citizenship: Feminism and Nationalism in Romania, 1880–1918," *Aspasia* 1 (2007): 84–102; Katja Mihurko Poniž, "Nation and Gender in the Writings of Slovene Women Writers, 1848–1918," *Aspasia* 2 (2008): 28–43; Voichița Năchescu, "The Visible Woman: Interwar Romanian Women's Writing, Modernity and the Gendered Public/Private Divide," *Aspasia* 2 (2008): 70–90; and Irina Gigova, "The Feminisation of Bulgarian Literature and the Club of Bulgarian Women Writers," *Aspasia* 2 (2008): 91–119.

their attempts to control the behavior of all Croats, for instance. The bodies of Croatian women were to be primarily manipulated through public policies aimed at reproductive control. The same can be said for Bulgaria, Greece and Romania. To be sure, men were also subjected to gender-specific biopolitical control. Whether forbidden from marrying a partner of a different ethnic or religious group, or pressured to assert their masculinity, men were also subjected to gender norms.

By the same token, one has to understand the specific ways in which women and men were able to use the gender-specific attributes of these ideas to craft new public roles for themselves. Croatian, Bulgarian and Romanian women sought to transform the focus on their bodies into an argument for becoming more powerful agents in the public sphere, if not in terms of direct political power or economic well-being, then at least symbolically and socially in their communities.[26] At the same time, male physicians arrogating the role of powerful policy-makers acted as both arbiters and also powerful embodiments of their masculine ideal. Their aggressive views on masculinity were in part a reflection of their self-image as men.[27] Much more needs to be explored in the complicated area of gender roles, especially in terms of actual practices in Southeastern Europe at this critical time. This volume therefore offers glimpses at the potential for innovative studies in this area.

## Comparative Insights

The most important contribution of these pieces as organized in the present volume rests in enabling the reader to think comparatively about basic assumptions about what public health was, or could become, in the late nineteenth century and first half of the twentieth century; the importance

26 In addition to the chapters by Dugac and Mircheva in this collection see also Bucur, *Eugenics and Modernization*; and Krassimira Daskalova, "A Life in History (Fani Popova-Mutafova)," *Gender and History* 14, 2 (2002): 321–339.

27 An interesting case in this regard is Iordache Făcăoaru, an aggressive, racist eugenicist in Romania who practiced what he preached by marrying a woman of Aryan background. The couple remained childless, however. On Făcăoaru's eugenics and Romanian eugenicist ideas regarding masculinity see Bucur, *Eugenics and Modernization*; Turda, "The Nation as Object"; and Marius Turda, *Eugenism şi antropologie rasială în România, 1874–1944* (Bucharest: Ed. Cuvântul, 2008).

of regional political, cultural, social and economic contexts for understanding how ideas about modernization, progress and health become translated into policies in Southeastern Europe; the diverse vantage points in framing a question in terms of the possible analytical payoff of a case study; and the need to be self-reflexive about what is left out of the scholarly gaze in which each of us individually engages. What is unnecessary detail for one scholar might be a crucial conceptual element for another (see gender analysis, for instance). In this regard, the manifest diversity of this collection is its greatest strength.

In the spirit of this great strength, my hope is that individual authors continue to move towards systematically acknowledging each other as participants in a larger dialogue. Promitzer, Mircheva and Popova write about contemporaneous developments in the same general milieu of Bulgarian public health reformers, but that larger common context is mostly background rather than an explicit comparative element. Likewise, Georgescu's Saxons are like a lone island in an unknown land (Romania), appearing to bear no relationship to the wider case described by Marius Turda. In fact, these two chapters on Romania add up to the realization that two vigorous, parallel and, in fact, oppositional eugenic movements could develop simultaneously in the same space, surely a particularity of Southeastern Europe in the larger international context.

On the other hand, explicit connections with international public health and especially eugenics movements in Europe and North America offer a clearer way to link these pieces. All authors make clear that Austrian, German, Swedish, French and American concepts of public health, welfare state institutions, authoritarian population legislation and overall medical training had paramount importance both directly (as many physicians from Southeastern Europe were trained in these centers) and indirectly (as examples in terms of eugenic theories on heredity or legislation regarding labor protection, for instance). The Rockefeller Foundation, for instance, makes its appearance (not surprisingly, based on scholarship focusing on other cases) as the embodiment of the Wilsonian vision of global health, one self-determining nation at a time. The kind of support offered by the Rockefeller Foundation in the Yugoslav case described by Dugac is very much in keeping with a larger vision of progress through centralized state institutions with coercive modernizing powers – a paradigmatic example of the Foucauldian definition of biopolitics. Such a vision naturally flew in the face of complicated ethno-politics on the ground, much like in Romania[28] or Greece. Such case studies are an excellent reminder to scholars of transnational movements in the early twentieth century —exemplified by the work of international foundations like the Rockefeller's in eradicating malaria— that Europe was as important as Africa or East Asia for shaping the subsequent institutions and power relations between policy makers (the Rockefeller's, the League of Nations, the United Nations, the World Bank and so on) and the "field." One can only hope that the paths opened by these scholars will continue to widen and diversify as these historians and others joining the field ask themselves and one another questions about the relationship between public health discourses and policies, on the one hand, and the shaping of modern states in Europe and beyond, on the other.

[28] See, for instance, Maria Bucur, "From Private Philanthropy to Public Institutions. The Rockefeller Foundation and Public Health in Interwar Romania," *Romanian Civilization* 4, 2 (1995): 47–60.

# CONTRIBUTORS

**Maria Bucur** is John V. Hill Chair in East European History, Indiana University, Bloomington. She is the author of *Eugenics and Modernization in Interwar Romania* (2002) and *Heroes and Victims: Remembering War in Twentieth-Century Romania* (2009), as well as co-editor with Nancy Wingfield of *Staging the Past: The Politics of Commemoration in Habsburg Central Europe, 1848 to the Present* (2001) and *Gender and War in Twentieth-Century Eastern Europe* (2006).

**Željko Dugac** is senior scientific collaborator at the Department for the History of Medicine, Institute of History and Philosophy of Science of the Croatian Academy of Sciences and Arts. His research interests are the history of public health, art and the cultural anthropology of medicine. He is the author of *Protiv bolesti i neznanja: Rockefellerova fondacija u međuratnoj Jugoslaviji* (2005).

**Brigitte Fuchs** is an independent scholar and lecturer in gender studies at the Department of Social Anthropology at Vienna University. She is the co-editor with Gabriele Habinger of *Rassismen and Feminismen. Differenzen, Machtverhältnisse und Solidaritat zwischen Frauen* (1996) and author of *'Rasse', 'Volk', Geschlecht. Anthropologische Diskurse in Österreich (1850–1960)* (2003).

**Katerina Gardikas** was a researcher at the Institute for Neohellenic Research at the National Hellenic Research Foundation. She is assistant professor at the Department of History, University of Athens since 2001, where she teaches modern Greek history. She has written on modern Greek political history, including a book (in Greek) entitled *Protection and Guarantees, 1821–1920* (1999). Her current interests are in the history of malaria in modern Greece.

**Tudor Georgescu** completed his PhD at Oxford Brookes University in 2009 on the emergence of eugenic and fascist movements among the Transylvanian Saxons. He is also the Outreach Officer for the Centre for Health, Medicine and Society at Oxford Brookes University and the Research Coordinator of the Working Group on the History of Race and Eugenics (HRE); an editor of 'Compass: Political Religion' published online by Wiley-Blackwell and a Director of the Pulse Project.

**Despina Karakatsani** is associate professor at the University of Peloponnesus. She is author (with Vassiliki Theodorou) of the book *Instructions of Hygiene. Health and Social Welfare for Children in the first decades of the 20th century* [*'Υγιεινής Παραγγέλματα': υγεία και κοινωνική πρόνοια για παιδιά τις πρώτες δεκαετίες του 20ού αιώνα*] (2010).

**Gergana Mircheva** graduated in law and cultural studies from the University of Sofia. Currently a doctoral student in History of Modern Bulgarian Culture at the Cultural Studies Department, University of Sofia, her research interests are in the field of cultural and social history of medicine and eugenics, modern Bulgarian culture and nationalism studies.

**Leda Papastefanaki** is a member of the Department of History and Archaeology, University of Ioannina. She is the author of *Labor, Technology and Gender in the Greek Industry. The Textile Industry of Piraeus (1870–1940)* [*Εργασία, τεχνολογία και φύλο στην ελληνική βιομηχανία. Η κλωστοϋφαντουργία του Πειραιά (1870–1940)*] (2009).

**Kristina Popova** is associate professor at the Chair for Bulgarian History and Archival Studies, South-West University "Neofit Rilski" in Blagoevgrad. She is member of the editorial board of the journal *Balkanistic Forum* and organizes the *International University Seminar for Balkan Studies and Specialization*. She is the author of *Natsionalnoto dete. Blagotvoritelnata i prosvetna na Sajua za zakrila na decata v Balgarija, 1925–1944* (1999).

**Christian Promitzer** is assistant professor at the Center for Southeast-European History at the Institute of History, University of Graz. His research interests are in the history of ethnic relations, medicine and racial anthropology in Bulgaria and former Yugoslavia. He is founding member of the Center for the Study of Balkan Societies and Cultures (CSBSC) in Graz and co-editor of *(Hidden) Minorities Language and Ethnic Identity between Central Europe and the Balkans* (2009).

**Vassiliki Theodorou** is associate professor of modern and contemporary Greek history at Democritus University of Thrace, Greece. She studied sociology and social history in Athens, Paris and Strasbourg. Her research interests lie in the fields of social history, history of childhood and public health in Greece during the first half of the 20th century. Her most recent publications include "Health policy in Interwar Greece: the intervention by the League of Nations Health Organization," *Dynamis* 28 (2008), 53–75.

**Sevasti Trubeta** is assistant professor at the University of the Aegean, Department of Sociology (Mytiline) and affiliated with the Free University of Berlin (Venia legendi candidate). She is the editor of *The Roma in the Modern Greek State* [*Οι Ρομά στο σύγχρονο ελληνικό κράτος. Συμβιώσεις, Αναιρέσεις, Απουσίες*] (2008), co-editor with Christian Voss of *Minorities in Greece-Historical Issues and new Perspectives* (*Jahrbücher für Geschichte und Kultur Südosteuropas*, Special Issue 5, 2003/2004) and author of *Die Konstitution von Minderheiten und die Ethnisierung sozialer und politischer Konflikte. Eine Untersuchung am Beispiel der im griechischen Thrakien ansässigen 'Moslemischen Minderheit'* (1999).

**Marius Turda** is Reader in 20th Century Central and Eastern European Biomedicine at Oxford Brookes University and the founder of the Working Group on the History of Race and Eugenics (HRE). He is the author of several books including, most recently, *Eugenism și antropologie rasială în România* (2008) and *Modernism and Eugenics* (2010).

**Paul Weindling** is Wellcome Trust Research Professor in the History of Medicine at Oxford Brookes University. He is the author of *Health, Race and German Politics* (1989), *Epidemics and Genocide in Eastern Europe* (2000), *Nazi Medicine and the Nuremberg Trials* (2004) and *John W. Thompson, Psychiatrist in the Shadow of the Holocaust* (2010). Together with Marius Turda he is director of a project on the reconstruction of life histories of victims of Nazi human experiments.

**Rory Yeomans** is an independent scholar affiliated with the Centre for the Study of South Eastern Europe, School of Slavonic and East European Studies, University College London. His first book (*Visions of Annihilation: The Croatian Ustasha Regime and the Cultural Politics of Mass Murder*) is due to be published by the University of Pittsburgh Press in 2011.

# INDEX

For Product Safety Concerns and Information please contact our EU representative GPSR@taylorandfrancis.com Taylor & Francis Verlag GmbH, Kaufingerstraße 24, 80331 München, Germany

**Batch number: 10725853**

Printed by Printforce, the Netherlands